ACOUSTICS

of

WOOD

Voichita Bucur

Institut National de la Recherche Agronomique
Centre de Recherches Forestières
Nancy, France

CRC Press
Taylor & Francis Group
Boca Raton London New York

CRC Press is an imprint of the
Taylor & Francis Group, an **informa** business

First published 1995 by CRC Press
Taylor & Francis Group
6000 Broken Sound Parkway NW, Suite 300
Boca Raton, FL 33487-2742

Reissued 2018 by CRC Press

© 1995 by Taylor & Francis
CRC Press is an imprint of Taylor & Francis Group, an Informa business

No claim to original U.S. Government works

A Library of Congress record exists under LC control number: 94036293

Publisher's Note
The publisher has gone to great lengths to ensure the quality of this reprint but points out that some imperfections in the original copies may be apparent.

Disclaimer
The publisher has made every effort to trace copyright holders and welcomes correspondence from those they have been unable to contact.

ISBN 13: 978-1-138-50647-3 (hbk)
ISBN 13: 978-1-138-56202-8 (pbk)
ISBN 13: 978-0-203-71012-8 (ebk)

Visit the Taylor & Francis Web site at http://www.taylorandfrancis.com and the
CRC Press Web site at http://www.crcpress.com

"Dubito, ergo cogito; cogito ergo sum"

René Descartes
(1596–1650)

CONTENTS

Chapter 10
Acoustic Emission

FOREWORD

The Hooke's law of elasticity

$$[\sigma_{ij}] = [C_{ijkl}][\varepsilon_{kl}]$$

appears in its general forms as Equation 4.1 at the beginning of this book, as it usually does in many texts of elasticity or acoustics. In order to thoroughly appreciate the spirit that inspired the author in writing the "Acoustics of Wood", one should have seen the very same formula projected by the same author on the screen of one of the Erice lecture halls during the presentation of an advanced research lecture, as the author herself quotes in the Preface to this volume, few, elegantly handwritten letters "$[\sigma_{ij}] = [C_{ijkl}][\varepsilon_{kl}]$" filled the whole screen in a symphony of pastel colours. The attention of the audience was gently captured. Science and art were locked together within a simple formula, as science and art link together in the author's life, as science and art frequently share common fate in wood history.

The making of violins, cellos, or pianos or other musical instruments has been an art long before being an object of scientific investigation. Architectural wood structures are artists' representations that rely on advanced achievements of mechanics. The scientific knowledge of wood properties and characteristics is a necessary step toward its best use in artistic representations. This might really be a rather personal interpretation of the reading of the book, but it could be really one of the ways to approach the reading.

Acoustics of wood deals with wood in all aspects that are of concern to acoustics: from sound barriers produced by forests and trees, to the use of wood in acoustical panels; from the crystallographic symmetry classes of different woods, to the surface wave propagation in wood structures; from the influence of aging and moisture in the elastic propagation in wood, to the chemical methods for improving acoustic properties; from the counting of the average ring width in violin tops, to the high Q properties of guitar wood for sustaining "sing" modes; from acoustic micrographs of acoustic microscopy techniques, to the acoustic emissions characteristics of different wood species.

Acoustics of wood, however, preliminarily needs information about wood, from seed germination to forest growth, through moisture, aging and anatomical properties. The intrinsic coordinate system of wood is a cylindrical system, that follows the axial direction of growth of the stem, the azimuthal and radial directions, the commonest case of wood materials presents an orthotropic symmetry, where three mutually perpendicular mirror planes of symmetry exist, related to the directions of growth. Velocity of ultrasonic waves present a wide spread of values, from 6000 m/s for longitudinal waves along the fiber direction to 400 m/s for shear waves in the radial-tangential plane.

An interesting general review of wave equations and solutions accompanies Part I devoted to material characterization, where elastic constant relations to technical constants is duely reviewed together with Christoffel equations and eigenvalue properties of the wave equation. That is the science part, as we said at the beginning, that matches with the technical part reported as the last section of the book, where probing of materials and common techniques of testing are also reviewed. And all is treated with meticulous care as to completeness and with careful attention to biographical sources, that are listed by subject at the end of each chapter. But art and style are blended with science, and that is materially achieved with a series of color plates properly selected to show grain and fiber structure in different samples.

Wood is technically studied because of its importance in the manufacture of musical instruments: what are the characteristics of a guitar plate or of a harp soundbox or a violin bow and what are the wood species to be used? Names with exotic charm like Manilkara, Mauritius ebony or Pernambuco wood, alternate with those of cultural Latin origin, like *Picea abies* or *Acer pseudoplatanus* and so on. And they are of interest because of their Young's modulus or Poisson ratio or high quality factor.

What was known about quality factor or Poisson ratio by the handcraft masters of the past? Why is wood the best material for many musical instruments, still not overtaken by the ubiquitous power of plastics? Perhaps Nature is science and art at the same time and we usually follow different routes to get to the target, only to discover at the end that it could have been achieved through either route.

Adriano Alippi
Istituto di Acustica "O. M. Corbino"
Rome, Italy

PREFACE

My involvement with the project that led to the publication of "Acoustics of Wood" began in 1985 when I first participated in a series of lectures organized by the International School of Physical Acoustics in a splendid and magical place called Erice in Sicily (Italy), on the subject of ultrasonic methods in evaluation of inhomogeneous materials. The interest of the participants in the subject and the subsequent invitations addressed to me by Professor Alippi to give "advanced research lectures" in the third and fourth courses at the School in Erice enhanced my idea that a book on wood acoustics could be helpful for scientifically educated persons wishing to know more about wood, as a natural anisotropic composite material.

All these ideas became a reality with the constant encouragement of Dr. Carleen M. Hutchins, fellow of the Acoustical Society of America and permanent secretary of The Catgut Acoustical Society, with whom I have worked very closely over the years on the subject of the acoustical properties of wood for violins and other musical instruments.

The aim of this book is to present a comprehensive account of progress and current knowledge in wood acoustics, presented in the specialized literature from the last 25 to 30 years. For earlier publications the reader is generally referred to books related to wood technology and wood physics.

This volume is divided into three main parts. The first part describes environmental acoustics, the second part presents acoustic methods for the characterization of the elastic behavior of wood and the third part deals with acoustic methods for wood quality assessment. To enhance the usefulness of the book, a cumulative index of subjects and authors is presented in the last chapter.

The reader is encouraged to thoroughly examine his subject from nearly 800 bibliographic references. The search of the bibliography in different databases (CAB Abstracts; COMPENDEX; INSPEC-Physics; ISMEC; NASA; PASCAL; CRIS USDA; etc) was made with the kind cooperation of Michel Dumas, librarian in our institute (Institut National de la Recherche Agronomique, Centre de Recherches Forestières de Nancy, France).

During the last 15 years, my colleague Pierre Gelhaye has drawn numerous figures for the slides I needed for my lectures at international conferences and symposia: almost all of them became figures in this book. It is through his generous help that the book was illustrated.

I am very much indebted to the following persons for reading the manuscript and making comments for the improvement of the comprehension of the exposed ideas and written text: Dr. Martin Ansell, University of Bath, United Kingdom; Professor I. Asano, University of Tokyo, Japan; Dr. Claire Barlow, University of Cambridge, United Kingdom; Dr. Ioan Facaoaru, RILEM and C.R.L. Comp. Vicenza, Italy; Dr. Daniel Haines, Catgut Acoustical Society, U.S.A.; M. Maurice Hancock, Catgut Acoustical Society, United Kingdom; Dr. Johannes Klumpers, Centre de Recherches de Nancy, France; Dr. Robert Ross, Forest Products Laboratory, Madison, U.S.A.; and Dr. John Wolf, Naval Research Institute, Washington, D.C., U.S.A.

Last, but not least, I would like to thank Professor Adriano Alippi, Università di Roma, Italy, for his enthusiastic support during obscure and doubtful moments that I spent when writing the book.

I am indebted to the "Institut National de la Recherche Agronomique—INRA" in France for providing the facilities required to complete this book, and particularly to my colleagues in the Forestry Research Center in Nancy, my students, and my professional friends mentioned in the bibliographic list.

Thanks are due to my sister Despina Spandonide, my family, and to all my friends from all over the world who followed the writing of this book with interest.

Finally, I would like to thank the editorial and production staff at CRC Press for their efficient contribution to the heavy task of transforming the manuscript into the finished book.

Voichita Bucur

To the memory of
Dr. R. W. B. Stephens,
a pioneer in ultrasonic activity
and an enthusiastic stimulator
of creative ideas in acoustics.

Chapter 1

Introduction

1.1. GENERAL REMARKS ON WOOD MATERIAL

Wood is a biologically renewable substance and one of the most fascinating materials due to its very complex structure and its wide, extensive uses. The *Concise Encyclopedia of Wood and Wood-Based Materials* defines wood as "the hard, fibrous tissue that comprises the major part of stems, branches and roots of trees, belonging to the plant groups known as the gymnosperms and the dicotyledonous angiosperms." Wood can be considered to be a biological composite that is produced by the living organisms of trees. Its organization can be observed on discrete scale levels, ranging from the molecular to the macroscopic.

Figure 1.1 gives an overview of the way in which the components of wood — from molecular to macroscopic levels — are held together by specific interactions, assuring high performance of the tree, without suffering the debilitation of difficult environmental conditions (wind, snow, rain, etc.). The complex assemblies reveal a hierarchical organization of the structure. For further information on the anatomic structure of wood, the reader is referred to: Côté (1965); Kollmann and Côté (1968); Bosshard (1974); Core et al. (1976); Jacquiot et al. (1976); Grosser (1977); Schweingruber (1978); Panshin and de Zeeuw (1980); Butterfield and Meylan (1980); Wilson and White (1986).

Before addressing the contemporary aspects of wood acoustics, let us consider the hierarchical structure of wood from the molecular to the macroscopic level, which provides background to understanding the behavior of wood.

The trunk of the tree is composed of millions of individual cells. The principal constituents of the cell wall are cellulose (50%), hemicellulose (35%), lignin (25%), and extractives (Fengel and Wegener, 1989). The proportion of these constituents varies between and within species, as well as between and within individual trees.

Cellulose is a polymer, containing repeated cellobiose segments, each of which is composed of two glucose units. The length of the cellobiose segment is 10.3 Å. Each crystalline cellulose unit contains two segments of cellobiose and is 8.35×7.9 Å in section. The average cellulosic chain length is 50.10^3 Å.

The hemicelluloses are low molecular weight polysaccharides, consisting of approximately 50 to 400 glucose units. Cellulose and hemicelluloses are present in highest concentration in the secondary wall of the cell. The cellulosic macromolecules aggregate during biosynthesis to form crystals that can be observed at the Angström scale level. The regions of oriented cellulose chains (about 600 Å in length, with a lateral dimension of 100×40 Å) are called crystallites. The crystallites alternate with relatively short, amorphous regions.

Lignin is a complex, amorphous polymer present in the middle lamella, between the walls of contiguous cells. Intramolecular covalent bonds and intermolecular van der Waals forces determine the specific arrangement of cellulosic crystals, which are embedded in an amorphous lignin matrix, in fibrils. The crystals have a diameter of 35 Å and contain approximately 40 cellulose chains. Aggregates of fibrils form microfibrils, having a diameter of 200 Å, containing approximately 20 micellar strands. The width of this unit is about 250 Å.

As the lignin system increases in complexity, the microfibrils are aggregated in macrofibrils. The macrofibrils are the basic building blocks of lamellae, which comprise the various layers of the cell wall. The lamellae have the following generally accepted abbreviations (Kollmann and Côté, 1968; Siau; 1971; Mark, 1967): M is the middle lamella; P is the primary wall of adjoining cells; S_1 is the outer layer of the secondary wall (1 μm thick in latewood); S_2 is the middle layer of the secondary wall (10 μm thick in latewood); S_3 is the inner layer of the secondary wall (1 μm thick in latewood); W is the warty membrane that lines the cell lumen.

Within each lamella the microfibrils are arranged in a typical parallel pattern and inclined with respect to the axis of the cells. This corresponds generally to the vertical growth direction of the tree. The pattern of the lamellae can be seen with an electron microscope. The cell wall, the structure of which is a layered composite, can be observed with an optical microscope.

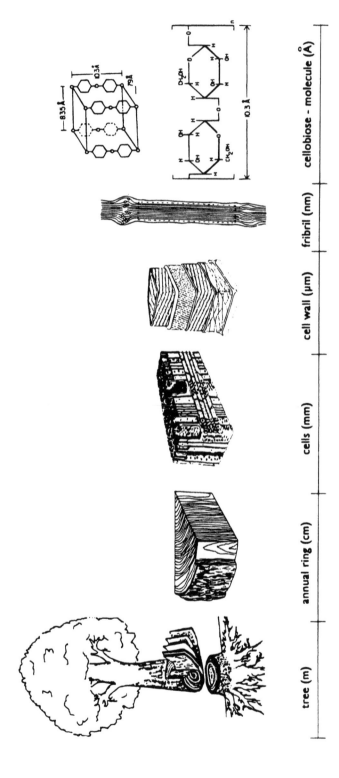

Figure 1.1. Hierarchical structure of wood. This figure is an improved presentation of the hierarchical structure of wood given by Baer, E., Hiltner, A., and Morgan, R. J. (1992).

Table 1.1. Cell dimensions of some wood species, average values

Parameters	Species		
	Softwoods		
	Abies alba	*Picea abies*	*Pinus sylvestris*
Density (kg/m³)	410	430	490
Cell dimensions: tracheids			
Length (mm)	4.3	2.9	3.1
Diameter (μm)	50	30	30
Cell percentage (%)			
Tracheids	90	95	93
Parenchyma	Scarce	1.4–5.8	1.4–5.8
Rays	9.6	4.7	5.5
	Hardwoods		
	Fagus sylvatica	*Quercus robur*	*Populus* spp.
Density (kg/m³)	680	650	400
Cell dimensions: vessels			
Length (mm)	3–7	1–4	5
Diameter (μm)	5–100	10–400	20–150
Cell percentage (%)			
Fibers	37	43–58	62
Vessels	31	40	27
Parenchyma	4.6	4.9	
Rays	27	16–29	11

From Fengel, D. and Wegener, G., *Wood: Chemistry, Ultrastructure, Reactions,* Walter de Gruyter, Berlin, 1989. With permission.

Using the millimetric scale it is possible to observe the main anatomic elements of wood: tracheids, fibers, vessels, rays, and parenchyma cells. A softwood tracheid is approximately 4 μm in diameter and 4 mm in length. The elements in hardwoods are shorter in length than those of the tracheids, but their diameter is wider. The cell dimensions of some wood species are given in Table 1.1, and the functions of these cells are noted in Table 1.2.

At the annual ring level the structure is again one of a layered composite built up with two layers corresponding to earlywood and latewood.

This hierarchical architecture of wood is responsible for its high anisotropic and viscoelastic behaviors. The anisotropy at the microscopic scale is related to the disposition of cells. The wood substance is also anisotropic down to the finest detail of its crystallographic and molecular elements.

Because the aim of our analysis is to consider solid wood as a product of a living organism, it is appropriate to observe that this material is characterized by both the high variability and the heterogeneity of its structural constituents.

The variability, as viewed from the microscopic scale, depends upon the relative proportion and distribution of different types of cells from species to species. The cells are variable in character, and the cell walls vary in chemical composition and in their organization at the molecular level. The morphological variability of cells within a tree is determined by the influence of crown elongation and cambium activity.

Table 1.2. Functions of the various cell types in wood

Species	Functions		
	Mechanical	**Conducting**	**Storing**
Softwoods	Latewood tracheids	Earlywood tracheids	Resin canal Parenchyma
Hardwoods	Fibers	Vessels	Rays Parenchyma

From Fengel, D. and Wegener, G., *Wood: Chemistry, Ultrastructure, Reactions,* Walter de Gruyter, Berlin, 1989. With permission.

More specifically, wood is also different from tree to tree, from the top to the bottom of the trunk itself, etc. The variability between trees is related to growth factors such as geographic location, site quality, soil type, and the availability of moisture.

The integrated effects of all these forms of heterogeneity and anisotropy must be taken into account when assessing the physical properties of solid wood.

To summarize, it is noted that wood is a complex, highly ordered material. The attention of the reader was focused on those anatomical features which could afford clear insight into the structure of wood for a better understanding of acoustic wave propagation phenomena in this material.

1.2. OUTLINE OF THE BOOK

A comprehensive understanding of wood behavior necessitates an interdisciplinary approach. This volume examines aspects related to the development of acoustic methods as an effective means for examining the physical properties of wood. The discussion is concerned particularly with the studies of short duration pulse methods and standing wave methods.

The chapters are organized into three sections: the acoustics of forests and forest products and wood in architectural acoustics, the methods for mechanical characterization of wood behavior, and quality assessment of wood products.

Part I, "Environmental Acoustics", presents a discussion of the physical phenomena associated with the propagation of acoustic waves in forests and studies the behavior of wood and wood composites as materials used in architectural acoustics.

Part II, "Material Characterization", was written in response to practical considerations about the uses of wood. Interest is increasing in the development of nondestructive techniques that predict the mechanical characteristics of wood. The methods based on acoustic energy are satisfying and practical. The challenges to the scientist or engineer interested in the development of acoustic nondestructive techniques are:

- To know what information is needed to fully characterize each wood product
- How to use this information to explain the behavior of wood
- To develop new or improved properties of wood
- To reduce costs

The material presented in the second section provides:

- An introductory understanding of basic aspects related to the theory of waves propagation in anisotropic solids
- Experimental methods for the acoustical characterization of solid wood and wood-based materials as they relate to the measurement of various parameters in ultrasonic and audible frequency ranges
- Procedures for global elastic characterization of the material as they relate to the determination of elastic constants in ultrasonic and audible frequency range
- Techniques for the local characterization of wood through acoustic microscopy and photoacoustics
- Examination of wood anisotropy using ultrasonic parameters

Part III, "Quality Assessment", is confined to the discussion of wood quality assessment. Wood used for musical instruments is considered to have the most remarkable qualities, with unique acoustic properties. In addition to the wood utilized for musical instruments, which is free of defects, the section also considers "common" wood, in which defects are always present. The ultrasonic velocity method is employed for the detection of natural defects such as knots, and to assess the deterioration or modification of the structure of the wood by parameters such as moisture content, temperature, and biological agents. The use of the acoustic emission technique is described for detecting defects in trees that are induced by cavitation phenomena or by biological agents and for monitoring different technological processes such as curing, drying, strength prediction of large structural elements, and wood machining. High-energy ultrasonic treatment for wood processing is presented in Chapter 11.

An extensive bibliography with subject and author indices concludes the book.

REFERENCES

Baer, E., Hiltner, A., and Morgan, R. J. (1992). Biological and synthetic hierarchical composites. *Phys. Today* 45(10), 60–67.

Bosshard, H. H. (1974). *Holzkunde II. Zur Biologie, Physik und Chemie des Holzes.* Birkhäuser, Basel.

Butterfield, B. G. and Meylan, B. A. (1980). *Three-Dimensional Structure of Wood. An Ultrastructural Approach.* Chapman & Hall, London.

Core, H. A., Côté, W. A., and Day, A. C. (1976). *Wood Structure and Identification.* Syracuse University Press, Syracuse, NY.

Côté, W. A. (Ed.) (1965). *Cellular Ultrastructure of Woody Plants.* Syracuse University Press, Syracuse, NY.

Fengel, D. and Wegener, G. (1989). *Wood: Chemistry, Ultrastructure, Reactions.* Walter de Gruyter, Berlin.

Grosser, D. (1977). *Die Hölzer Mitteleuropas.* Springer-Verlag, Berlin.

Jacquiot, C., Trenard, Y., and Dirol, D. (1973). *Atlas d'Anatomie des Bois des Angiospermes.* Centre Technique du Bois, Paris.

Kollmann, F. F. P. and Côté, W. A. (1968). *Principles of Wood Science and Technology.* Springer-Verlag, Berlin.

Mark, R. E. (1967). *Cell Wall Mechanics of Tracheids.* Yale University Press, New Haven, CT.

Panshin, A. J. and de Zeeuw, C. (1980). *Textbook of Wood Technology,* 4th ed., Vol. 1. McGraw-Hill, New York.

Schniewind, A. P. (Ed.) (1989). *Concise Encyclopedia of Wood and Wood-Based Materials.* Pergamon Press, Oxford.

Schweingruber, F. H. (1978). *Mikroskopische Holzanatomie.* Zürcher AG, Zurich.

Siau, J. (1971). *Flow in Wood.* Syracuse University Press, Syracuse, NY.

Wilson, K. and White, D. J. B. (1986). *The Anatomy of Wood: Its Diversity and Variability.* Stobart & Son, London.

Part I
Environmental Acoustics

Chapter 2

Acoustics of Forests and Acoustic Quality Control of Some Forest Products

2.1. ACOUSTICS OF FORESTS AND FOREST PRODUCTS

Trees and different kinds of vegetation belts (forest floor, grass, lawn, etc.) are of interest to acousticians because of the general belief in the ability of forests or of plantations to attenuate environmental noise and to create an inexpensive and pleasant microclimate. Measurements were reported from field and reverberant rooms with a view to establishing the influence of vegetation on the attenuation of noise (Eyring, 1946; Embleton, 1963; Beranek, 1971; Burns, 1979; Leschnik, 1980; Price et al., 1988; Attenborough, 1982, 1988; Rogers and Lee, 1989; Rogers et al., 1990). This section examines first the field results related to the attenuation of sound by forests, plantations, and shelter vegetation belts, and second, the results of measurements in reverberant rooms.

As noted by Bullen and Fricke (1982), the main phenomena directly related to sound attenuation in forests are

- the interference between direct and ground-reflected sound
- the scattering by tree trunks and branches, the ground, and (possibly) air turbulence
- the absorption by trees (mainly the bark and the foliage), the ground, and the air

Climatic conditions such as wind and temperature (Brown, 1987; Naz et al., 1992) have a minor effect on sound attenuation. The relative humidity of the air has an important influence on traffic noise (Delany, 1974). These measurements are also influenced by the parameters of the equipment: source and receiver height, the time of the response, etc.

Huisman and Attenborough (1991) reported measurements of attenuation of environmental noise, within a range of 50 to 6000 Hz, on a plantation of Austrian pines (*Pinus nigra,* 29 years old, 160-mm diameter, 11.2-m height, and a density of 0.19 trees per square meter) located on a polder (flat ground), in The Netherlands. The litter layer was covered with decaying needles, moss, and herbs. The vertical profile of the trees was divided into three sections: the canopy with living branches, the upper trunk with dead branches, and the stem (Figure 2.1). The corresponding field arrangement is presented

Figure 2.1. Vegetation profile in a pine plantation 15 m wide and 2 m deep. (a) Living canopy; (b) dead branches; (c) stem; (d) litter covered with decaying needles, moss, herbs, and branches. (From Huisman, W. H. T. and Attenborough, K., *J. Acoust. Soc. Am.,* 90(5), 2664, 1991. With permission.)

a) **canopy**   c) **stem**

b) **dead branches**   d) **litter**

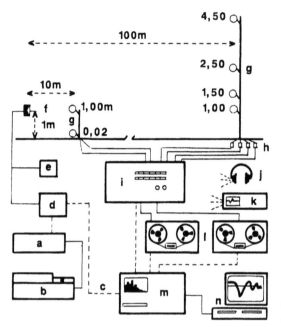

Figure 2.2. Field acoustical arrangement for the measurements of the attenuation of environmental noise. Signal production: (a) sine generator, (b) frequency sweep controller, (c) pink noise generator, (d) amplifier, (e) loudspeaker current monitoring unit, (f) single cone loudspeaker. Signal recording and analysis: (g) microphone and preamplifiers, (h) field preamplifier + 20 dB, (i) interface unit, (j) headphones, (k) oscilloscope, (l) stereo tape recorders, (m) 1/3 oct-band real-time analyzer, (n) computer. (From Huisman, W. H. T. and Attenborough, K., *J. Acoust. Soc. Am.*, 90(5), 2664, 1991. With permission.)

in Figure 2.2. The source was fixed at 1 m above the ground and the microphones were placed on the litter, either at the stem level or the canopy level.

Because the soft forest floor has an important effect on low frequency sound propagation, the pine plantation (when compared to pasture or free field) produced lower emission levels from road traffic noise at all frequencies (Figure 2.3).

Fricke (1984) analyzed the influence that age, density, and the diameter of the plantation trees have on sound attenuation for in-field measurements. Measurements were performed using a very intense noise source ("a gas scare gun" with a peak level of 150 dB at 10 m), with a microphone located in plantations of *Pinus radiata* of different densities and maturities, on very porous ground.

Three characteristic plantations were chosen: two were relatively old, having 1500 and 400 trees per hectare, with individuals of 160 mm in diameter and a height of 13.5 m, and the other was a young plantation, having 1350 trees per hectare that were 110 mm in diameter and 8 m high. Sound attenuation was strongly related to the frequency range of measurements. The older, denser plantation had the highest attenuation for frequencies, >2000 Hz, and the lowest attenuation for frequencies <125 Hz. It seems that at high frequency the scattering and the absorption of the trees play an important role, whereas at low frequencies the effect of the ground is more important in sound attenuation.

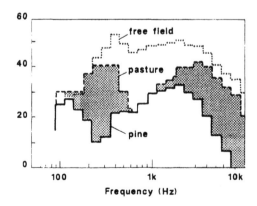

Figure 2.3. Traffic noise spectra for pine plantation, pasture, and free field. (From Huisman, W.H.T. and Attenborough, K., *J. Acoust. Soc. Am.*, 90(5), 2664, 1991. With permission.)

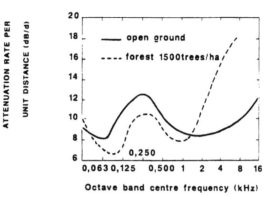

Figure 2.4. Attenuation rate measured in a *Pinus radiata forest* (1500 trees per hectar, 160 mm diameter) and over open ground. Measurements: the source height at 0.6 m; the microphone height at 1.2 m and 90% relative humidity. (From Fricke, F., *J. Sound Vibr.,* 92(1) 149, 1984. With permission.

Table 2.1. Ground characteristic impedance data for sound propagation from a stationary jet engine

Soil type	Measured impedance
Sand (15 cm)	0.270
Sandy soil	0.269
Lawn	0.420
Meadow with grass 8–10 cm high	0.073
Forest floor	0.394
Meadow with low vegetation	0.280

Note: "The normalized characteristic impedance was defined as the ratio of the pressure and normal velocity at the surface of a semi-infinite medium divided by the characteristic impedance (product of density and velocity) for air."

From Attenborough, K., *Appl. Acoust.,* 24, 289, 1988. With permission.

Figure 2.4 graphs the attenuation rates through a forest of approximately 1500 trees per hectare (about 160 mm diameter at the breast height) and those through open ground. Vegetation has an important effect on the sound propagation for all frequencies up to 2 kHz. At high frequencies absorption could be considered the most important phenomena. It was suggested that ground scattering is more important than vegetation scattering because the attenuation in the forest is not very different from that over the open ground at frequencies <1 kHz.

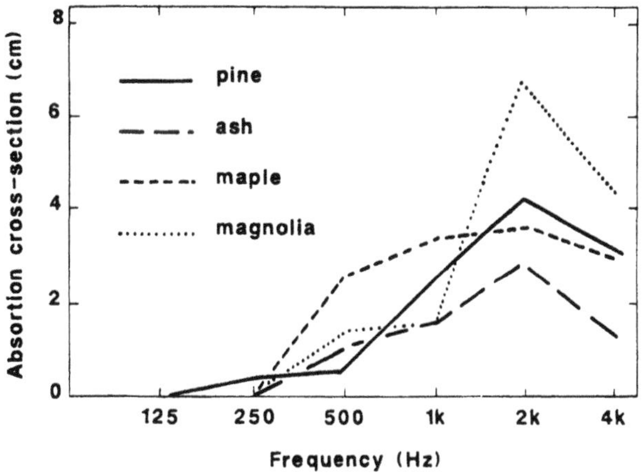

Figure 2.5. Absorption cross-section of various species vs. frequency. Measurements on 2 m high plants in a reverberant room. (From Bullen, S. H. and Fricke, F., *J. Sound Vibr.,* 80(1)11, 1981. With permission.)

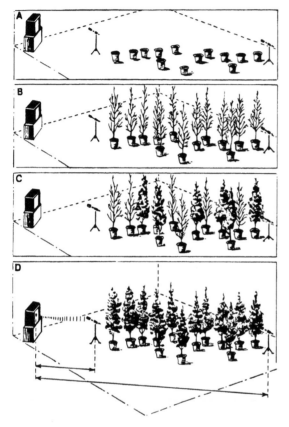

Figure 2.6. Measurements of noise spectra in an anechoic chamber for different experimental configurations. (A) birch trees, all sawn down; (B) birch trees (*Betula* spp.) fully defoliated, stems, branches and twigs present; (C) 46 birch trees of which 23 are defoliated; (D) 46 fully foliated birch trees. (From Martens, M. J. M., *J. Acoust. Soc. Am.*, 67, 66, 1980. With permission.)

Kragh (1979) noted the influence of barrier (shelter) belts of trees and vegetation on the attenuation of the noise produced by passing trains over various, but not very different terrain configurations. The noise was recorded at a section of railroad track (as reference) and at the shelter belt position. Two types of dense vegetation with leaf covering were studied:

In site 1 (400 m long × 50 m wide): 50-year-old birches and elms mixed with 15-year-old beeches and various conifers and bushes

In site 2 (1200 m long × 25 m wide): 20-year-old oaks mixed with hornbeams, poplar, and silver fir, 10-year-old larch and bushes

The difference between attenuation measurements near the track and behind a 50-m wide shelter belt was 6 to 7 dB for the first site and 8 to 9 dB behind a 25-m wide shelter belt for the second site. The extra attenuation obtained in the second studied configuration was due to a combined effect of vegetation and minor terrain variations. It was impossible to measure the attenuation caused solely by the shelter belt itself, but Kragh stated that similar belts of trees and bushes could provide a means of practical noise reduction.

The nature, shape, and length of the ground have important effects on the attenuation spectrum of environmental noise. Theoretical models for the prediction of sound propagation near the absorbing ground and the relationship of the acoustical properties to various physical parameters of the ground surface were reviewed by Attenborough (1988). Data were taken under neutral conditions, when the temperature difference between monitoring points at 1.2 and 12.2 m above the ground was <0.3°C and the vector wind speed was <1.52 m/s. Several types of ground are described by their characteristic impedance in Table 2.1. Lawn and forest floor exhibited the highest coefficients of characteristic impedance data.

In order to avoid the ground effect and the influence of the meteorological conditions on attenuation, measurements in reverberation room were performed. Only the scattering effect of the vegetation is

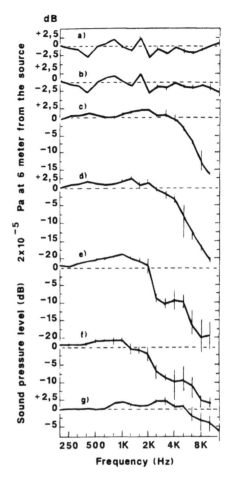

Figure 2.7. Noise spectra in an anechoic chamber measured for the experimental configuration from Figure 2.6. The reference spectrum considered at 0 dB sound pressure level indicates the noise field generated by the source in the empty anechoic chamber, measured on the microphone at 6 m. The vertical bars indicate the spreading of three measurements. (a) Birch trees, all sawn down; (b) birch trees (*Betula* spp.) fully defoliated, stems, branches and twigs present; (c) 46 birch trees of which 23 are defoliated; (d) 46 fully foliated birch trees; (e) 25 fully foliated hazeltrees (*Corylus avellana*); (f) 26 fully foliated tropical plants; (g) 12 fully foliated privet (*Ligustrum vulgare*). (From Martens, M. J. M., *J. Acoust. Soc. Am.*, 67, 66, 1980. With permission.)

analyzed in such cases. The sound absorption of plants (pine, maple, ash, and magnolia), 2 m high, in tubs filled with earth was measured (Figure 2.5). The foliage seems to play an important role near the 2-kHz frequency (Bullen and Fricke, 1982).

Martens (1980) investigated sound transmission through vegetational foliage in an anechoic chamber. A wide range of species originating from temperate and tropical zones were studied. Species selection was based on the biomass quantity and on the dimensions of leaves (length × width from 20 × 5 mm to 750 × 350 mm). Measurements were performed at 21°C and 55 to 62% relative humidity. The distance between the noise source and microphone was 6 m (Figure 2.6). The noise spectrum was recorded in foliage, with and without leaves, planted in earthenware flower pots. The spectra for the experimental conditions are presented in Figure 2.7 and the results are summarized in Table 2.2. Martens concluded that the foliage acts as a filter for the noise. In midfrequency range (<2 kHz) the noise is slightly amplified and in high frequencies the noise is attenuated. The sound pressure level is decreased by the plant foliage. The traffic noise spectrum could be changed in its pitch, and the tree foliage could reduce the environmental noise produced by the traffic.

2.2. ULTRASOUND FOR THE GERMINABILITY DETECTION OF ACORNS

A great deal of interest has arisen recently in developing nondestructive techniques to characterize the germinability of forest seeds. In Scandinavian countries and elsewhere in the world, various aspects of the application of X-ray analysis to coniferous tree seeds have been considered. Specific radiographic techniques, i.e., tomography, fluoroscopy, and X-ray spectroscopy, are being developed and refined

Table 2.2. Sound pressure level (SPL) measured in anechoic chamber for different species

No. plants	Biomass (kg)	Total SPL (dB) at 6 m		Attenuation of SPL	Frequency drop (kHz)
		Empty room	Plants foliated		
		Birch trees foliated			
46	8.4	94	92.8	−0.2	2–4
		Birch trees defoliated			
46	5.9	94	93.0	0.0	8–10
		Tropical plants			
26	11.5	95	92.0	−2.7	1–1.2
		Hazel trees			
25	3.6	96	94.1	−0.9	2–2.5
		Privets			
12	2.5	96	96.0	+1.2	5–6.4

From Martens, M.J.M., *J. Acoust. Soc. Am.*, 67, 66, 1980. With permission.

constantly to meet specific problems (Gustafsson and Simak, 1956; Paci and Perulli, 1983; Vozzo and Linebaugh, 1974).

It has been proved that the X-ray technique is useful in the evaluation of seed maturity and quality, moisture content, seed ripeness, diagnosis of old and dead seeds, physiological changes, insect infestation, and mechanical damage. The interpretation of data using this technique relies on an adequate understanding of the limitations involved. At present it is impossible to distinguish between nonviable seeds or seeds having a conspicuous delay in germination by means of X-ray photography alone. Seed material with well-developed embryos and endosperm and normal X-ray absorption may be dead due to unsuitable storage, heat treatment of some kind, or aging without this fact being discovered on the photographic plate.

Other serious limitations of the X-ray method arise from data collection as well as from the difficulty of adapting the method to a mechanized test system for all seeds under consideration. Because of the growing interest in the ultrasonic characterization of biological tissue, attention was focused on the measurement of ultrasonic propagation parameters on acorns (Bucur and Muller, 1988). Two aspects were explored:

1. The determination of the relationships between the morphological characteristics of acorns described in Figure 2.8 and the physical parameters (ultrasonic velocity, impedance, stiffness, density) presented in Table 2.3. The statistical analysis of data (Figure 2.9) emphasizes the influence of density on the

Table 2.3. Morphological and physical characteristics of acorns

	Germinated acorns		Non-viable acorns	
	Average	Coefficient of variation (%)	Average	Coefficient of variation (%)
Morphological parameters				
Axis D(mm)	25.10	15.18	23.50	20.50
Axis d(mm)	15.92	19.27	14.86	14.72
D/d	1.59	13.66	1.58	16.21
Physical parameters				
Density (kg/m^3)	1085	19.86	858	37.59
Velocity (m/s)				
on D, V_D	1184	26.72	1139	29.10
on d, V_d	800	23.52	700	26.46

From Bucur, V. and Muller, C., *Ultrasonics*, 26, 224, 1988. With permission

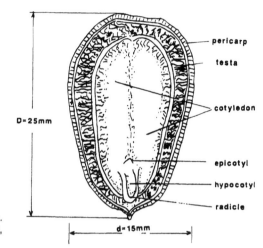

Figure 2.8. Longitudinal section of acorn. (From Bucur V. and Muller, C., *Ultrasonics* 26, 224, 1988. With permission.)

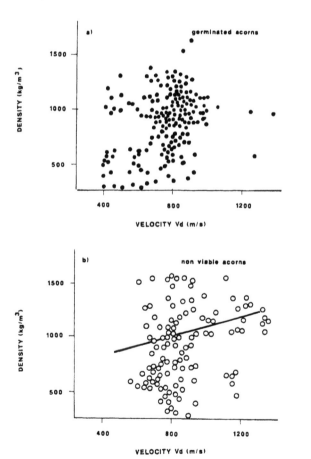

Figure 2.9. Experimental relationship between physical and morphological parameters of acorns. Relationships between density and velocity for (a) germinated acorns (r = N.S., or nonsignificant) and (b) nonviable acorns (r = 0.319***).

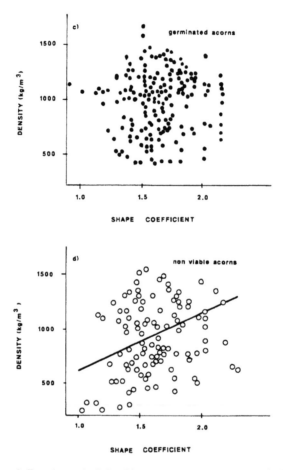

Figure 2.9. (*Continued*) Experimental relationship between physical and morphological parameters of acorns. Relationships between density and velocity for (c) germinated acorns (*r* = N.S.) and (d) nonviable acorns (*r* = 0.346***). (From Bucur, V. and Muller, C., *Utrasonics, 26,* 224, 1988. With permission.)

ultrasonic velocity and on the morphological coefficient of acorn shape. In contrast to the strong correlation found for nonviable acorns no significant correlation coefficient was obtained for the germinated acorns. The lack of correlation for germinated acorns reveals a distinction between the two populations and emphasizes the difference in character resulting from the germinative process. A possible explanation for this difference may be the low variability in the germinated acorn population.

2. The assessment of the results for practical acorn testing purposes (using principal components analysis), to observe whether the parameters under consideration are sufficiently sensitive in their measurement capability so that they may form the basis of a quality control system for industrial production processes.

The swarm of observation points and vectors corresponding to density and velocity are presented in Figure 2.10. Germinated acorns are located mostly around their proper inertial center in the first quadrant, whereas the nonviable acorns are structured around another center in the third quadrant. For acorns of the same size and weight, velocity may be a discriminant parameter able to detect delamination between successive layers of the acorn envelope. Often delamination can take place between the pericarp and the testa. This phenomenon arises from moisture content loss, fungus attacks, or mechanical injuries. Delamination can be considered to be an impedance discontinuity in the propagation medium of the ultrasonic waves of acorns, therefore, delaminated acorns should be characterized by lower velocities.

It is interesting to speculate that for practical acorn testing purposes it could be useful to measure the weight and velocity for each specimen, to calculate the coordinates in the principal components

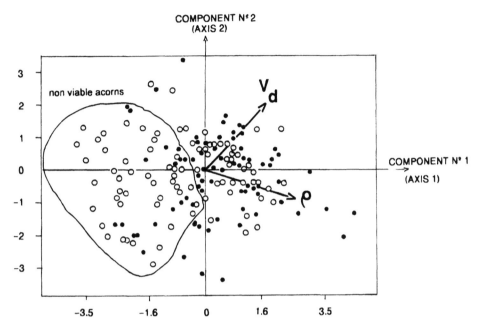

Figure 2.10. The swarm of experimental points in the principal components plane. The germinated acorns (x) are located mostly around their inertial center in the first quadrant. The nonviable acorns (o) are clustered around another center in the third quadrant. (From Buchur V. and Muller, C., *Ultrasonics*, 26, 224, 1988. With permission.

plane, and to show each point on the reference graph using the procedure outlined previously. This could be the basis of a rather sophisticated method of selecting germinative acorns. The interpretation of any data using this technique relies on an adequate understanding of the limitations involved. Theses arise from data collection and computation, the characteristics of the transducers used, the specimen, and from the real acoustic situation, and require the development of a theory or a measurement method to be evaluated.

Ultrasonic velocity measurements complemented by other propagation parameters of the ultrasonic waves (absorption, attenuation, etc.) could be applied to other forest seeds, and may provide a useful characterization technique for the evaluation of germinability and the detection of defective or poorly developed acorns (Muller, 1980, 1986). Ultrasonic measurements could aid in some fundamental studies of germinative processes for the understanding of the genetic behavior of agricultural and forest seeds.

REFERENCES

Attenborough, K. (1982). Predicted ground effect for highway noise. *J. Sound. Vibr.* 81(3), 413–424.
Attenborough, K. (1988). Review of ground effects on outdoor sound propagation from continuous broadband sources. *Appl. Acoust.* 24, 289–319.
Beranek, L. L. (1971). *Noise and Vibration Control.* McGraw-Hill, New York.
Brown, E. H. (1987). Atmospheric acoustics, in *Encyclopedia of Physical Science and Technology,* Vol. 2, Academic Press, New York, pp. 147–163.
Bucur, V. and Muller, C. (1988). Non-destructive approach for analysing the germinability of acorns. *Ultrasonics.* 26, 224–228.
Bullen, R. and Fricke, F. (1981). Sound propagation through vegetation. *J. Sound. Vibr.* 80(1), 11–23.
Burns, S. H. (1979). The absorption of sound by pine trees. *J. Acoust. Soc. Am.* 65, 658–661.
Delany, M. E. (1974). Traffic noise, in *Acoustics and Vibration Progress,* Stephens, R. W. B. and Leventhall, H. G., Eds., Chapman & Hall, London, pp. 1–48.

Embleton, T. F. N. (1963). Sound propagation in homogeneous deciduous evergreen woods. *J. Acoust. Soc. Am.* 35, 137–146.

Eyring, C. F. (1946). Jungle acoustics. *J. Acoust. Soc. Am.* 18, 257–270.

Fricke, F. (1984). Sound attenuation in forests. *J. Sound. Vibr.* 92(1), 149–158.

Gustafsson, A. and Simak, M. (1956). X-ray diagnostics and seed quality in forestry, in Proc. 12th IUFRO Congr. Vol. 1, pp. 398–413.

Huisman, W. H. T. and Attenborough, K. (1991). Reverberation and attenuation in pine forest. *J. Acoust. Soc. Am.* 90(5), 2664–2677.

Kragh, J. (1979). Pilot study on railway noise attenuation by belt of trees. *J. Sound. Vibr.* 66(3), 407–415.

Leschnik, W. (1980). Sound propagation in urban and forest areas. *Acustica,* 44, 14–22.

Martens, M. J. M. (1980). Foliage as a low-pass filter: experiments with model forests in an anechoic chamber. *J. Acoust. Soc. Am.* 67, 66–72.

Muller, C. (1980). Long term storage of fir seeds and its influence on the behaviour of seed in the nursery. *Seed Sci. Technol.* 8, 103–118.

Muller, C. (1986). Le point sur la conservation des semences forestières et la levée de dormance. *Rev. Forest. Fr.* 38(3), 200–204.

Naz, P., Parmentier, G., and Canard-Caruana, S. (1992). Propagation acoustique à grande distance dans les basses couches de l'atmosphère. *J. Phys.* 4(2), C1-745–C1-748.

Paci, M. and Perulli, D. (1983). Imagini preliminari sull'impiego del methodo radiografico per valutare la vitalita del seme de *Pinus nigra. Ital. Forest Mont.* 38(6), 317–330.

Price, M. A., Attenborough, K., and Heap, N. W. (1988). Sound attenuation through trees: measurements and models. *J. Acoust. Soc. Am.* 84, 1836–1844.

Rogers, J. C., Shen, W., and Lee, S. M. (1990). Acoustic scattering: measurements on model trees and their extension to natural forests. *Proc. I.O.A.* 12(1), 827–833.

Rogers, J. C. and Lee, S. M. (1989). Sound scattering from forests: considerations and initial measurements. *13th Int. Conf. Acoustics, Belgrade,* 3(7), 277–281.

Vozzo, J. A. and Linebaugh, S. (1974). Tomography in seed research. *Proc. Assoc. Off. Seed Anal.* 64, 94–96.

Chapter 3

Wood and Wood-Based Materials in Architectural Acoustics

Owing to the interaction between acoustics and architecture, it may be helpful to survey the field of human reaction to sounds before considering the capacity of wood and wood-based materials as acoustic insulators. The threshold of human hearing is limited at 0 dB, the equivalent of 2×10^{-3} μb of pressure. The limit of human pain induced by noise is 120 dB or 2×10^2 μb. Thus, the range of sound intensity tolerated by humans is very large.

Buildings constructed with satisfactory acoustics are characterized by the control of the transmission loss of sound through the construction elements, the absorption of sound within a space, and the separation of noise sources from quiet spaces. Solid wood and some wood-based composites can be considered acoustic materials because of their ability to absorb a significant amount of incident sound in order to reduce the sound pressure level or the reverberation time in a room. Wood materials are applied to walls and ceiling surfaces or to floor platforms and are occasionally suspended within the room depending on its performance requirements,—for speech and music listening, offices, industrial buildings, homes, etc. It was demonstrated by Cremer and Muller (1982) that "it is possible to accomplish some predetermined acoustical design objectives by selecting the enclosure surfaces to absorb, reflect or transmit the incident wave. How well this objective is accomplished will depend upon the designer's knowledge and skill in the selection and use of materials." Similar statements were advanced by Beranek (1960, 1962).

Sound absorption and sound reflection efficiency over the audible spectrum are strongly related to the internal structure of the material, the surface treatment, the type of mounting, the geometry, etc. For example, plywood or particleboard provides sound absorption in the lower-frequency region of the audible spectrum (<500 Hz) and porous artificial materials are remarkably efficient absorbants in mid- and high frequencies (2000 to 4000 Hz) (Beranek, 1960).

This chapter attempts to cover the following points: the influence of the structural organization of wood on its sound absorption characteristics, wood-based materials as sound insulators, and some aspects related to the utilization of wood and wood-based composites in room acoustics.

3.1. INFLUENCE OF ANATOMICAL STRUCTURE OF WOOD ON SOUND ABSORPTION CHARACTERISTICS

The acoustic efficiency of walls constructed with wood depends on the methods of installation and on the basic properties of the material. A deeper understanding of the very complex phenomena related to the sound insulation of walls must consider the sound absorption of different wood species. As cited by Kollamann and Côté (1968), data on sound absorption of wood (fir, 20-mm thickness) as a function of frequency were first published by Sabine in 1927. The absorption coefficient was estimated at 10%. (The coefficient of sound absorption for an open window is considered to be 1 or 100%.) The values of sound absorption coefficients are affected by the experimental configuration (thickness of the specimen, rear space from the specimen to the rigid wall of the Kundt tube, species, etc.) and by the frequency range. Studies on the relationship between the sound absorption coefficient and the anatomic structure of different species were published by Watanabe et al. (1967) and by Hayashi (1984).

Standing waves were used to determine the absorption coefficients and acoustic impedance of different tree species (Japanese cedar, *Cryptomeria japonica*; Saghalin fir, *Abies sachalinensis*; maple, *Acer* spp.; and willow, *Salix* spp.). Measurements were performed on disks of 31 to 101 mm in diameter and 2 to 11 mm thick, in a frequency range of 90 to 6500 Hz. The specimens were cut in principal axes and at different angles α vs. the principal directions and were submitted to the sonic field. Figure 3.1 shows the variation of absorption coefficient vs. frequency on several specimens of maple, the first in *LR* plane at α = 0°, the second in the same plane at α = 90°, and three others for intermediate angles: α = 23°, α = 45°, and α = 67°. Measurements of the specimen at α = 0° are less influenced by the frequency range than are the measurements of the specimen at α = 90°. For the latter specimen the maximum of the absorption coefficient was measured at 1.5 kHz. This behavior is explained by the motion of the air contained in the cavities of wood cells. Watanabe et al. (1967) noted that when

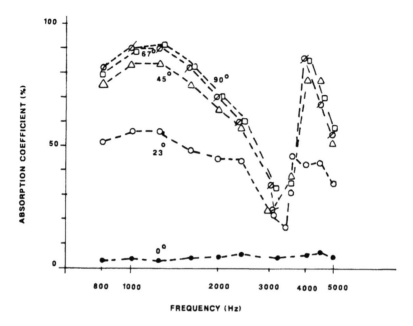

Figure 3.1. Absorption coefficient (%) as a function of frequency for maple (*Acer mono* Maxim). (From Watanabe et al., *Mokuzai Gakkaishi,* 13(5), 177, 1967. With permission.)

thin specimens (at a maximum of 10 mm) are exposed to the incident sonic energy in the anisotropic planes of wood *LR* or *LT*, the absorption coefficient is small and quite constant for all frequencies for all species analyzed.

3.2. WOOD MATERIALS AS ACOUSTIC INSULATORS

Wood is a basic building material. A wide range of structural applications of wooden members in foundations, light frame construction, beams, columns, and bridges, was cited by Freas (1989). Increased attention was given recently to the utilization of solid wood and wood-based composites as acoustical insulators in floors, ceilings, and walls.

Great variability exists in the requirements for noise control in various types of buildings (industrial buildings, homes, etc.). The discussion in this section is confined houses, the reduction of indoor and outdoor noise sources of which is of major concern. Outdoor noise sources are principally produced by vehicular and pedestrian traffic and commercial or industrial activities, among others. Indoor noise sources include automatic home appliances; heating, air conditioning, and sanitary systems; entertainment devices; musical instruments; conversations, floor-impact noise (produced by walking, falling objects, etc.), the activities of children and adults, etc. A schematic representation of the transmission of sound between two rooms is seen in Figure 3.2a. Three ways of sound transmission are possible: the sound is transmitted through the adjacent wall (1), the ceiling (2), and or the floor (3). Figure 3.2b shows the relationships between the incident acoustic energy (Ee), the absorbed energy by the wall (Ea), the reflected energy (Er), and the transmitted energy (Ed). If the acoustic insulation is good enough, the transmitted energy through the wall is very small.

The acoustic capacities of a wall separating two rooms are expressed by two factors, transmission loss and noise reduction. The transmission loss factor (R) is defined (Braune, 1960) as the log of the ratio of the acoustic incident energy to the acoustic energy transmitted through the wall:

$$R = 10 \log Ee/Ed \ [\text{dB}] \tag{3.1}$$

The noise reduction factor is a ratio of pressures, the difference in sound pressure level on the two sides of the wall and the incident sound pressure, and can be calculated as:

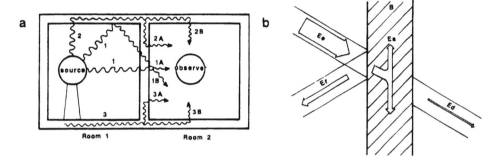

Figure 3.2. Transmission of sound through panels. (a) Schematic representation of sound propagation between adjoining rooms. (Beranek, 1962.) (b) Schematic representation of the incident acoustic energy (Ee) on a wall and the corresponding transmitted (Ed), absorbed (Ea), and reflected (Er) energies. (Braune, 1960.)

$$A = (Ee - Er)/Ee \times 100 \; [\%] \qquad (3.2)$$

For the single wall, the sound transmission loss is governed by its weight (Braune, 1960).

As an example we cite the measurements of a wood panel, 2 cm thick, and a panel of glass fibers, also 2 cm thick (Braune, 1960). The transmission loss of the wood panel is 22 dB and the acoustical absorption is 3%, whereas the corresponding parameters for the glass fiber panel are 3 dB and 65%. This behavior in two different walls of the same thickness is roughly determined by the mass density of the constitutive materials.

The transmission loss of airborne sound through a single panel of wood-based material vs. frequency is given in Figure 3.3, in which three types of materials are analyzed: plywood, particleboard, and hardboard. The measured values of the transmission loss factor are compared to the values deduced from the law of mass incidence. A reduction of 20 to 30% at low to middle frequency was observed, probably produced by the mechanical impedance of the wall and by the energy dissipation properties of the constitutive materials.

It is useful to emphasize that for the acoustic insulation of walls the relevant constants are the surface mass or the mass density, the critical frequency*, and the damping factor.

Cremer et al. (1973) and Craik and Barry (1992) reported results for specimens that were tested using the impedance head method in order to determine for each material the loss factor, mechanical impedance, and sound velocity (Table 3.1). Wood is characterized by low density, high velocity, and relatively low loss factor when compared to other materials having the same density.

The critical frequency for different building materials is given in Table 3.2. Note that for wood panels 2 to 3 cm thick, the critical frequency is in the low range (<1000 Hz).

Another important parameter for acoustic insulators is the noise reduction coefficient, defined as the average of the sound absorption coefficients at 250, 500, 1000, 2000, and 4000 Hz. Some example data are given in Table 3.3. The highest sound absorption coefficient for all frequencies was observed for a floor covered with wool carpeting and the smallest coefficient was measured on a brick wall.

Recent studies (Takahashi et al., 1981 and 1987a, b) reported on the control of sounds traveling from one room to another through the floor to ceiling path. Figure 3.4 reproduces the most common impact noise sources in a wooden house (children running and jumping, adults walking, an object dropped) and the corresponding measuring systems. To reproduce the impact noise several standards (ISO, JIS, SIS, NF, etc.) proposed the utilization of a tapping machine and the dropping of a tire. Takahashi et al. (1987b) noted the similarities between the profile of frequency spectra for the floor impact noise produced by the tapping machine and by walking in women's high-heeled shoes and also

---

*Critical frequency is defined as the frequency measured when the velocity of the bending wave in the wall is equal to the velocity of the oblique incident wave. The frequency is deduced from the relation: $fc = 2.05 \times 10^4 \times 1/d(ro/E)^{1/2}$, where ro is the density of the wall, d is the thickness of the wall, and E is the Young's modulus of the wall material (Braune, 1960).

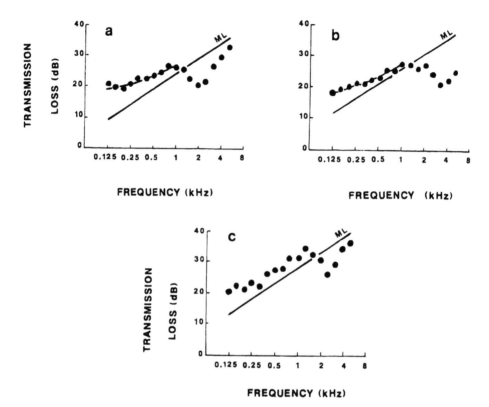

Figure 3.3. Sound transmission loss (dB) vs. frequency in different materials: (a) plywood; (b) particle board; (c) hard board. (From Suzuki et al., *Mokuzai Gakkaishi,* 32(3), 155, 1986. With permission.)

Table 3.1. Mechanical properties of building materials under standard conditions (Cremer et al., 1973).

Material	Density (10³ kg/m³)	Young's modulus (10⁸ N/m²)	Sound Velocity (m/s)	Loss factor
Asphalt	1.8–2.3	77	1900	.380
Brick	1.9–2.2	160	3000	0.020
Cork	0.12–0.25	25	4300	0.170
Dense concrete	2.3	260	3400	0.008
Dry sand	1.5	—	170	0.120
Glass	2.5	60	4900	0.002
Gypsum board	1.2	70	2400	0.006
Light concrete	1.3	38	1700	0.015
Oak, solid wood	0.7–1.0	80	3500	0.010
Plywood	0.6	54	3000	0.013
Porous concrete	0.6	20	1700	0.010

by the noise produced by dropping a tire and by a barefoot male walking. The authors stated that the profiles of the impact noise frequency spectra for tapping machine impact and tire drop impact were very different, but that their radiation factor spectra are almost the same.

Nakao et al. (1989) proposed some improvements in the reduction of floor impact sound levels in the receiving room. The floor-ceiling-wall system was remodeled using high rigidity panels, and a reduction of 10 dB occurred in the tire dropping test. Measured and calculated values of floor impact sound levels for different floor systems are given in Table 3.4.

Table 3.2. Critical frequencies for different building materials as a function of thickness (Braune, 1960).

	Thickness of wall				Density (kg/m³)
	5 cm	10 cm	15 cm	20 cm	
Brick wall	470	235	160	120	1,500
Concrete wall	—	165	110	82	2,400

Panels in Wood and Wood-Based Materials

	Thickness of panels				Density (kg/m³)
	2 mm	5 mm	2 cm	3 cm	
Fir	—	—	500	330	500
Beech	—	2,400	600	400	750
Plywood	7,000	2,800	700	470	550
Particleboard	—	4,300	1,090	725	660
PVC	20,200	8,100	2,020	1,350	1,400

Panels in Metal and Glass

	Thickness of panel				Density (kg/m³)
	0.5 mm	3 mm	5 mm	10 mm	
Iron	26,000	4,350	2,600	1,300	7,600
Glass	24,000	4,000	2,400	1,200	2,500
Lead	106,000	17,700	10,600	5,300	11,300

Table 3.3. Average sound absorption coefficients of some building materials

Material	Thickness (cm)	Coefficient at frequency (Hz)					
		125	250	500	1000	2000	4000
Brick wall	46	0.02	0.02	0.03	0.04	0.05	0.05
Plaster/wood wall	2	0.04	0.30	0.20	0.15	0.10	0.10
Wood, polished	5	0.1	—	0.05	—	0.04	0.04
Wood platform	—	0.40	0.30	0.20	0.17	0.15	0.10
Glass	—	0.04	0.04	0.03	0.03	0.02	0.02
Wood floor, pitch pine		0.05	0.03	0.06	0.09	0.10	0.22
Wool carpets	1.5	0.20	0.25	0.35	0.40	0.50	0.75

From Beranek, L., *Acoustics*, 2nd. ed., American Institute of Physics, Cambridge, MA, 1986. With permission.

Sound insulation of wood joist floors and procedures for the reduction of noise nuisance from footsteps were also analyzed in many European countries (Fothergill and Royle, 1991; Fothergill and Savage, 1987; Centre Scientifique et Technique du Bâtiment, 1989; Utley and Cappelen, 1978; Utley, 1979; Compien, 1977) and in the U.S. (Grantham and Heebink, 1971 and 1973; Dickerhoff and Lawrence, 1971).

A considerable number of experiments have been conducted on timber platform floating floors (Fothergill and Royle, 1991). Laboratory and field measurements were related to the insulation capacities of the materials used for the resilient layer of the floor, to the absorbing materials between the joists, and to the sound transmission up and down through the supporting walls of the floor. A typical section through a timber platform floating floor is shown in Figure 3.5. Typical insulation values, determined in agreement with ISO 717-1982 (BS 5821: 1984), were 54 dB against airborne and 61 dB against impact sounds.

Windows are important for the reduction of outdoor noise entering a house. The attenuation of windows can be improved using thick glass and the proper joint (nearly hermetically sealed). Some architectural details are given by Braune (1960).

3.3. WOOD AND THE ACOUSTICS OF CONCERT HALLS

This section discusses some aspects related to the utilization of wood material as an insulator in the acoustics of concert halls. Room acoustics are greatly improved when wood material is used, although we recognize that the acoustic quality of a room is subjective. The design and construction of a concert hall can be defined as an art in the classic sense of the word. The scientific knowledge in this field is advanced (Muller, 1986). The sound field in a room is very complicated and is not open to precise mathematical treatment. The large number of vibrational modes of the sound field as well as the total possible data require the introduction of average functions and the statistical treatment of data.

Two points are relevant to the fundamental aspects of room acoustics: the generation and propagation of the sound in an enclosure and the physiological and psychological factors that provide clues about the good or poor acoustics of concert halls, opera houses, lecture rooms, churches, restaurants, offices,

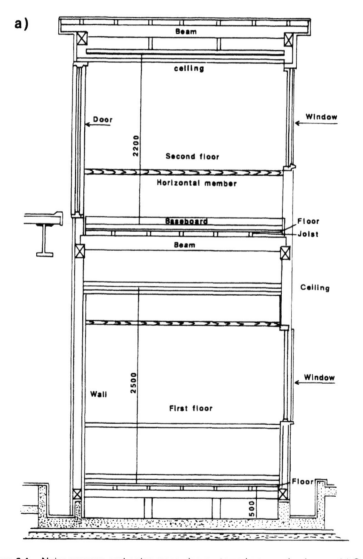

Figure 3.4. Noise sources and noise measuring systems in a wooden house. (a) Section.

b)

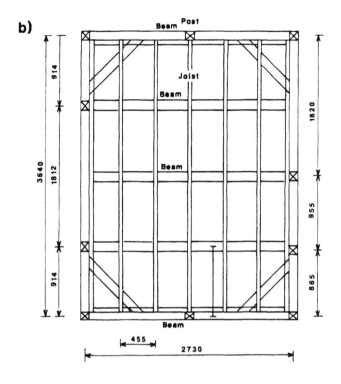

c)

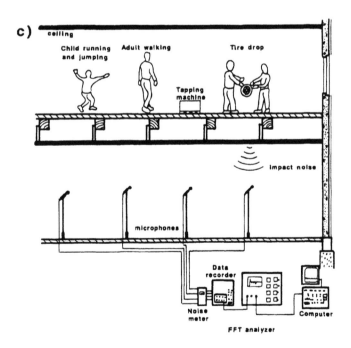

Figure 3.4. (*Continued*) Noise sources and noise measuring systems in a wooden house. (b) plan view of floor system on the second floor of the experimental wooden structure. (c) noise sources and noise measuring systems. (From Takahashi et al., *Mokuzai Gakkaishi,* 33(12), 950, 1987a. With permission.)

Table 3.4. Measured and calculated values of floor-impact sound levels for different floor systems

Floor system		Sound level (dB) at	
		63 Hz	125 Hz
Standard floor without ceiling	Measured	101.0	92.0
	Calculated	103.4	92.5
Standard floor + nonsuspended ceiling	Measured	89.8	76.7
	Calculated	91.5	74.6
High rigidity floor + nonsuspended ceiling	Measured	88.0	74.8
	Calculated	89.3	72.5

From Nakas et al., *Mokuzai Gakkaishi*, 35(2), 85, 1989. With permission.

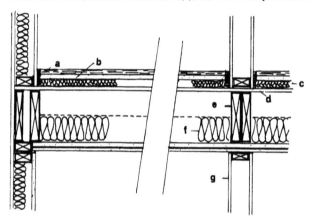

Figure 3.5. Section through a timber platform floating floor. (a) 18 mm tongue and groove chipboard; (b) 19 mm plasterboard; (c) 25 mm mineral wool 64 kg/m³; (d) 12 mm tongue and grooved plywood; (e) floor joist; (f) 10 mm mineral wool 12 kg/m³; (g) 12.5 mm plasterboard. (From Fothergill, L. C. and Royle, P., *Appl. Acoust.*, 33, 249–261, 1991. By permission of the Controller of HMSO: British Crown copyright).

and the like. The first aspect is related to the physical phenomenon of wave propagation and sound field description. The second aspect is related to a subjective perception of sound by the listener(s).

A newly built room has several requirements:

1. An exact definition of the practical purposes of the room (drama, musical events, sports events, etc.), which must be related to the values of sound field parameters such as reverberation time, local or directional distribution of sound, and limitation and peculiarities of subjective listening abilities.
2. The architectural plan or design of the hall: the shape and dimensions, position of the sound sources, stage enclosure, arrangement of the audience and seats, walls, ceiling, and floor. The position of the latter elements are essential to the frequency spectrum of the reflected sound remaining similar to that of the direct sound.
3. The materials used for the construction. Wooden-plated panels in front of an air cushion are used for absorbing low frequencies. Wooden plates act as resonators, wherein the basic resonance frequency is related to the mass per square unit and to the stiffness of the air cushion behind. Wooden linings lead to a bright sound because of low frequency absorption. Other systems such as Helmholtz resonators or thin gypsum plates can be used for this purpose, with some success. The absorption of high frequencies in normal auditoria is caused by the audience (the effects of clothing fabrics, etc.) and the volume of the air. It is interesting to note Beranek's (1988) statement, "The absorbing power of a seated audience, orchestra and chorus in a large hall for music increases in proportion to the floor area occupied, nearly independent of the number of seated persons in those areas." The acoustic quality of halls is strongly dependent on the initial time delay gap (<20 ms) and is defined as "the difference in the time of arrival at a listener's ear of the first of the reflected waves and the direct sound wave".
4. The subjective preference of the sound in a concert hall is related to the psychoacoustic parameters as defined by Ando (1988): the preferred initial time delay gap, the preferred listening level, the preferred reverberation time subsequent to the arrival of early reflections, the magnitude of the preferred measured interaural cross-correlation. "As different music is performed in a hall, the total preference value changes according to the auto-correlation function of the music."

It is difficult to cite only one hall design as having the most excellent acoustics. Beranek (1992) points to four basic concert hall design types: rectangular (Grosser Musikvereinsaal, Vienna), large, fan-shaped (Koussevitzky Music Shed, at Tanglewood Lenox, MA, U.S.), segmented nonsymmetrical audience arrangement of "vineyard" type (Berlin's Philharmonie Hall), and oval shaped with extensive use of multiple upper sidewall reflectors (Christchurch Town Hall, New Zealand).

The oldest is the rectangular hall (also known as the "shoe box"). Two examples of halls with very good acoustics built in the last century are Grosser Musikvereinsaal, Vienna and Symphony Hall in Boston. Both are characterized by a rectangular shape and by sound-diffusing surfaces on interior ceiling and walls.

The first large concert hall built after World War II in Europe (1951) was the Royal Festival Hall, London, followed by the prestigious Beethoven Hall,Bonn, 1959; Berliner Philarmonie, 1964, Barbican Concert Hall, London, 1982, Gasteig Philarmonie, Munich, 1985, Opéra Bastille, Paris, 1989, and Teatro Carlo Felice, Genova, 1991. etc.). Around the same time in North America construction began on Avery Fisher Hall and The Metropolitan Opera, New York, Joseph Meyerhoff Symphony Hall, Baltimore, The Orange Country Performing Arts Center, Southern California, Meyerson Symphony Center in Dallas, Tanglewood Music Shed, Lenox, MA, and Roy Thompson Hall, Toronto, etc.), in South America on the Municipal Theatre, Lima, Peru, in Australia on the Sydney Opera House, and in Japan on Bunka Kaikan Hall, Tokyo.

Several physical parameters can assist in judging the quality of concert hall acoustics:

- Whether a good connection exists between the orchestra, the musicians, and the listeners
- Sufficient reverberation time to give tonality to the music
- Reasonable balance between strings, woodwinds, and percussion in the orchestra

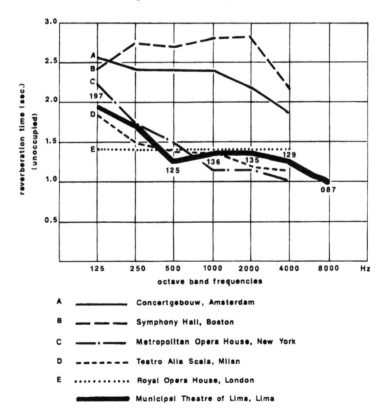

Figure 3.6. Reverbaration time vs. frequency in several concert halls. (Jimenez-D. Carlos, *Appl. Acoust.*, 35, 153, 1992 with permission.)

• Sufficiently loud sound, without distortion, echoes, or undesirable noise
• Musicians are able to hear themselves and the other players

The acoustic response of a concert hall can be deduced from the corresponding "signature". This occurs when excitation is produced via a loud, impulsive sound such as pistol shots, or via an omnidirectional dodecaphonic loudspeaker system, giving 1 ms pulses (Muller, 1986).

ISO 3382-1975(E), "Measurement of Reverberation Time in Auditoria", allows comparison of the acoustic quality of different halls expressed in reverberation time. Figure 3.6 shows the reverberation time vs. frequency in several halls that were unoccupied at the time of testing. At 500 Hz the reverberation time varies between 1.25 and 2.6 s. Timbre-related effects, gauged on the spectra of sound levels in three concert halls (Amsterdam, Boston, and Vienna), were studied by Bradley (1991).

In all concert halls wood was used as acoustic material in walls, ceiling, chairs, and flooring (GADE, 1989), as seen in Table 3.5. Table 3.6 shows dimensions of various concert halls.

Figure 3.7 presents the complex structure of a coffered ceiling. Maximum absorption was obtained at 200 Hz. This maximum was due probably to the mass of the plates and the stiffness of the air cushions; it also may be due to the resonance frequencies of the wooden elements. This type of wooden structure shows the "favorable influence on acoustics of wooden panelling, wooden stages and vibrating wooden floors. The corresponding absorption at low frequencies is a welcome complement to the medium- and high-frequency absorption of the audience." (Cremer and Muller, 1982). A more recently built concert hall is the Arsenal in Metz, France, built in 1987–1989, seating 1548 (Table 3.7). The main acoustical material is the wood used for the ceiling, floor, and walls (Commins, 1992). This hall is well considered by both the musicians and the listeners for classical symphonic music and recitals.

Typical diffusing elements are shown in Figure 3.8. Reverberation time over a wide frequency range is given in Table 3.7. Differences in reverberation time between unoccupied and and occupied halls are more important in frequencies ranging <2000 Hz.

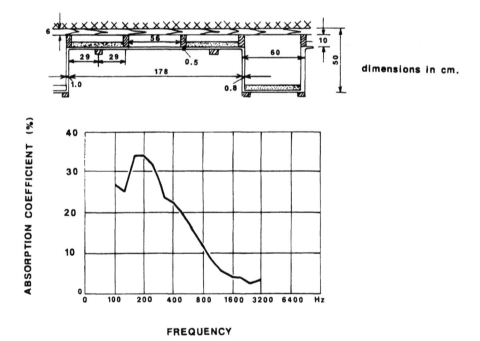

Figure 3.7. Absorption coefficient for a coffered ceiling, measured in a reverberant room. (From Cremer L. and Muller, H. A., *Applied Science*, London, 1982. With permission.)

Table 3.5. Wood material used as an insulator in different European concert halls

Gasteig Philharmonie, Munich, Germany

Function: symphonic concerts, 85%; drama and opera, 5%; rock jazz, pop concerts, 5%; miscellaneous, 5%
Inaugurated: 1985
Geometrical data: volume 30,000 m^3; platform area 300 m^2; seating area 1500 m^2; number of seats 2,387
Acoustical data: reverberation time 2.2 s (nonoccupied), 2.1 s (occupied)
Ceiling: suspended convex and concave elements of 60 mm wood
Walls: 38 mm veneered wood fiberboard in front of concrete; wooden reflectors on major sidewall areas
Floor: parquet on concrete; platform floor of 44 mm wood over air space with a very flexible hydraulic riser system supplemented with loose wooden riser elements
Chairs: fixed, wooden folding chairs with 8 cm upholstery on seat and brackrest; the rear side of the backrests are made of plywood

Grosses Festspielhaus, Salzburg, Austria

Function: symphonic concerts 70%; drama and opera 30%
Inaugurated: 1960, major modifications to orchestra shell, 1979
Geometrical data: volume 15,500 m^3; platform area 260 m^2; seating area 1,050 m^2; number of seats 2,168
Acoustical data: reverberation time 2.0 s (nonoccupied); 1.5 s (occupied)
Ceiling: painted plaster on reeds
Walls: plaster on reeds covered with a thin layer of wood; convex wooden panels of 40 mm
Floor: linoleum on wood with carpet on main aisles
Platform floor: linoleum on wood over air space with moveable risers of 22 mm wood
Chairs: fixed, wooden folding chairs with upholstered seats and thinly upholstered backrests

Barbican Concert Hall, London, England

Function: Symphonic concerts 60%; recitals and chamber music 10%; rock and jazz concerts 10%; miscellaneous 20%
Inaugurated: 1982
Geometrical data: volume 17,750 m^3; platform area 200 m^2; seating area 1,050 m^2; number of seats 2,026
Acoustical data: reverberation time 2.0 s (nonoccupied), 1.7 s (occupied)
Ceiling: concrete with exposed concrete beams and ventilation ducts
Walls: wood panels in front of concrete
Floor: parquet on hard surface
Platform floor: 22 mm parquet on 22 mm plywood and gypsum over air space; a wooden canopy is suspended over the platform
Chairs: fixed wooden chairs with upholstered seats and backrests

Liederhalle, Stuttgart, Germany

Function: symphonic concerts 60%; chamber music 15%; rock, jazz, pop concerts 15%; drama and opera 5%; miscellaneous 5%
Inaugurated: 1956
Geometrical data: volume 15,000 m^3; platform area 178 m^2; seating area 1,150 m^2; number of seats 1,994
Acoustical data: reverberation time 2.1 s (nonoccupied); 1.6 s (occupied)
Ceiling: plaster on metal lath in the central part and plane and slotted fiberboard near the walls
Walls: plywood of varying thickness in front of air spaces of varying depths
Floor: parquet on main floor
Chairs: wooden folding chairs with upholstered seats and backrests

Musikverin, Vienna, Austria

Function: symphonic concerts 75%; recitals and chamber music 25%
Inaugurated: 1870
Geometrical data: volume 15,000 m^3; platform area 125 m^2; seating area 620 m^2; number of seats 1,600
Acoustical data: reverberation time 3.2 s (nonoccupied); 2.1 s (occupied)
Ceiling: gilded and painted plaster on wood
Walls: plaster on brick, wooden doors; wooden panelling around the platform; balcony fronts are plaster on wood
Floor: linoleum on wood with carpet; platform floor is wood over air space with steep, fixed risers
Chairs: fixed wooden chairs with upholstered seats

Table 3.5. *Continued*

Concertgebouw, Amsterdam, The Netherlands

Function: symphonic music 75%; recitals and chamber music 14%; miscellaneous 11%
Inaugurated: 1888
Geometrical data: volume 18,700 m³; platform area 150 m²; seating area 900 m²; number of seats 2,040
Acoustical data: reverberation time 2.5 s (nonoccupied), 2 s (occupied)
Ceiling: coffered, 40 mm plaster reeds
Walls: plaster on bricks at floor level; plaster on reeds above balcony
Floor: wood over 70 mm air space filled with sand; carpets
Chairs: fixed wooden folding chairs with upholstered seats and backrests

From Gade, A. C., Acoustical Survey of Eleven European Concert Halls, Rep. # 44, Technical University of Denmark, Copenhagen, 1989. With permission.

Table 3.6. Dimensions of various concert halls

Gasteig (Munich)	Grosses Festspielhaus (Salzburg)	Barbican (London)	Liederhalle (Stuttgart)	Musikvereinsaal (Vienna)	Concertgebouw (Amsterdam)
Audience Area					
Distance from platform front to rearmost seat (m)					
42	29	33	36	40	26
Mean width between side walls (m)					
55	34	40	35	19.5	28.5
Mean ceiling height (m)					
15	14	15	12.6	15	17.5
Angle between side walls (m)					
72	30	60	50	0	0
Floor slope angle (degrees)					
20	15	15	5	5	0
Ratio of width: height					
3.666	2.429	2.667	2.778	1.114	1.629
Platform Area					
Distance from platform front to rear wall (m)					
17	14	12.5	10	9	13
Mean width between side walls (m)					
20	23	17	19	19.5	28.5
Mean ceiling height (m)					
15	8.5	9	9	15	15.5
Angle between side walls (degrees)					
12	70	60	12	0	0
Platform volume (m³)					
5100	2737	1912	1700	2633	7068
Height of platform over main floor (cm)					
60	90	91	125	71	150

Data from Gade, A. C., Acoustical Survey of Eleven European Concert Halls, Rep. # 44, Technical University of Denmark, Copenhagen, 1989. With permission.

Table 3.7. Reverberation time (seconds) vs. frequency (Hertz) in the Arsenal Concert Hall, in Metz, France (Commins, 1992).

Frequency (Hz)	125	250	500	1000	2000	4000
Reverberation time (s)						
Occupied hall	1.5	1.8	2.1	2.4	2.4	2.1
Nonoccupied hall	1.7	2.1	2.3	2.7	2.4	2.1

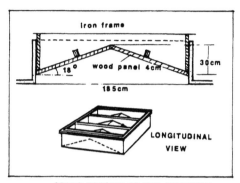

CROSS SECTION OF THE DIFFUSING
ELEMENTS ABOVE THE STAGE

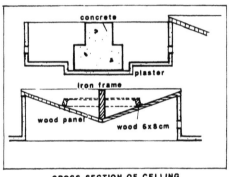

CROSS SECTION OF CEILING
DIFFUSING ELEMENTS

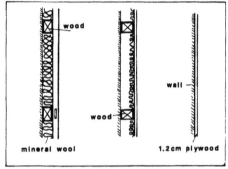

WALL PANEL SYSTEM

Figure 3.8. Diffusing elements in the Arsenal Concert Hall in Metz, France. (From Commins, 1992. With permission.)

REFERENCES

Ando, Y. (1985). *Concert Hall Acoustics.* Springer-Verlag, Berlin.

Auzou, S. (1976). Etude des caractéristiques acoustiques de matériaux et d'équipements. *Cah. Cent. Sci. Tech. Bati.* no. 173.

Barron, M. (1988). Subjective study of British Symphony concert halls. *Acustica.* 66, 1–14.

Beranek, L. (1960). *Noise Reduction.* McGraw-Hill, New York.

Beranek, L. (1962). *Music, Acoustics and Architecture.* John Wiley & Sons, New York.

Beranek, L. (1986). *Acoustics.* 2nd ed. American Institute of Physics, Cambridge, MA.

Beranek, L. (1988). Concert hall acoustics, 25 years of experience. *Proc. Inst. Acoust. U.K.* 10(2), 75–77.

Beranek, L. (1992). Concert hall acoustics. *J. Acoust. Soc. Am.* 92(1), 1–39.

Bradley, J. S. (1991). A comparison of three classical concert halls. *J. Acoust. Soc. Am.* 89(3), 1176–1192.

Braune, B. (1960). *Documentation Bois. Acoustique.* Lignum, Union Suisse en Faveur du Bois, Zurich.

Centre Scientifique et Technique du Bâtiment (1989). Construction de maisons et bâtiments à ossature en bois. *Doc. Tech. Unif.* no. 31.2, Paris.

Commins, D. (1992). Acoustics of the "Arsenal" Concert Hall in Metz (France). Paper presented at Acoustics, Architecture and Auditoria Meet., May 19, 1992 (RIBA; IOA, U.K.) Birmingham, England.

Commins, D. (1987). Acoustique. L'auditorium du Musée d'Orsay. Des caractéristiques équilibrées pour une petite salle destinée à plusieurs usages: musique, conférences et cinéma. *Le Moniteur.* 30 April, 63–64.

Compien, B. (1977). *Le Bois et ses Dérivés dans l'Isolation et la Correction Acoustique.* Union Nationale Francaise des Chambres Syndicales des Charpantes, Menuiserie et Parquets, Paris.

32

Craik, R. J. M. and Barry, P. J. (1992). The internal damping of building materials. *Appl. Acoust.* 35, 139–148.

Cremer, L. and Muller, H. A. (1982). *Principles and Applications of Room Acoustics,* Vol. 1 and 2. Applied Science, London.

Cremer, L., Heckel, M., and Ungar, E. E. (1973). *Structure-Borne Sound.* Springer-Verlag, London.

Dickerhoff, H. E. and Lawrence, J. D. (1971). Wood floor system is cost competitive with concrete slab. *For. Prod. J.* 21(2), 13–18.

Fothergill, L. C. and Royle, P. (1991). The sound insulation of timber platform floating floors in the laboratory and field. *Appl. Acoust.* 33, 249–261.

Fothergill, L. C. and Savage, J. E. (1987). Reduction of noise nuisance caused by banging doors. *Appl. Acoust.* 21, 39–52.

Freas, A. D. (1971). Wood products and their use in construction. *Unasylva.* 25 (101–3), 53–68.

Freas, A. D. (1989). Building with wood, in *Concise Encyclopedia of Wood and Wood-Based Materials.* Schniewind, A. P., (Ed.). Pergamon Press, Oxford, 31–37.

Gade, A. C. (1989). Acoustical Survey of Eleven European Concert Halls. Rep. No. 44, Acoustics Laboratory, Technical University of Denmark, Copenhagen.

Grantham, J. B. and Heebink, T. B. (1971). Field measured sound insulation of wood-framed floors. *For. Prod. J.* 21(5), 33–38.

Grantham, J. B. and Heebink, T. B. (1973). Sound attenuation provided by several wood-framed floor-ceiling assemblies with troweled floor toppings. *J. Acoust. Soc. Am.* 54(2), 353–360.

Hase, N., Yamada, Y., Arima, T., Suzuki, S., and Ono, K. (1988). The relationships between the classification of impact-sound insulation and the vibrational characteristics of composite, wooden vibration damping floorings. *Mokuzai Gakkaishi.* 34(6), 500–507.

Hayashi, H. (1984). Sound Absorption and Anatomical structure of Japanese Cedar, Saghalin Fir, Maple and Willow, Paper presented at Proc. Pac. Regional Wood Anatomy Conf., Tsukuba, Japan, 22–24.

ISO 3382-1975(E), Measurement of Reverberation Time in Auditoria.

Jimenez-D., C. (1992). Acoustical evaluation of the Municipal Theatre of Lima, Peru. *Appl. Acoust.* 35, 153–166.

Josse, R. (1972). *Notions d'Acoustique à l'Usage des Architectes, Ingénieurs, Urbanistes.* Editions Eyrolles, Paris.

Kollman, F. F. P. and Côté, W. A. (1968). *Principles of Wood Science and Technology.* Springer-Verlag, Berlin.

Kuttruff, H. (1981). *Room Acoustics,* 2nd ed., Update, New York.

Muller, H. A. (1986). Room acoustical criteria: prediction and measurements. Proc. 12th Int. Conf. Acoustics, Toronto, Paper E 4–11.

Nakao, T., Takahashi, A., and Tanaka, C. (1988). Characteristics of footstep noise on wood-based floors. *Mokuzai Gakkaishi.* 34(1), 14–20.

Nakao, T., Takahashi, A., Nanba, M., Itoh, H., and Tanaka, C. (1989). Sound proofing against floor impact sounds in wooden houses. *Mokuzai Gakkaishi.* 35(2), 85–89.

Raes, A. C. (1964). *Isolation Sonore et Acoustique Architecturale.* Editions Chiron, Paris.

Smith, W. R. (1989). Acoustic properties of wood. Architectural acoustics, in *Concise Encyclopedia of Wood and Wood-Based Materials.* Schniewind, A. P. (Ed.) Pergamon Press, Oxford, 7–8.

Suzuki, M., Kojima, Y., and Koyasu, M. (1986). Relationship between sound transmission loss of wood based panels at low frequencies and resistive term. *Mokuzai Gakkaishi.* 32(3), 155–162.

Takahashi, A., Nanba, M., Nakao, T., and Tanaka, C. (1987a). Soundproofability against floor-impact noise in wooden houses. The effect of ceiling construction and separated soundproof walls. *Mokuzai Gakkaishi.* 33(12), 950–956.

Takahashi, A., Tanaka, C., Nakao, T., Iwashige, H., Minamizawa, A., and Schniewind, A. P. (1987b). The characteristics of impact sounds in wood-floor systems. *Mokuzai Gakkaishi.* 33(12), 941–949.

Takahashi, A., Tanaka, C., and Shiota, Y. (1981). Solid-borne noise from wood-joist floors and floating-floor variations of the traditional Japanese wooden-house construction. *Mokuzai Gakkaishi.* 27(12), 833–844.

Utley, W. A. (1979). Methods for improving the sound insulation of existing simple wood joist floors. *Appl. Acoust.* 12, 349–360.

Utley, W. A. and Cappelen, P. (1978). The sound insulation of wood joist floors in timber frame constructions. *Appl. Acoust.* 11(2), 147–164.

Watanabe, H., Matsumoto, T., Kinoshita, N., and Hayashi, H. (1967). Acoustical study of wood products. I. On the normal absorption coefficient of wood. *Mokuzai Gakkaishi.* 13(5), 177–182.

Part II
Material
Characterization

Chapter 4

Theory and Experimental Methods for the Acoustic Characterization of Wood

This chapter highlights the potential uses of vibrational methods in (ultrasonic and audible frequency ranges) for the characterization of the mechanical behavior of solids in general and of wood in particular. The analysis of mechanical wave propagation in media of various complexities enables us to focus our attention on the theoretical basis of the techniques (elastic moduli, Poisson's ratios, and tensile or shear yield strengths) used for the measurement of elastic properties in solid wood or wood-based composites. The behavior of an acoustically vibrating body was analyzed in such fundamental reference books as those of Harris and Crede (1961); Hearmon (1961); Snowdon (1968); Musgrave (1970); Auld (1973); Achenbach (1973); and Dieulesaint and Royer (1974).

The elastic properties of solids can be defined by generalizing Hooke's law, which relates the volume average of stress $[\sigma_{ij}]$ to the volume average of the strains $[\epsilon_{kl}]$ by the elastic constants $[C_{ijkl}]$:

$$[\sigma_{ij}] = [C_{ijkl}][\epsilon_{kl}] \qquad (4.1)$$

or

$$[\epsilon_{kl}] = [S_{ijkl}][\sigma_{kl}] \qquad (4.2)$$

where $[C_{ijkl}]$ are termed elastic stiffnesses, $[S_{ijkl}]$ are the elastic compliances, and $i,j,k,l = 1$, 2, or 3. Stiffnesses and compliances are fourth-rank tensors. Hearmon (1961) noted that "the use of the symbols [S] for compliances and [C] for stiffness is now almost invariably followed". This is the notation that is used hereafter.

Following the general convention on matrix notation, $[C_{ijkl}]$ could be written as $[C_{ij}]$, in terms of two-suffix stiffnesses, or symbolically as $[C]$. Similarly, the $[S_{ijkl}]$ could be written as $[S_{ij}]$ or $[S]$. In many applications it is much simpler to write Equations (4.1) and (4.2) in the following condensed form:

$$[\sigma] = [C] \cdot [\epsilon] \qquad (4.1')$$

and

$$[\epsilon] = [S] \cdot [\sigma] \qquad (4.2')$$

It is apparent that the stiffness matrix $[C]$ is the inverse of the compliance matrix $[S]$, as $[C] = [S]^{-1}$ and $[S] = [C]^{-1}$. Experimentally, the terms of the $[C_{ij}]$ matrix may be determined from ultrasonic measurements, whereas the terms of the $[S_{ij}]$ matrix are determined from static tests.

From Equation (4.1) it can be deduced that because strain is dimensionless the stiffnesses have the same dimensions as the stresses (N/m², MPa, or GPa), while the compliances (Equation (4.2)) are in m²/N or in MPa⁻¹. As an example, the case of spruce in longitudinal direction, for which $C_{11} = 150 \times 10^8$ N/m² = 15,000 MPa.

For solids of different symmetries, isotropic, transverse isotropic, or orthotropic, the stiffness matrix can be turned into a compliance matrix, following a specific procedure (Bodig and Jayne, 1982). Symmetry features of solids are introduced by the microstructural elements of the material, e.g., orientation of the fibers.

4.1. ELASTIC SYMMETRY OF PROPAGATION MEDIA

Because this chapter aims to provide the theoretical basis needed to understand wave propagation phenomena in solids, we first analyze the elastic symmetry of media of varying complexities. For solid wood and for wood-based composites, orthotropic and transverse isotropic symmetries are the most frequently observed. For simplicity, we begin by analyzing the isotropic solid. (Homogeneity of solids is assumed here.)

4.1.1. ISOTROPIC SOLIDS

The simplest elastic symmetry is that of an isotropic solid, which has only two independent constants, the stiffness moduli C_{11} and C_{12}. Lamé's constants, v and μ, can be defined as

$$\mu = E/2 \, (1 + v) \tag{4.3}$$

$$v = E/(1 - 2v)(1 + v) \tag{4.4}$$

$$G = E/2 \, (1 + v) = \mu \tag{4.5}$$

where E is the Young's modulus, G is the shear modulus, and is Poisson's ratio. For an isotropic solid the stiffnesses are

$$C_{11} = v + 2\mu \tag{4.6}$$

$$C_{12} = \lambda \tag{4.7}$$

$$C_{44} = \mu \tag{4.8}$$

The velocity of the propagation of a bulk longitudinal wave in an isotropic solid is related to the elastic constants by:

$$V_L = [(\lambda + 2\mu)/\rho]^{1/2} \tag{4.9}$$

where ρ is the density. The velocity of the propagation of a transverse wave is related to the elastic constants by:

$$V_T = (\mu/\rho)^{1/2} \tag{4.10}$$

4.1.2. ANISOTROPIC SOLIDS

The origin of anisotropy, perceived as the variation of material response in the direction of the applied stress, lies in the preferred organization of the material's internal structure. The structure may be, for example, the atomic array in monocrystals, the morphological texture in polycrystalline aggregates such as metals, rocks, sand, etc., the orientation of fibers in composites and human tissues, or the orientation of layers in laminated plastics, plywood, etc.

One instance of complex elastic symmetry is that of an orthotropic solid, because constants are influenced by three mutually perpendicular planes of elastic symmetry. The corresponding stiffness matrix contains nine independent constants: six diagonal terms (C_{11}, C_{22}, C_{33}, C_{44}, C_{55}, C_{66}) and three off-diagonal terms (C_{12}, C_{13}, C_{23}), as can be seen in Equation (4.11).

For transverse isotropy, the material may possess an axis of symmetry in the sense that all directions at right angles to this axis are equivalent. The corresponding stiffness matrix (Equation (4.12)) contains five independent constants (C_{11}, C_{22}, C_{55}) four diagonal terms, and two off-diagonal terms (C_{12}, C_{13}). The transverse isotropy is a particular case of an orthotropic solid.

$$\begin{bmatrix} C_{11} & C_{12} & C_{23} & & & \\ & C_{22} & C_{13} & & & \\ & & C_{33} & & & \\ & & & C_{44} & & \\ & & & & C_{55} & \\ & & & & & C_{66} \end{bmatrix} = [C]_{\text{orthotropic}} \tag{4.11}$$

Orthotropic material: nine independent constants

$$
\begin{bmatrix}
C_{11} & C_{12} & C_{12} & & & \\
 & C_{22} & C_{23} & & & \\
 & & C_{22} & & & \\
 & & & (C_{11} - C_{12})/2 & & \\
 & & & & C_{55} & \\
 & & & & & C_{55}
\end{bmatrix} = [C]_{\text{transverse}} \qquad (4.12)
$$

Transverse isotropic material: five independent constants

$$
\begin{bmatrix}
C_{11} & C_{12} & C_{12} & & & \\
 & C_{11} & C_{12} & & & \\
 & & C_{11} & & & \\
 & & & (C_{11} - C_{12})/2 & & \\
 & & & & (C_{11} - C_{12})/2 & \\
 & & & & & (C_{11} - C_{12})/2
\end{bmatrix} = [C]_{\text{isotropic}} \qquad (4.13)
$$

Isotropic material: two independent constants

The following is a brief discussion of the relationships between the engineering elastic parameters and the terms of stiffness and compliance matrices for solid wood (considered to be an orthotropic material) and for composites of wood-based materials (plywood, flakeboards, fiberboards, etc.) which are expected to exhibit transverse isotropy (Jayne, 1972; Bodig and Jayne, 1982; Guitard, 1985). The discussion is based on the fundamental data of Love (1944), Green and Zerna (1968), and Hearmon (1961).

First we consider the orthotropic symmetry of solid wood. The terms of the compliance matrix are given by:

$$
[S] = \begin{bmatrix}
S_{11} & S_{12} & S_{13} & & & \\
 & S_{22} & S_{23} & & & \\
 & & S_{33} & & & \\
 & & & S_{44} & & \\
 & & & & S_{55} & \\
 & & & & & S_{66}
\end{bmatrix} \qquad (4.14)
$$

The physical significance of the compliances can be understood when expanding Equation (4.2):

$$
\epsilon_{11} = S_{1111}\sigma_{11} + S_{1112} \quad \sigma_{12} + S_{1133} \quad \sigma_{13} + S_{1114} \quad \sigma_{14} + S_{1115} \quad \sigma_{15} + S_{1116} \quad \sigma_{16} \qquad (4.15)
$$

with five similar equations for $\epsilon_{22} \ldots \epsilon_{66}$. As noted by Hearmon (1961), several interesting points can be generalized:

1. $S_{iiii} (\equiv S_{qq})$ $(q = 1, 2, 3)$ relates an extensional stress to an extensional strain, both in the same direction, and $S_{qq} = 1/Eq$. For the particular symmetry of solid wood this relation gives Young's moduli E_L, E_R, E_T.
2. $S_{iijj} (\equiv S_{qr})$ $(q \neq r;\ q, r = 1, 2, 3)$ relates an extensional strain to a perpendicular extensional stress and $S_{qr} = -vrq/E_q = -vqr/E_r$. The six Poisson's ratios can be calculated in this way.
3. $S_{ijij} (\equiv S_{qq})$ $(q = 4, 5, 6)$ relates a shear strain to a shear stress in the same plane, and $S_{qq} = 1/G_q$. For solid wood the shear terms deduced are C_{44}, C_{55}, and C_{66}, corresponding to the planes 23, 13, and 12.

4. The other terms, S_{iijk}, S_{ijjk}, S_{iiij}, relate an extensional strain to a shear stress in the same plane or in a perpendicular plane. As they are not integral to the present discussion, they are not included here.

The relationships between the stiffness terms and the compliance terms for the orthotropic solid are

$$C_{11} = [S_{22}S_{33} - (S_{23})^2] \cdot /S/^{-1}$$

$$C_{22} = [S_{11}S_{33} - (S_{13})^2] \cdot /S/^{-1}$$

$$C_{33} = [S_{11}S_{22} - (S_{12})^2] \cdot /S/^{-1}$$

$$C_{12} = [S_{21}S_{33} - S_{23}S_{31}] \cdot /S/^{-1}$$

$$C_{13} = [S_{31}S_{22} - S_{21}S_{32}] \cdot /S/^{-1} \qquad (4.16)$$

$$C_{23} = [S_{31}S_{12} - S_{11}S_{32}] \cdot /S/^{-1}$$

$$C_{44} = 1/S_{44}$$

$$C_{55} = 1/S_{55}$$

$$C_{66} = 1/S_{66}$$

where

$$/S/ = S_{11}\,S_{22}\,S_{33} + 2S_{12}\,S_{23}\,S_{31} - S_{11}(S_{23})^2 - S_{22}(S_{13})^2 - S_{33}(S_{12})^2$$

The terms of the compliance matrix are related to the terms of the stiffness matrix, with the terms S replaced by the terms C. When the axes are labeled 1, 2, 3, engineering constants are related to the compliances in the following way:

$$[S] = \begin{bmatrix} 1/E_1 & -v_{12}/E_2 & -v_{13}/E_3 & & & \\ -v_{21}/E_1 & 1/E_2 & -v_{23}/E_3 & & & \\ -v_{31}/E_1 & -v_{32}/E_2 & 1/E_3 & & & \\ & & & 1/G_{23} & & \\ & & & & 1/G_{13} & \\ & & & & & 1/G_{12} \end{bmatrix} \qquad (4.17)$$

Finally, it is worth repeating the relationships between the terms of the stiffness matrix and the technical constants:

$$C_{11} = (1 - v_{23}\,v_{32})[E_2\,E_3\,/S/]^{-1}$$

$$C_{22} = (1 - v_{31}\,v_{13})[E_1\,E_3\,/S/]^{-1}$$

$$C_{33} = (1 - v_{21}\,v_{12})[E_1\,E_2\,/S/]^{-1}$$

$$C_{12} = (v_{21} + v_{23}\,v_{31})[E_2\,E_3\,/S/]^{-1}$$

$$C_{23} = (v_{32} + v_{31}\,v_{12})[E_3\,E_1\,/S/]^{-1} \qquad (4.18)$$

$$C_{13} = (v_{13} + v_{12}\,v_{23})[E_1\,E_2\,/S/]^{-1}$$

$$C_{44} = G_{23} \qquad C_{55} = G_{13} \qquad C_{66} = G_{12}$$

and

$$/S/ = [1 - \nu_{12}\nu_{21} - \nu_{23}\nu_{32} - \nu_{13}\nu_{31} - 2\nu_{21}\nu_{32}\nu_{31}](E_1 E_2 E_3)^{-1} \tag{4.18'}$$

Accurate measurement of the set of orthotropic constants is not an easy task. Among these constants, Poisson's ratios are perhaps the most difficult to measure. For a more thorough understanding of the problem, the realistic boundaries of Poisson's ratios must be considered. The reasons that underlie this observation follow, first by analyzing the isotropic solids and then the orthotropic solids.

For an isotropic solid, the relationships (Green and Zerna, 1968) between the Poisson's ratios (defined as the quotient "lateral constriction/longitudinal extension") for a specimen under tension and the elastic constants are

$$K = E/3(1 - 2\nu) \tag{4.19}$$

for the bulk modulus K and

$$G = E/2(1 + \nu) \tag{4.20}$$

for the shear modulus.

The strain energy function is positive definite for a homogeneous isotropic elastic continuum. This means that $K > 0$ and $G > 0$. Consequently, $E > 0$ and $(1 - 2\nu) > 0$ or $(1 - \nu) > 0$. The boundary conditions for Poisson's ratios are

$$-1 \langle \nu \rangle 1/2 \tag{4.21}$$

For an orthotropic solid, the question is more complex due to the correspondence of the six Poisson's ratios to the three symmetry planes.

Bearing in mind that the strain energy function W must be defined as positive, we form:

$$W = 1/2 \, C_{ijkl}\epsilon_{ij}\epsilon_{kl} > 0 \tag{4.22}$$

and similarly,

$$W = 1/2 \, S_{ijkl}\sigma_{ij}\sigma_{kl} > 0 \tag{4.23}$$

Consequently, $C_{ijkl} > 0$ and $S_{ijkl} > 0$, meaning that all the terms of the stiffness and compliance matrices must be positive; in other words:

$$C_{11}, C_{22}, C_{33}, C_{44}, C_{55}, C_{66}, C_{12}, C_{13}, C_{23} > 0 \tag{4.24}$$

$$S_{11}, S_{22}, S_{33}, S_{44}, S_{55}, S_{66}, S_{12}, S_{13}, S_{23} > 0$$

For a real material, the Young's and shear moduli must be also positive definite:

$$E_1, E_2, E_3, G_{12}, G_{13}, G_{23} > 0 \tag{4.25}$$

Considering the relationships between the terms of $[C]$ and $[S]$ matrices and the engineering constants we can deduce the boundary conditions for all Poisson's ratios of an orthotropic solid. From Equation (4.18) we can establish the simultaneous relationships between all six Poisson's numbers:

$$1 - \nu_{12}\nu_{21} - \nu_{23}\nu_{32} - \nu_{13}\nu_{31} - 2\nu_{21}\nu_{32}\nu_{31} > 0 \tag{4.26}$$

The relationships between two Poisson's ratios, corresponding to a well-defined symmetry plane, are deduced from Equation (4.18) when the terms C_{11}, C_{22}, and C_{33} are considered:

Table 4.1. Engineering parameters of solid wood (Hearmon, 1948).

Species	Density (kg/m³)	Young's moduli (× 10⁸ N/m²)			Shear moduli		
		E_L	E_R	E_T	G_{TR}	G_{LT}	G_{LR}
Balsa	200	63	3.0	1.1	0.3	2.0	3.1
Yellow poplar	380	97	8.9	4.1	1.1	6.7	7.2
Birch	620	163	11.1	6.2	1.9	9.2	11.8
Oak	660	53	21.4	9.7	3.9	7.6	12.9
Ash	670	158	15.1	8.0	2.7	8.9	13.4
Beech	750	137	22.4	11.4	4.6	10.6	16.1
Sitka spruce	390	116	9.0	5.0	0.39	7.2	7.5
Spruce	440	159	6.9	3.9	0.36	7.7	7.5
Douglas fir	450	157	10.6	7.8	0.88	8.8	8.8
Fir	450	127	9.3	4.8	1.40	7.5	9.3
Scotch pine	550	163	11.0	5.7	0.66	6.8	11.6

Species	Density (kg/m³)	Poisson's ratio					
		ν_{RT}	ν_{TR}	ν_{RL}	ν_{LR}	ν_{TL}	ν_{LT}
Balsa	200	0.66	0.24	0.018	0.23	0.009	0.49
Yellow poplar	380	0.70	0.33	0.030	0.32	0.019	0.39
Birch	620	0.78	0.38	0.034	0.49	0.018	0.43
Oak	660	0.64	0.30	0.130	0.33	0.086	0.50
Ash	670	0.71	0.36	0.051	0.46	0.030	0.51
Beech	750	0.75	0.36	0.073	0.45	0.044	0.51
Sitka spruce	390	0.43	0.25	0.029	0.37	0.020	0.47
Spruce	440	0.47	0.25	0.028	0.44	0.013	0.38
Douglas fir	450	0.39	0.37	0.020	0.29	0.022	0.45
Fir	450	0.60	0.35	0.030	0.45	0.020	0.50
Scotch pine	550	0.68	0.31	0.038	0.42	0.015	0.51

$$(1 - \nu_{23}\nu_{32}) > 0 \qquad (1 - \nu_{12}\nu_{32}) > 0 \qquad (1 - \nu_{12}\nu_{21}) > 0 \qquad (4.27)$$

From these equations we recognize that the corresponding in-plane ratios ν_{rq} and ν_{qr} could have the same sign (+) or (−). On the other hand, the relationships between the Poisson's ratios and the Young's moduli $\nu_{rq} = \nu_{qr}E_q/E_r$ imply:

$$\nu_{qr}^2 > E_r/E_q. \qquad (4.28)$$

However, for anisotropic solids it is possible to have $E_r > E_q$ and therefore, $\nu_{qr} > 1$.

Indeed, negative values of Poisson's ratios or values >1 may contradict intuition if the majority of experience is in isotropic solids, but recently such data were reported for composite materials (Jones, 1975) and for foam materials (Lipsett and Beltzer, 1988).

In their excellent review of methods used for measuring mechanical properties McIntyre and Woodhouse (1985–1986) suggested that idealized two-dimensional honeycomb patterns of transverse wood structure could produce a Poisson's ratio ν_{RT} in the range −1 to ∞. Referring to the analysis above, the assumption of orthotropy suggests that nine independent stiffnesses or compliances characterize the elastic behavior of solid wood analyzed in a rectangular coordinate system. As consequence we find 12 engineering parameters: three Young's moduli, three shear moduli, and six Poisson's ratios. Table 4.1 supplies some values of the 12 solid wood engineering parameters.

For a wide range of European, American, and tropical species, Bodig and Goodmann (1973) and Guitard (1987) deduced statistical regression models that were able to predict the terms of the compliance matrix as a function of density. These data may be used by modelers in finite element calculations, or with nondestructively tested lumber when the elasticity moduli are required. In engineering practice, however, the elastic constants of solid wood can be used for the accurate estimation of the elastic properties of plywood. Gerhard (1987) defined a homogeneous "equivalent orthotropic material" that

Table 4.2. Dynamic moduli of plywood and corresponding solid wood for birch and Sitka spruce

Elastic constants ($\times$ 10^8 N/m^2)					
E_1	E_2	E_3	G_{12}	G_{13}	G_{23}
Birch Three Plywood					
96.6	54.1	27.0	10.6	8.78	6.04
95.1	53.3	22.4	10.6	8.78	6.04
Sitka Spruce Three Plywood					
79.6	42.3	9.9	7.2	10.6	0.57
79.0	42.0	9.0	7.2	10.6	0.57
Birch-Solid Wood (620 kg/m^3)					
163	11.0	6.2	9.1	11.8	1.9
Sitka Spruce-Solid Wood (390 kg/m^3)					
116	9.0	5.0	7.2	7.5	0.4

From Gerhards, C. C., *Wood Sci. Tech.*, 21, 335, 1987. With permission.

enables conventional analytical methods to be applied to the elastic characterization of plywood. Gerhard's approach is deduced from the strain energy method. The properties of the proposed material are compared to those of an equivalent material deduced from the "law of mixtures" proposed by Bodig and Jayne in 1982. The values of Young's and shear moduli for the "equivalent plywood" and for solid wood are given in Table 4.2. Plywood exhibits fewer anisotropic mechanical properties than does solid wood. Young's moduli E_2 and E_3 as well as the shear modulus G_{23} for plywood are strongly increased as compared to the same properties of solid wood.

Another interesting example of an orthotropic wood composite is that of machine-made paper. Mann et al. (1980) describe the measurement of nine elastic constants on a heavy milk carton stock (780 kg/m^3) using a transmission technique. The engineering constants are presented in Table 4.3. These constants indicate that the paperboard is highly anisotropic. The Poisson's ratios corresponding to the planes, which include axis 3, the axis normal to the thickness, are remarkably high, undoubtedly connected to the misalignment of fibers in the plane of the sheet.

For wood composites or some tropical wood species that exhibit plane isotropy the terms of the stiffness matrix may be reduced, bearing in mind that:

Table 4.3. Dynamic elastic constants ($\times$ 10^8 N/m^2) of machine-made, heavily bleached Kraft milk carton stock

Dynamic elastic constants ($\times$ 10^8 N/m^2)					
Young's moduli			Shear moduli		
E_1	E_2	E_3	G_{44}	G_{55}	G_{66}
74.4	34.7	0.39	0.99	1.37	20.4
Poisson's ratios					
ν_{12}	ν_{21}	ν_{13}	ν_{31}	ν_{23}	ν_{32}
0.15	0.32	0.008	1.52	0.021	1.84

Note: The three principal axes are (1) the machining direction; (2) the cross-machine direction; and (3) the thickness direction.

From Mann et al., *Tappi J.*, 63(2), 1980. With permission.

$$C_{11} = C_{22} \qquad C_{21} = C_{21} \qquad C_{66} = (C_{11} - C_{12})/2 \qquad (4.29)$$

For a solid having transverse anisotropy, the corresponding relationships between the terms of the stiffness matrix and the engineering constants are

$$E_1 = [(C_{11} - C_{13}^2/C_{33})^2 - (C_{12} - C_{13}^2/C_{33})^2]/(C_{11} - C_{13}^2/C_{33})$$

$$E_3 = C_{33} - 2C_{13}^2/(C_{11} + C_{12})$$

$$\nu_{12} = \nu_{21} = (C_{12}C_{33} - C_{13}^2)/(C_{11}C_{33} - C_{13}^2)$$

$$\nu_{13} = [C_{13}(C_{11} - C_{12})]/(C_{11}C_{33} - C_{13}^2) \qquad (4.30)$$

$$\nu_{31} = C_{13}/(C_{11} + C_{12})$$

$$C_{55} = G_{13}$$

$$C_{66} = G_{12}$$

The corresponding relationships between the terms of the compliance matrix and the engineering terms are given by Equation (4.31), if $E_1 = E_2 = E$; $\nu_{12} = \nu_{21} = \nu$; and $G_{12} = G = E/2 (1 + \nu)$.

$$[S] = \begin{bmatrix} 1/E & -\nu/E & -\nu_{31}/E_3 \\ -\nu/E & 1/E & -\nu_{32}/E_3 \\ -\nu_{13}/E & -\nu_{23}/E & 1/E_3 \\ & & & 1/G_{23} \\ & & & & 1/G_{13} \\ & & & & & E/2(1 + \nu) \end{bmatrix} \qquad (4.31)$$

Note that seven is the total number of independent stiffnesses or compliances derived from the particular form of the Hooke's law for plane isotropic solids. Correspondingly, the number of engineering elastic parameters is nine: two Young's moduli, two shear moduli, and five Poisson's ratios.

Using transverse isotropic hypothesis for the structure of a standing tree, Archer (1987) presented a procedure for growth strain estimation. The same symmetry was used by Baum and Bornhoeft (1979) for the estimation of Poisson's ratios in paper (69, 42, and 26 lb Kraft linerboard).

4.2. WAVE PROPAGATION IN ANISOTROPIC MEDIA

Hearmon (1961), Musgrave (1970), Green (1973), Auld (1973), Dieulesaint and Royer (1974), and Alippi and Mayer (1987) have previously discussed the propagation of waves in ansiotropic solids. We consider first the case of an isotropic solid in which bulk waves are propagating. When the motion of a particle is along the propagation direction, this is a longitudinal wave. When the motion of a particle is perpendicular to the propagation direction, this is a shear wave or a transverse wave. In anisotropic materials both longitudinal and transverse waves can propagate either along or out of the principal symmetry directions. Figure 4.1 shows the bulk velocities in an orthotropic solid. Surface waves can propagate in any direction on any isotropic or anisotropic substrate. Recent applications for surface waves involve the characterization of elastically anisotropic solids having piezoelectric properties and that of layered solids (Edmonds, 1981).

Some theoretical considerations are presented in this section relationing to the propagation phenomena of ultrasonic waves in orthotropic solids. This symmetry was chosen because of the interest in the cartesian orthotropic wood structure model. As seen here, the most rapid way to obtain stiffness is via the ultrasonic velocity method.

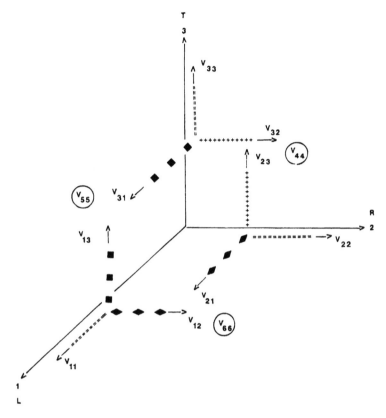

Figure 4.1. Ultrasonic velocities in an orthotropic solid. Longitudinal waves: $V_{11} = V_{LL}$; $V_{22} = V_{RR}$; $V_{33} = V_{TT}$. Transverse waves: $V_{44} = V_{RT}$, deduced from V_{23} and V_{32}; $V_{55} = V_{LT}$, deduced from V_{13} and V_{31}; $V_{66} = V_{LR}$, deduced from V_{12} and V_{21}.

The following notations are used in this section:

$[\sigma]$	stress tensor
$[\epsilon]$	strain tensor
ρ	density
$\ddot{u}_j$	displacement
$[C_{ijkl}]$	stiffness tensor
$[S_{ijkl}]$	compliance tensor
$[\Gamma_{ij}]$	Christoffel tensor
A_o	amplitude of the wave
p_j	unit displacement polarization vector
k_m	wave vector component along the x_m direction
x_m	position vector
ω	angular frequency
P_m	polarization vector
n_k	direction cosines
α	angle of unit wave vector from the symmetry direction
β	displacement angle
δ_{ij}	Kronecker tensor
D_{ij}	flexural rigidities in plates
ν_{ij}	Poisson's ratios
V	phase velocity
v	group velocity
a_1 to a_4	coefficients depending on the supported conditions of a plate

4.2.1. ULTRASONIC BULK WAVE PROPAGATION IN ORTHOTROPIC MEDIA

The equation of wave motion in anisotropic media is

$$\partial \sigma_{ij}/\partial x_i = \rho \ddot{u}_j \tag{4.32}$$

For an elastic medium under a small deformation, the constitutive relation (Hooke's law) and the strain displacement relation are given by Equation (4.1) ($[\sigma_{ij}] = [C_{ijkl}][\epsilon_{kl}]$) and the related strain displacement is given by

$$\epsilon_{ij} = 1/2 \left(\partial u_i/\partial x_j + \partial u_j/\partial x_i \right) \tag{4.33}$$

From the constitutive relation and the strain displacement relation (4.33), the equation of motion (4.32) can be written as:

$$C_{ijkl} \partial^2 u_k/\partial x_i \partial x_l = \rho \ddot{u}_j \tag{4.34}$$

We assume a plane harmonic wave solution of the above equation:

$$u_j(x_m, t) = A_o P_j \exp\left[i(k_m x_m - \omega t)\right] \tag{4.35}$$

After substituting Equation (4.35) into Equation (4.34), the equation of motion takes the form of an eigenvalue equation for the polarization vector P_m:

$$(C_{ijkl} n_i n_j - \delta_{ik} \rho V^2) P_m = 0 \tag{4.36}$$

These are Christoffel's equations, valid for the most general kind of anisotropic solids. The second order tensor Γ_{ij}, Christoffel's tensor, is given by Equation (4.37) and has the coefficients (4.38):

$$\Gamma_{ij} = C_{ijkl}\, n_i n_j \tag{4.37}$$

	$n_1^2\ n_2^2\ n_3^2$	$2n_2 n_3$	$2n_1 n_3$	$2n_1 n_2$	
$\Gamma_{11} =$	$C_{11}\ C_{66}\ C_{55}$	C_{56}	C_{15}	C_{16}	
$\Gamma_{22} =$	$C_{66}\ C_{22}\ C_{44}$	C_{24}	C_{46}	C_{26}	
$\Gamma_{33} =$	$C_{55}\ C_{44}\ C_{33}$	C_{34}	C_{35}	C_{45}	
$\Gamma_{12} =$	$C_{16}\ C_{26}\ C_{45}$	$1/2(C_{25} + C_{46})$	$1/2(C_{14} + C_{56})$	$1/2(C_{12} + C_{66})$	(4.38)
$\Gamma_{13} =$	$C_{15}\ C_{46}\ C_{35}$	$1/2(C_{36} + C_{45})$	$1/2(C_{13} + C_{55})$	$1/2(C_{14} + C_{56})$	
$\Gamma_{23} =$	$C_{56}\ C_{24}\ C_{34}$	$1/2(C_{23} + C_{44})$	$1/2(C_{36} + C_{45})$	$1/2(C_{25} + C_{46})$	

For example, in symmetry plane 12, for an orthotropic solid:

$$n_1 = \cos \alpha \qquad n_2 = \sin \alpha \qquad n_3 = 0$$

$$\Gamma_{11} = C_{11} n_1^2 + C_{66} n_2^2$$

$$\Gamma_{22} = C_{22} n_2^2 + C_{66} n_1^2$$

$$C_{55} = 0 \qquad C_{56} = 0 \qquad C_{15} = 0 \qquad C_{16} = 0$$

Christoffel's equations supply the relations between the elastic constants C_{ijkl} and the phase velocity V of ultrasonic waves propagating in the medium.

4.2.1.1. Velocities and Stiffnesses, The Eigenvalues of Christoffel's Equations

The eigenvalues and the eigenvectors of Christoffel's equations can be calculated for specific anisotropic materials. The non-zero values of P_m are obtained as characteristic eigenvectors that correspond to the characteristic eigenvalues. These are the roots of Equation (4.36).

$$
\begin{bmatrix} \Gamma_{11} - \rho V^2 & \Gamma_{12} & \Gamma_{13} \\ \Gamma_{12} & \Gamma_{22} - \rho V^2 & \Gamma_{23} \\ \Gamma_{13} & \Gamma_{23} & \Gamma_{33} - \rho V^2 \end{bmatrix} \begin{bmatrix} p_1 \\ p_2 \\ p_3 \end{bmatrix} = 0 \qquad (4.38')
$$

This equation is a cubic polynomial in phase velocity squared.

From Equation (4.38') the first issue addressed is the determination of the elastic constants (Γ_{ij}) of a given material, when the phase velocity is known. Equations (4.36) form a set of simultaneous equations in P_m. For their unique solution:

$$
[\Gamma_{ij} - \delta_{ik}\rho V^2] = 0 \qquad (4.39)
$$

If this equation is written for wave propagation along the symmetry axes for an orthotropic solid, three solutions are obtained:

$$
\begin{bmatrix} \Gamma_{11} - \rho V^2 & 0 & 0 \\ 0 & \Gamma_{22} - \rho V^2 & 0 \\ 0 & 0 & \Gamma_{33} - \rho V^2 \end{bmatrix} = 0 \qquad (4.40)
$$

These solutions show that along every axis it is possible to have three types of waves, one longitudinal and two transverse, as can be seen from Equations (4.41):

$$
\begin{cases} \Gamma_{11} - \rho V^2 = 0 \text{ or, } \rho V^2 = C_{11}, \text{ longitudinal wave} \\ \Gamma_{22} - \rho V^2 = 0 \text{ or, } \rho V^2 = C_{66}, \text{ transverse wave} \\ \Gamma_{33} - \rho V^2 = 0 \text{ or, } \rho V^2 = C_{55}, \text{ transverse wave} \end{cases} \qquad (4.41)
$$

Such solutions enable us to calculate the six diagonal terms of stiffness matrix $[C]$ by a relation which may be presented in the general form

$$
C_{ii} = V^2 \rho, \quad \text{where } i = 1, 2, 3 \cdots 6 \qquad (4.42)
$$

The three off-diagonal stiffness components can be calculated when the propagation is out of the principal axes of symmetry of the solid as, for example, in the 12 plane:

$$
\begin{bmatrix} \Gamma_{11} - \rho V^2 & \Gamma_{12} & 0 \\ \Gamma_{12} & \Gamma_{22} - \rho V^2 & 0 \\ 0 & 0 & \Gamma_{33} - \rho V^2 \end{bmatrix} = 0 \qquad (4.43)
$$

Then

$$
\Gamma_{12}^2 = (\Gamma_{11} - \rho V^2)(\Gamma_{22} - \rho V^2) \qquad (4.44)
$$

We can calculate, for example, Γ_{12} or C_{12}:

$$
(C_{12} + C_{66})n_1 n_2 = \pm[(C_{11}n_1^2 + C_{66}n_2^2 - \rho V^2)(C_{66}n_1^2 + C_{22}n_2^2 - \rho V^2)]^{1/2} \qquad (4.45)
$$

where V depends on the angle of propagation α, of quasi-longitudinal or quasitransverse bulk waves, in infinite solids.

Table 4.4. Propagation of bulk waves in an orthotropic solid

Propagation direction	Wave normal component	Polarization vector	Type of wave	Velocity
Propagation Along Principal Directions				
Axis X_1	$n_1 = 1$	X_1	L	$V_{11}^2\rho = C_{11}$
	$n_2 = 0$	X_2	T	$V_{66}^2\rho = C_{66}$
	$n_3 = 0$	X_3	T	$V_{55}^2\rho = C_{55}$
Axis X_2	$n_1 = 0$	X_1	T	$V_{66}^2\rho = C_{66}$
	$n_2 = 1$	X_2	L	$V_{22}^2\rho = C_{22}$
	$n_3 = 0$	X_3	T	$V_{44}^2\rho = C_{44}$
Axis X_3	$n_1 = 0$	X_1	T	$V_{55}^2\rho = C_{55}$
	$n_2 = 0$	X_2	T	$V_{44}^2\rho = C_{44}$
	$n_3 = 1$	X_3	L	$V_{33}^2\rho = C_{33}$
Propagation Out of Principal Directions				
Plane X_1X_2	n_1, n_2	$p_1/p_2 = \Gamma_{12}/(\rho V^2 - \Gamma_{11})$ $= \Gamma_{12}/(\rho V^2 - \Gamma_{22})$	Q_L, Q_T	$2V_{QLQT}^2\rho = (\Gamma_{11} + \Gamma_{22})$ $\pm [(\Gamma_{11} - \Gamma_{22})^2 + 4\Gamma_{12}^2)]^{1/2}$
	$n_3 = 0$	X_3	T	$2V_T^2\rho = C_{55}n_1^2 + C_{44}n_2^2$
Plane X_1X_3	n_1, n_3	$p_1/p_3 = \Gamma_{13}/(\rho V^2 - \Gamma_{11})$ $= \Gamma_{13}/(\rho V^2 - \Gamma_{33})$	Q_L, Q_T	$2V_{QLQT}^2\rho = (\Gamma_{11} + \Gamma_{33})$ $\pm [(\Gamma_{11} - \Gamma_{33})^2 + 4\Gamma_{13}^2)]^{1/2}$
	$n_2 = 0$	X_2	T	$2V_T^2\rho = C_{66}n_1^2 + C_{44}n_3^2$
Plane X_2X_3	n_2, n_3	$p_2/p_3 = \Gamma_{23}/(\rho V^2 - \Gamma_{22})$ $= \Gamma_{12}/(\rho V^2 - \Gamma_{22})$	Q_L, Q_T	$2V_{QLQT}^2\rho = (\Gamma_{22} + \Gamma_{33})$ $\pm [(\Gamma_{22} - \Gamma_{33})^2 + 4\Gamma_{23}^2)]^{1/2}$
	$n_1 = 0$	X_1	T	$2V_T^2\rho = C_{55}n_3^2 + C_{66}n_2^2$

Note: L = longitudinal wave; T = transverse wave; QL = quasilongitudinal wave; QT = quasitransverse wave.
$\pm$ For QL calculation see velocity expressions with +; for QT calculation see velocity expressions with −.

By permutations of indices we obtain the corresponding expression for C_{13} and C_{23}. Details of the calculation are given in Table 4.4.

If we admit that the matrix $[C] > 0$, and consequently $C_{ij} > 0$, then for the propagation angle α:

$$0 \langle\alpha\rangle \pi/2 \quad \text{or} \quad \pi \langle\alpha\rangle 3\pi/2$$

the expression under the square root (Equation (4.45)) must be considered with the sign (+). For $\pi/2 \langle\alpha\rangle \pi$ or $3\pi/2 \langle\alpha\rangle 2\pi$, the expression under the square root must be considered with the sign (−).

In the interest of clarity, calculation of the nondiagonal terms of the stiffness matrix requires the value of the velocity of a quasilongitudinal or a quasitransverse wave, or both. Also note that those values epend on the propagation vector and consequently on the orientation of the specimen (angle $\alpha°$).

It is well known that the physical properties of wood depend strongly on the orientation of reference coordinates, i.e., they depend on the angle α. This directional dependency of wood constants renders conventional averaging techniques inapplicable when measurements are taken with specimens at different angles α. For this reason Chapter 5 provides rational procedures for data averaging or optimization of directionally dependent measurements.

Having now obtained optimized values for all nine terms of the stiffness matrix $[C]$, the calculation of elastic constants can be done easily. The matrix $[C]$ is inverted to obtain compliance terms of $[S]$ and subsequently, Young's moduli and Poisson's ratios, using simple relations as seen previously in the chapter. When an optimization procedure for off-diagonal term was used, the values of the $[C]$ matrix could be employed for the calculation of the characteristic velocity or slowness surfaces.

The velocity surface is the locus of the radius vector, which has a length proportional to the velocity in the direction of the vector. The slowness surface is formed with the radius proportional to $1/V$. The

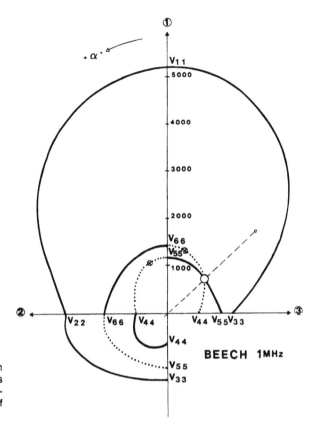

Figure 4.2. Velocity surface of beech deduced from experimental values measured using broadband transducers that have a central frequency of 1 MHz.

normal to the slowness surface coincides with the direction of the flux energy. The wave surface is the polar reciprocal of the slowness surface. Both the slowness and wave surfaces of isotropic materials are spheres.

The velocity surface of orthotropic solid (Musgrave, 1970) is formed by the intersection of three separate surfaces (sheets) — quasilongitudinal (QL), quasitransverse (QT), and transverse (T) — and are calculated using the corresponding velocities. As an example, consider them in the 12 plane with following equations:

$$2\rho V_{QL}^2 = (\Gamma_{11} + \Gamma_{22}) + [(\Gamma_{11} - \Gamma_{22})^2 + 4\Gamma_{12}^2]^{1/2} \quad \text{(QL wave)}$$

$$2\rho V_{QT}^2 = (\Gamma_{11} + \Gamma_{22}) - [(\Gamma_{11} - \Gamma_{22})^2 + 4\Gamma_{12}^2]^{1/2} \quad \text{(QT wave)} \quad \textbf{(4.46)}$$

$$\rho V_T^2 = C_{55}n_1^2 + C_{44}n_2^2 \quad \text{(T wave)}$$

Figure 4.2 shows the velocity surface for an orthotropic solid (beech). The curves help establish the discrepancies between the theoretical and experimental values of velocities and help to quantify the anisotropy of materials.

Musgrave (1970) demonstrated that wave propagation in any medium can be represented by the velocity surface, the slowness surface (the inverse of velocity), and the wave surface. Figure 4.3 shows, for historic interest, the slowness and wave surfaces of solids with orthorhombic symmetry (uranium and spruce). Several intersection points can be observed on the slowness surface between sheets, demonstrating the mode conversion phenomena. To identify the displacement vectors, the conditions of cusps must be studied on the wave surface. The cusps are associated with the departure of the slowness surface from an elliptical shape.

48

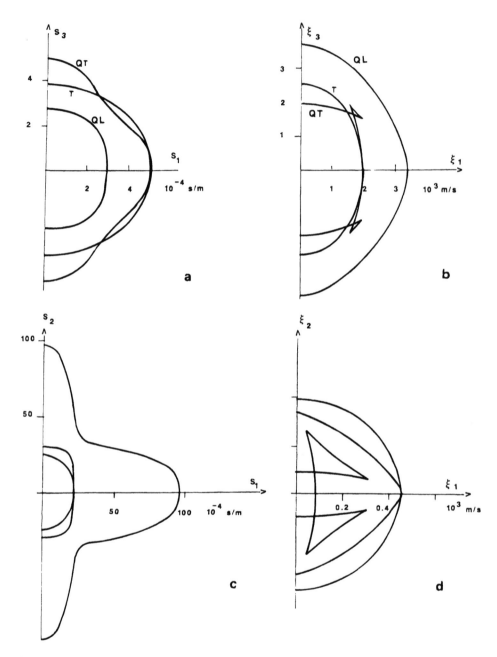

Figure 4.3. Slowness surface and wave surface in solids with orthotropic symmetry (uranium and spruce). (a) Uranium, slowness surface in 1–3 anisotropic plane; (b) uranium, wave surface in 1–3 anisotropic plane; (c) spruce, slowness surface in 1–2 anisotropic plane; (d) spruce, wave surface in 1–2 anisotropic plane. *QL*, quasilongitudinal wave; *QT*, quasitransverse wave; *T*, transverse wave. In the figures for spruce, the intercept of all the curves with coordinate axes should be at right angles. In this case, it is therefore incorrect to assign the labels *T* and *QL*; *T* is always associated with eliptic sections. We also have to bear in mind that wood is not a perfect crystal and some deviations from theoretical approaches may be expected. (Musgrave, M. J. P., *Crystal Acoustics*, Holden-Day, San Francisco, 1970. With permission of the author.)

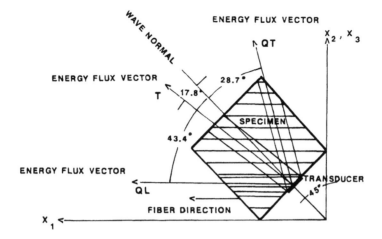

Figure 4.4. Energy-flux deviation in the X_1–X_2 plane in a transverse isotropic graphite fiber material (Kriz and Stinchcomb, 1979).

Moreover, the flux energy in anisotropic media deviates from the wave normal. Figure 4.4 gives an example of energy-flux deviation in transversely isotropic graphite fibers (Kriz and Stinchcomb, 1979). The deviation of energy flux from the wave normal can generate the conical refraction of waves phenomena.

4.2.1.2. The Eigenvectors of Christoffel's Equations

Eigenvectors can be obtained very simply, using Equation (4.38), if two off-diagonal components of the tensor are zero. The linear equation for the displacement $P_m(p_1, p_2, 0)$, associated with the quadratic factor of Equation (4.39) for $(n_1, n_2, 0)$, is

$$\begin{cases} (\Gamma_{11} - \rho V^2)p_1 + \Gamma_{12}p_2 = 0 \\ \Gamma_{12}p_1 + (\Gamma_{22} - \rho V^2)p_2 = 0 \end{cases} \tag{4.47}$$

where

$$p_1/p_2 = \Gamma_{12}/(\rho V^2 - \Gamma_{11}) = (\rho V^2 - \Gamma_{22})/\Gamma_{12} \tag{4.48}$$

If the polarization is allowed to correspond to the same sign ($p_1 = \sin \beta$ and $p_2 = \cos \beta$, $\beta =$ displacement angle), we obtain:

$$tg\ \beta = \Gamma_{12}/(\rho V^2 - \Gamma_{11}) \tag{4.49}$$

and

$$p_2 = \cos \beta = (\rho V^2 - \Gamma_{11})/[(\rho V^2 - \Gamma_{11})^2 + \Gamma_{12}^2]^{1/2} \tag{4.50}$$

Using Equation (4.50), the particle velocity polarization is expressed in terms of phase velocity, propagation direction, and stiffness constants of the solid for each plane of symmetry.

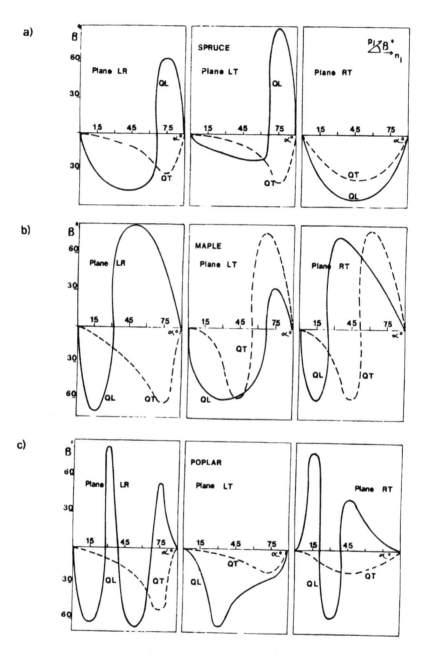

Figure 4.5. Dependence of β(α), the deviation from wave normal of quasilongitudinal (*QL*) and quasitransverse (*QT*) displacement for spruce (a), maple (b), and poplar (c) (Bucur, 1987).

The dependence of β(α), which relates the deviation of QL and QT displacements from the wave normal, is plotted in Figure 4.5. The results indicate how displacements excited in the interior of poplar wood propagate within the medium. It is appropriate to observe that the results concerning poplar, maple, and spruce, reported here are primarily of mathematical significance in relation to the methodology adopted for the calculation of the optimum C_{ij}. We note with considerable interest the calculated

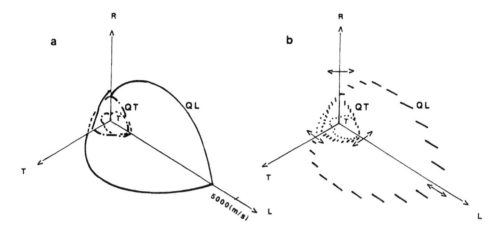

Figure 4.6. Velocity surfaces in all anisotropic planes and calculated displacement fields for spruce. (From Kriz, R. D. and Ledbetter, H. M., in *Rheology of Anisotropic Materials,* Huet C., Bourgoin, D., and Richemond, S., (Eds.), Editions Cepadues, Toulouse, France, 1986. With permission.)

displacement fields and the particular geometrical features of QL and QT curves on the velocity surface (Figure 4.6). On the spruce velocity surface longitudinal-transverse mode conversion was observed in 13 and 23 planes. This means that both longitudinal and transversal modes exist on the same wave surface. Thus, one must take care to identify the displacement vectors associated with different regions of the velocity or slowness sheets.

4.2.2. MECHANICAL VIBRATIONS IN THE ACOUSTIC FREQUENCY RANGE

The common audio frequency acoustic methods for wood testing use frequencies below 20,000 Hz. Steady-state or transient (impact) excitation can be used in dynamic tests at resonance vibration, when elastic moduli are to be determined.

The direct, accurate measurement of engineering constants of wood (the three Young's moduli, the three shear moduli, and the six Poisson's ratios) is important to engineering and to product design. The most convenient technique for precise measurement of these parameters depends upon measuring the resonance frequencies of longitudinal, flexural, or torsional resonant modes of both a bar-shaped sample of circular or rectangular cross-section, or a plate sample. The fact that the technique is resonant ensures that frequency measurements will be highly precise. Self-consistency of data, related to Young's moduli, for example, can be checked back using the same bar-shaped sample with longitudinal and flexural resonant modes. The technique can be extended to measure the internal friction, if the quality factor Q or the logarithmic decrement are measured in addition to resonance frequency.

Experimental studies on the elasticity of solid wood and wood-based composites are extensive and a large number of techniques have been developed. Free oscillation methods and methods with forced vibrations (resonance methods) have been used for measurements over a wide range of frequency, from 10^2 to 10^4 Hz. The main disadvantage of the resonance technique is related to the shape of the specimen, rod, or plate. It is well known that it is easy to obtain rods or plates in the LR or LT planes for solid wood, but it is very difficult in the RT plane. Moreover, care must be taken to ensure that losses through the suspension system, which is used to support the specimen, are not significant as compared to those intrinsic to the specimen (Kollmann and Krech, 1960; Becker and Noack, 1968; Kataoka and Ono, 1975–1976).

4.2.2.1. Resonance Vibration Modes in Rods and Plates

This section discusses the determination of the engineering constants of wood by resonance methods when the specimen is a rod or a plate. The most common vibratory resonant motions in a rod are longitudinal vibrations, flexural vibrations, and torsional vibrations. These vibrations are the dynamic counterpart of static tension, static bending, and static torsion.

We saw in Chapter 4.1 that for full characterization of a wood, nine elastic constants and nine damping constants are required. On a rod cut in the principal direction of the material only one Young's modulus and one shear modulus, and their corresponding damping constants, can be determined. On a thin quarter-cut plate four elastic constants together with four damping constants can be measured. These statements demonstrate the limitation of the technique. It is worth noting that the principal advantage of frequency resonance methods is related to the direct access to Young's moduli, shear moduli, and Poisson's ratios via measurement, as mentioned in several standard reference texts (Harris and Crede, 1961; Hearmon, 1961; Snowdon, 1968; Cremer and Heckl, 1973; Bodig and Jayne, 1982; Read and Dean, 1978; Vinh, 1982).

Because the theoretical considerations related to the vibrational resonances in bars and plates were analyzed in the books cited above and are well known, only the final relationships between engineering constants and resonance frequencies of different types of specimens are discussed.

The Young's modulus of a bar specimen from a homogeneous and nondispersive medium can be deduced from

$$E = 4\rho L^2 (f_{L,n}/n)^2 \qquad (4.51)$$

where ρ is the density, L is the length of the specimen, and $f_{L,n}$ is the frequency of the longitudinal mode ($n = 1$). For other modes ($n = 2, 3, \ldots$) this relation must be corrected with specific coefficients.

Young's modulus can also be deduced from a flexural mode by:

$$E = [4\pi^2 L^4 \rho (f_{F,n})^2 A]/I k_i^4 \qquad (4.52)$$

where A is the area of cross-section of the bar. Thus, we have for circular sections, $I = \pi r^4/4$, where r is the radius of the cross-section; for square cross-section, $I = a^4/12$, where a is the width; and for rectangular cross-section, $I = bh^3/12$, where b is the width and h is the height. The correction coefficient k_i is related to n mode by:

Mode	k_i
1	4.730
2	7.853
3	10.996
4	14.137
5	17.279
6	20.420

The shear modulus can be deduced from the torsional modes of the bars of circular section, using

$$G = 4\rho l^2 (f_n^T/n)^2 \qquad (4.53)$$

where f_n^T is the torsional frequency of mode n ($n = 1, 2, 3 \ldots$).

Flexural modes at higher frequencies can be used for the simultaneous estimation of Young's and shear moduli, when the Timoshenko bar theory is applied.

Figure 4.7 presents the displacements for different vibrational modes of bars.

When the vibration theory is employed on a flat, thin quarter-cut plate, four wood elastic constants can be determined on a single specimen. The corresponding theory was developed by McIntyre and Woodhouse (1978, 1985–1986, 1988) and by Caldersmith (1984).

4.2.2.2. Engineering Constants

Table 4.5 gives the Young's moduli of several species determined from the L, R, and T directions from longitudinal modes. The parameter measured was the resonance frequency. The anisotropy of wood can be deduced from the values noted in Table 4.5, and may be expressed as the ratios of E_R/E_L, which lie between 7/100 and 16/100, or of E_T/E_L, which lie between 5/100 and 9/100, and $E_L \gg E_R > E_T$, as expected.

Longitudinal modes of vibration of bar specimens were also used for purposes other than the characterization of elastical behavior:

- Estimating the elastic parameters of the fine structure of the cellular wall (Sobue and Asano, 1976; Tonosaki et al., 1983).
- Varying the moisture content in wood (James, 1986; Olszewski and Struk, 1983; Rebic and Srepcic, 1988; Sasaki et al., 1988; Suzuki, 1980; Tang and Hsu, 1972).

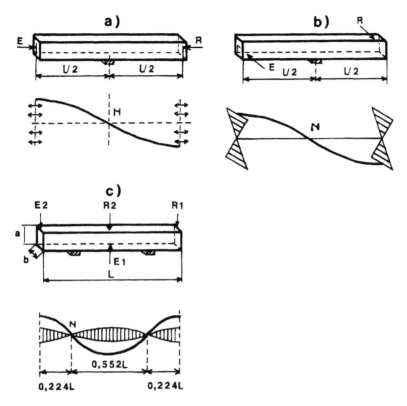

Figure 4.7. Displacements at different vibrations modes in bars: (a) Longitudinal vibrations in bars; (b) torsional vibrations in bars; (c) flexural vibrations in bars. (N) nodal point; (E) emission; (R) receiver. Note: for each mode the first figure shows the support position during testing and the second figure shows the displacement at $n = 1$. (From Facaoaru, I. and Jones, R., *Nondestructive Testing of Concrete,* Ed. Technica, Bucharest, 1971. With permission.)

- Chemical modifications induced by acetylation and formaldehyde cross-linking (Norimoto et al., 1988; Akitsu et al., 1991).
- Genetic variations observed as differences in tree clones (Fujisawa et al., 1992; Takada et al., 1989).

Longitudinal vibrations also can be produced by the stress-wave method (Dunlop, 1978, 1980; Gerhards, 1982), in which velocity is the measured parameter. When a frequency analysis is performed on the power spectrum of the stress-wave signal, it is possible to determine the resonance frequencies of different modes and, consequently, to compute instantaneously the Young's modulus on small clear specimens and on commercial-sized lumber (the maximum reported dimensions are 20 cm × 20 cm × 6m and 114 kg). This very elegant approach was developed by Sobue (1986a, b) and involves tapping the specimen with a hammer and receiving the signal with a wireless microphone. A microcomputer connected to the system permits the instantaneous treatment of data.

Flexural modes have been extensively used for the measurements of E on small, clear specimens, especially for the characterization of wood used for musical instruments. This approach is discussed in Chapter 7. Other interesting applications of flexural mode testing were reported in studying the anisotropy of plywood as it relates to the disposition of veneer sheets (Sobue, 1983; Sobue and Yvasaki, 1981a), the effect of adhesives on laminated lumber (Sobue and Yvasaki, 1981b), and the mechanical properties of epoxy-poplar composite materials (Moore et al., 1983).

Regarding testing purposes, the flexural mode offers the advantage of a relatively low frequency range ($n = 4$ or 5) on specimens of reasonable length (30 cm). Table 4.6 gives values of Young's

Table 4.5. Young's moduli in principal directions of wood deduced from longitudinal vibrations at 11% moisture content (frequency range 4 to 21 kHz)

Direction	Density (kg/m³)	E (10⁸ N/m²)	Anisotropy
	Sitka Spruce (*Picea sitchensis*)		
L	460	128	100
R	449	9.09	7.42
T	454	6.24	5.19
	Lauan (*Pentacme contorts*)		
L	481	118	100
R	489	13.8	11.7
T	478	6.38	5.40
	Makore (*Tieghemella heckelii*)		
L	669	136	100
R	673	21.58	15.8
T	670	12.3	9.07
	Matoa (*Pometia pinnata*)		
L	795	194	100
R	700	19.3	9.97
T	674	11.4	5.58
	Mizunara (*Quercus mongolica*)		
L	630	120	100
R	654	15.8	13.2
T	620	9.69	8.09
	Yachidamo (*Fraxinus mandshurica*)		
L	570	128	100
R	548	13.6	10.6
T	517	7.48	5.84

From Ono. T. and Norimoto, M., *Japan J. Appl. Phys.*, 24(8)960, 1985. With permission.

Table 4.6. Young's moduli for different modes of vibration, determined at 12% moisture content (Hearmon 1948).

Mode of vibration	E_L (10⁸ N/m²)
Beech (630 kg/m³)	
Bending of clamped specimen with additional mass	112
Longitudinal mode free-free	117
Beech (710 kg/m³)	
Bending of clamped specimen with additional mass	121
Longitudinal mode free-free	123
Maritime Pine (570 kg/m³)	
Bending of clamped specimen with additional mass	75
Longitudinal mode free-free	82

moduli as deduced by Hearmon (1948) from different testing configurations. Dynamic measurements with longitudinal modes give values that are a few percent higher than those from flexural modes. Brenndörfer (1972) and Radu and Brenndörfer (1976) indicated differences of the same order between Young's moduli deduced from longitudinal and bending modes in principal directions. When measurements were performed out of principal directions (Table 4.7), the moduli determined from longitudinal modes were between 1.02 and 1.16 higher than those measured from flexural modes.

Flexural testing modes afford a very useful opportunity to observe the coupling effect between shear and extensional stress in wood. Dynamic methods to determine elastic constants, deduced from the flexural vibrations of beams, were developed by Sobue (1986b) on small, clear specimens and on

Table 4.7. Dynamic Young's moduli determined by longitudinal and flexural resonance method in *LR* plane at different angles for Sitka spruce

Specimen angle	Longitudinal test (ᴸ)		Flexural test (ᶠ)		
	f_L (Hz)	E_L^L (10^8 N/m²)	f_F (Hs)	E_L^F (10^8 N/m²)	$E_L^L{:}E_L^F$
0°	9531	141	581	138	1.02
15°	9530	68	943	58	1.17
30°	6194	15	612	25	0.60
45°	4906	16	472	15	1.06
60°	4269	18	410	11	1.63
75°	4057	19	382	10	1.90
90°	3797	19	359	9	2.11

Note: 0 degrees = specimen in *L* direction; 90 degrees = specimen in *R* direction.

From Tonosaki et al., *Mokuzai Gakkashi*, 29(9), 547, 1983. With permission.

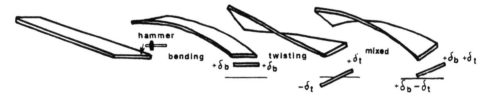

Figure 4.8. Complex vibration of a plank struck by a hammer (Sobue, 1987).

commercial-sized samples such as logs (Sobue, 1990; Arima et al., 1991) and timber (Sobue, 1988; Chui, 1991). Figures 4.8 and 4.9 present the complex vibrations emanating from a plank struck with a hammer and the corresponding displacements for different modes. Specimens were tapped with the hammer at the edge, at one end, and at a point located one quarter of the length from the end, to enhance the peaks of the higher harmonics. The tap tone was detected with two condenser microphones (Figure 4.10). A very sophisticated signal treatment, with a fast Fourier transform analyzer (Figure 4.11), enabled the separation of peaks corresponding to flexural and twisting vibrations. The value of the frequency corresponding to each peak was determined precisely, and consequently the values of the elastic constants were deduced.

Among the resonance methods, the technique utilizing the torsional mode with bar specimens (Radu and Brenndörfer, 1976; Morze et al., 1979; Olszewski et al., 1983; Becker and Noack, 1968) is less used, probably due to practical difficulties related to the torsional pendulum and to the very small frequency band obtained during testing. Some experimental difficulties could be overcome if we bear in mind that the temperature and frequency are equivalent parameters and that a relatively low resonance frequency may be obtained by increasing the temperature of the specimen.

Becker and Noack (1968) published an interesting study on beech, which gave the relationship between shear modulus, temperature (20° to 100°C), and moisture content (5 to 30%) for the characterization of the viscoelastic behavior of wood (Figure 4.12).

Another way to determine the elastic constants of wood is to use plate vibration tests on which complex Young's moduli, shear moduli, and Poisson's ratios can be determined (Caldersmith and Rossing, 1983; Nakao et al., 1985; Tonosaki and Okano, 1985; Sobue and Katoh, 1990; McIntyre and

Figure 4.9. The displacement of a beam seen in transverse section at different vibration modes. (a) Flexural mode; (b) twisting mode; (c) flexural mode and twisting mode. (From Sobue, N., *Mokuzai Gakkaishi*, 34(8), 652, 1988. With permission.)

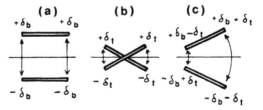

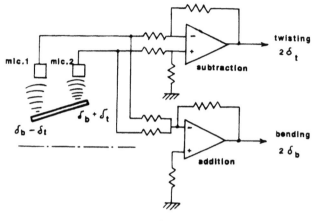

Figure 4.10. Device used for the addition and subtraction of signals obtained from a plank struck by a hammer. (From Sobue, N., *Mokuzai Gakkaishi*, 34(8), 1988. With permission.)

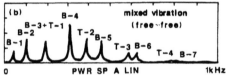

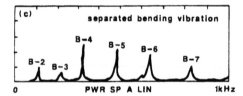

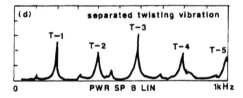

Figure 4.11. Power spectra deduced from a piece of dimension lumber (50 mm × 100 mm × 3 m). (From Sobue, N., *Mokuzai Gakkaishi* annual report, 1987. With permission.)

Woodhouse, 1988; Schumacher, 1988; Molin and Jansson, 1989; Sobue and Kitazumi, 1991). On orthotropic plates the resonance frequencies $f_{r(i,j)}$ are related to the elastic parameters by:

$$f_{r(i,j)} = 1/2\pi \cdot [S'/\rho h]^{1/2} \tag{4.54}$$

and

$$S' = D_{11}a_{1(ij)}a^{-4} + D_{22}a_{2(ij)}b^{-4} + 2\,D_{12}a_{3(ij)}a^{-2}b^{-2} + 4\,D_{66}a_{4(ij)}a^{-2}b^{-2} \tag{4.55}$$

where ρ is the density of the material, h is height, a is the length and b is the width of the plate, and a_1 to a_4 are coefficients depending on the supported condition of the plate (clamped, free, or simply supported). Table 4.8 gives the coefficients for the most common experimental condition, the free vibration mode. D_{11}, D_{22}, and D_{12} are flexural rigidities and D_{66} is the torsional rigidity.

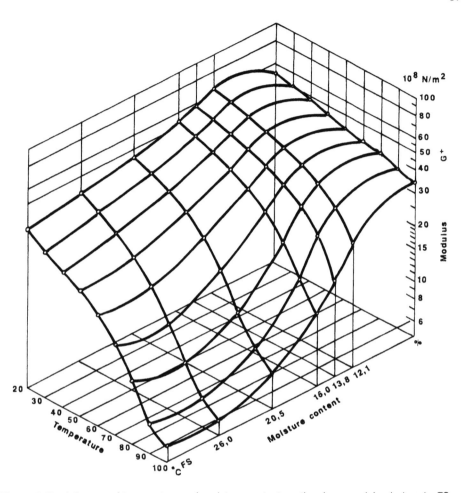

Figure 4.12. Influence of temperature and moisture content on the shear modulus in beech. FS = moisture content at fiber saturation point. (From Becker, H. and Noack, D., *Wood Sci. Tech.*, 2, 213, 1968. With permission.)

Table 4.8. Values of coefficients a_1 to a_4[a] for a completely free vibration of a rectangular orthotropic plate (Sobue and Kitazumi, 1991).

Vibration modes		Coefficients			
m	n	a_1	a_2	a_3	a_4
1	1	0	0	0	144
0	2	0	500.6	0	0
0	3, 4, etc.	0	Y^4	0	0
2	0	500.6	0	0	0
3, 4, etc.		X^4	0	0	0
1	2	0	500.6	0	593.76
1	3,4, etc.	0	Y^4	0	$12.3Y(Y+6)$
2	1	500.6	0	0	593.76
2	2	500.6	500.6	151.3	2448.3

Note: $X = (m - 0.5)\pi$; $Y = (n - 0.5)\pi$.

[a] From Equation 4.55.

58

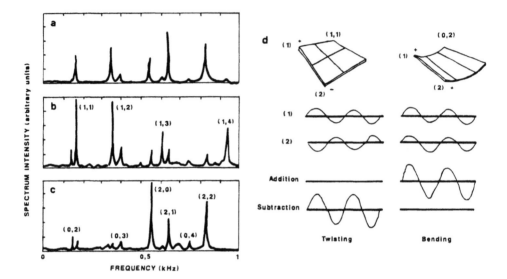

Figure 4.13. Power spectra on a rectangular freely vibrating Western red cedar plate. (a) Detected by the microphone; (b) separated by subtraction procedure for twisting modes; (c) separated by addition procedure for flexural modes; (d) vibrational modes of the plate. (Sobue, N. and Kitazumi, M., *Mokuzai Gakkaishi*, 37(1), 9, 1991. With permission.)

The engineering constants E_1, E_2, and G_{12} and Poisson's ratio in the corresponding anisotropic plane can be calculated by the relations:

$$E_1 = 12h^{-3}D_{11}\mu \tag{4.56}$$

$$E_2 = 12h^{-3}D_{22}\mu \tag{4.57}$$

$$G_{12} = 12h^{-3}D_{66}\mu \tag{4.58}$$

$$\nu_{12} = D_{11}/D_{22} \tag{4.59}$$

where:

$$\mu = 1 - D_{12}^2/D_{11}D_{22} \tag{4.60}$$

Based on these theoretical assumptions, Sobue and Kitazumi (1991) precisely determined the resonance frequency of a rectangular free vibrating plate, identifying the corresponding peaks from the power spectra of different vibration modes (i.e., [0,2], [1,1], [2,0], [2,2]). Figure 4.13 gives the typical power spectrum of a Western red cedar plate (300 × 300 × 10mm). The misidentification of peaks corresponding to flexural or torsional modes was avoided because the deflection phase was considered. The proposed procedure was used for the automatic identification of resonance peaks of the power spectrum and consequently for simultaneous, routine measurement of engineering constants E_1, E_2, G_{12}.

4.3. VELOCITY OF ULTRASONIC WAVES IN WOOD

The measurement of ultrasonic wave velocities in wood (considered an orthotropic material) is the basis of the nondestructive evaluation of its elastic or viscoelastic properties. The fundamentals of the propagation of ultrasound in homogeneous solids are given by McSkimin (1964) and Papadakis (1990) in polycrystalline media. Given that these are excellent references, the emphasis in this section is on theoretical aspects that are related directly to the measurement techniques that are appropriate for wood.

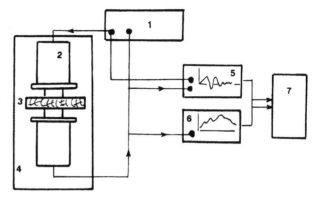

Figure 4.14. Block diagram for velocity and attenuation measurements, using direct transmission technique in wood. (1) Ultrasonic generator; (2) transducer; (3) specimen; (4) mechanical device; (5) oscilloscope; (6) spectrum analyzer; (7) computer. (From Bucur, V. and Böhnke, I., *Ultrasonics*, 32(5), 385, 1994. With permission.)

The principal wave types used for measuring wood properties are bulk waves (longitudinal or transverse-shear) and surface waves (Rayleigh, Lamb, and Love waves). The waves are characterized by the direction of propagation and by the particle motion. More specifically, for longitudinal waves the particle trajectory is in the direction of propagation; for transverse waves the particle motion is perpendicular to the direction of propagation; for Rayleigh waves the particle trajectory is elliptical in the plane which is perpendicular to the tested surface and parallel to the direction of propagation.

In dispersive media the ultrasonic velocity is dependent on frequency, and both phase and group velocities (or the velocity of the wave packet) can be measured.

The relationships between group velocity v and phase velocity V were extensively commented upon by Guilbot, 1992 and can be summarized in the relation:

$$v = V/(1 - f/V \, dV/df) \tag{4.61}$$

The phase velocity is

$$V = \omega/k = f \cdot \lambda \tag{4.62}$$

where $\omega = 2\pi f$ and f is the frequency, the propagation constant is $k = 2\pi/\lambda$, λ is the wavelength. When $k = k(\omega)$ and when V is a function of frequency, the medium is dispersive. Dispersion may be induced by the geometry of the specimen, the nature of the material, the scattering produced by the inhomogeneities of the structure, the absorption of wave energy by the material during propagation, etc. Experimental methods for the determination of phase and group velocities in dispersive solids are given by Sachse and Pao (1978) and by Sahay et al. (1992).

4.3.1. MEASUREMENT SYSTEM

Ultrasonic velocity measurements can be made with broadband pulses or with narrowband bursts, and are related to the time of flight and length of specimen measurements.

Either the immersion technique (when the specimen is immersed in a liquid) or the direct transmission technique (when the specimen is in contact with the transducer) can be used for wood material. Immersion is more appropriate for laboratory testing while direct transmission is convenient for both laboratory and field measurements. The principal advantage of either techniques is the flexibility in measuring velocity and attenuation of ultrasonic waves.

Wood material which is to be sensed and probed with ultrasonic waves may be most conveniently divided into three main groups: trees and logs, small clear specimens of solid wood, and wood-based composites. The direct transmission technique seems to be the most generally appropriate technique for ultrasonic testing of all these types of wood.

4.3.1.1. Devices

The most common block diagram for ultrasonic velocity and attenuation measurements is presented in Figure 4.14. The electric signal is transmitted from the generator to the emitter, transducer $-E$, and transformed in ultrasonic pulse. This pulse travels through the specimen and is received by the receiver,

Table 4.9. Acoustic and piezoelectric parameters of piezoelectric materials for ultrasonic transducers

Symbol	Definition
d	Transmission constant—(strain out/field in)
g	Receiving constant—(field out/stress in)
ρ	Density
v	Ultrasonic velocity in a particular direction $[(C^E/\rho)^{1/2}]$
Z_0	Characteristic acoustic impedance ($\rho \times v$)
ϵ^T	Free dielectric constant (unclamped)
k_T	Electromechanical coupling efficiency ($k_T^2 = e^2/\epsilon^s C^E$)
Qm	Mechanical quality factor

From O'Donnell et al., *Methods of Experimental Physics: Ultrasonics*, Edmonds, P. D. (Ed.), Academic Press, NY., 1981. With permission.

transducer −R, and is transformed again in an electric signal, which is visualized on an oscilloscope. This allows the measurement of the elapsed time between emission and reception. The time delay is measured on the oscilloscope, over the path length of ultrasonic signals. The technique is very simple and the time measurements very accurate (<1%). A spectrum analyzer can be used for more sophisticated measurements related to attenuation or phase.

4.3.1.2. Transducers

The physical basis for the generation and detection of ultrasonic signals with piezoelectric transducers has been extensively commented upon in the physical acoustics texts of Sachse and Hsu (1979), Hutchins and Hayward (1990), O'Donnell et al. (1981), and Lynnworth (1989). In the ultrasonic generation mode the transducer incorporates a piezoelectric element which converts electrical signals into mechanical vibration. The inverse effect is used for the detection of ultrasonic waves traveling through the specimen.

The active element in a piezoelectric transducer is a disk, made from ceramic materials such as barium titanate, lead zirconate-titanate, lead metaniobate, all of which have a favorable combination of mechanical, electrical, and piezoelectric properties. Some data on the acoustic and piezoelectric parameters of materials used for ultrasonic transducers are given in Tables 4.9 and 4.10. The mode of vibration of piezoelectric ceramics (longitudinal shear, etc.) is determined by the orientation of the disk, relative to an axis that is imposed on the material during the sintering process. Polarizing the structure is achieved by first raising the temperature of the ceramics above the Curie point and then by cooling the ceramics in a strong electric field.

Dimensions and shapes of several transducers commonly used in the 1980s for velocity and attenuation measurements are given in Figure 4.15. The time and frequency domain responses of a 1-MHz broadband transducer is given in Figure 4.16.

Transducer performance is related to constructional parameters such as radiation surface area, mechanical damping, characteristics of the piezoelectric and backing materials, and the connection of electrical and acoustical components of the system. Selection of the piezoelectric material is dictated by the specific application required by the transducers (Tables 4.9 and 4.10). The efficiency of the transducer as a transmitter is related to a large "d" constant, whereas its sensitivity as a receiver depends more on a large "g" constant.

The basic requirements of an ultrasonic transducer are good sensitivity and resolution, controlled beam pattern and reproducible performance under various testing conditions, and high signal to noise ratio.

The source of the experimental limitations of piezoelectric transducers is the coupling medium with the specimen under test. Very often the ultrasonic signal is disturbed, and interference phenomena, phase shift, and attenuation of the signal are associated with the propagation in the coupling layer.

4.3.2. SPECIMENS FOR ULTRASONIC TESTING

It is appropriate to observe that measurements of ultrasonic velocity and attenuation are influenced by requirements related to sample preparation, coupling of the transducer to the sample, and to the refine signal processing. This section reviews some specific requirements for testing trees, small, clear specimens, and wood composites.

Table 4.10. Piezoelectric material properties for longitudinal and transverse waves

	Longitudinal Waves					
	Quartz*	PZT-4	PZT-5	PZT-5H	PbNb$_2$O$_6$	BaTiO$_3$
d (10^{-12} m/V)	2	289	374	593	75	149
g (10^{-3} Vm/N)	50	26	25	20	35	14
ρ (kg/m^3)	2,650	7,600	7,500	7,500	5,900	5,700
v (m/s)	5,650	3,950	3,870	4,000	2,700	4,390
Z_0 (10^6 kg/m^2 s)	15	30	29	30	16	25
ϵ^T/ϵ^o	4	1,300	1,700	3,400	240	1,700
k_T (%)	11	70	70	75	40	48
Qm	>25,000	<500	<75	<65	<5	<400

	Transverse Waves					
	Quartz[1]	Quartz[2]	PZT-4	PZT-5	PZT-5H	BaTiO$_3$
d (10^{-12} m/V)	4.4	3.4	496	584	741	260
g (10^{-3} Vm/N)	110	80	38	38	27	20
ρ (kg/m^3)	2,650	2,650	7,600	7,500	7,500	5,700
v (m/s)	3,850	3,320	1,850	1,680	1,770	2,725
Z_0 (10^6 kg/m^2 s)	10	9	14	13	13	16
ϵ^T/ϵ^o	5	5	1,475	1,730	3,130	1,450
k_T (%)	14	9	71	68	65	50
Qm	>25,000	>25,000	<500	<75	<75	<300

Note: Quartz* = quartz 0° X-cut, quartz[1] = 0° Y-cut; quartz[2] = AT-cut; X = thickness expander; AT = thickness shear cut when the quartz plate orientation is considered. (The mode of vibration of crystalline piezoelectrics is determined by the orientation of the plate relative to the crystalline axes.)

From O'Donnell et al., in *Methods of Experimental Physics: Ultrasonics*, Edmonds, P.D. (Ed.), Academic Press, N.Y., 1981. With permission.

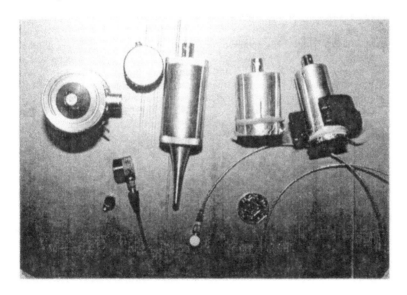

Figure 4.15. Sample broadband ultrasonic transducers for ultrasonic velocity and attenuation measurements using direct transmission technique in wood. (Courtesy of Panametrics Inc., Waltham, MA)

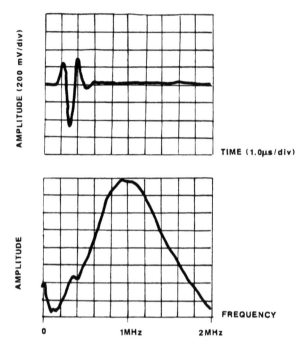

Figure 4.16. Time and frequency domain response characteristics of a 1-MHz broadband transducer. (Courtesy of Panametrics, Inc., Waltham, MA)

4.3.2.1. Preparation of Samples

The term "samples" includes all wood specimens such as trees, small, clear specimens of solid wood, and wood composite specimens. The specific differences in the requirements for preparation of particular samples is discussed later.

Ultrasonic measurements of trees may be performed on the periphery of the trunk, without bark. Measurements of logs can be done on the transverse sections at both ends. In this case no specific care is needed, although the surfaces must be parallel and as flat as possible.

The potential of the ultrasonic velocity method was demonstrated in the nondestructive measurement of the slope of the grain of living trees and when qualitative parameters for round wood were established (see Chapter 8). The range of ultrasonic velocities measured in wood at 12% moisture content at 1 MHz is between 6000 m/s for longitudinal waves in the fiber direction and 400 m/s for shear waves in the radial-tangential plane. The values of the attenuation coefficients are roughly 2 dB/cm for longitudinal waves in the direction of the fiber and 15 dB/cm for shear waves in the transverse anisotropic plane.

The conditions for satisfactory specimen preparation depend essentially on the magnitude of attenuation of ultrasonic waves in the wood species under test. Generally, the higher the attenuation, the greater the requirements concerning the flatness and parallelism of the specimen surfaces. In addition, the samples must be accurately perpendicular to the principal direction or to the required direction out of the main symmetry axes along which the ultrasonic waves propagate. The geometric shapes of typical small, clear wood specimens used for acoustic measurements may be rectangular parallelepipeds, cubes, rods, disks, or plates. The purpose of the research determines the choice of shape. Generally speaking, use of the ultrasonic immersion technique necessitates plates, while any specimen shape is suitable for the direct contact technique. If wave velocities are of interest along only a restricted range of directions (most commonly, the longitudinal direction), then the minimum requirement for specimen preparation is two plane, parallel faces. The larger the specimen dimensions, the longer the propagation time of the ultrasonic signal in wood, and the higher the accuracy of measurement.

At relatively high frequencies the requirements for parallelness of the opposite faces of wood specimens become more severe, and the bond between the transducers and the sample becomes more critical. Velocity measurements along the longitudinal direction are reported in the wood science literature for a large dimensional range, varying in scale from millimeter (Bucur, 1983) to meter (McDonald, 1978).

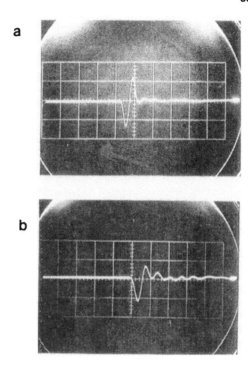

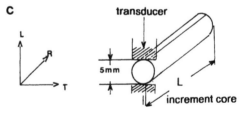

Figure 4.17. Pulse shape displayed when propagation of a 1-MHz signal in a 5-mm douglas fir increment core occurs in the longitudinal direction. (a) Signal with transducers in contact; (b) signal with increment core (T/2 μs); (c) increment core for testing. (Bucur, 1984b)

Figure 4.17 shows the pulse shape displayed for propagation of a 1-MHz signal for a 5-mm thickness Douglas fir increment core in the longitudinal direction. In such an extreme case attention should be paid to the excitation of the ultrasonic pulse (and if necessary, to the repetition rate) as well as to the excitation pulse length.

Errors may be introduced into the velocity measurements if the arrival times of the distorted and undistorted pulses are compared when the probes are in contact. The misalignment of transducers and specimens is another important introduced error. The error becomes larger when the sample is not accurately aligned with the transducer. The procedure used to diminish this effect is to maximize the amplitude of the signal from the interface.

If all elastic stiffnesses of a wood species are required, the sample size and shape constraints are more severe. For laboratory measurements it is necessary to manipulate rather small specimens in order to limit the effect of any spatial inhomogeneity of wood induced by its anatomical structure and to allow neglect of the annual ring curvatures related to the T direction. However, the specimen proposed for measuring the nondiagonal terms of the stiffness matrix does allow the propagation of quasilongitudinal and quasishear waves out of the principal symmetry axes. The specimen and the ultrasonic beam must be rotated relative to each other. This may be done in two ways:

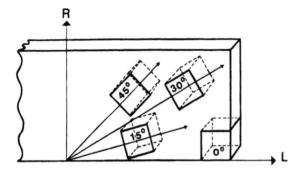

Figure 4.18. Rotation of cubic specimen in *LR* orthotropic plane. (Bucur, 1984a.)

1. The ultrasonic beam may be rotated with respect to a fixed sample which has edges that coincide with the material symmetry axes.
2. The ultrasonic beam is held in a fixed position while the specimen cutoff axis is rotated.

The first method is typically used in the immersion technique, which uses plate-type specimens (45 × 45 × 10 mm) (Preziosa et al., 1981). Commonly, three plates, one corresponding to each anisotropic plane, are employed for the characterization of a wood species. The second method employs cubic specimens or disks together with a direct contact technique. Figure 4.18 shows the rotation of cubic specimen in *LT* plane. Cubes with 16-mm sides were cut from the same board at 0°, 15°, 30°, and 45° to the main orthotropic axis. The same cubes were used to obtain velocities at 105°, 120°, and 135°. In all, ten cubes were necessary for the determination of the terms of the stiffness matrix of one wood species (Bucur and Archer, 1984). The cubic shape of the specimens did not present technical difficulties in the cutting process, but the great number of samples caused some practical difficulties.

To avoid this very tedious procedure and to restrict the natural variability of specimens, the multifaced disk sample may be used (Figure 4.19). The diameter of the disk is 35 mm. The faces are cut at 15°, 30°, 45°, and 75°, as well as at 0° and 90° (Bucur and Perrin, 1988). To achieve experimental efficiency, economy of sample material, and economy time for specimen production, the disks seem to be an ideal sample type for ultrasonic measurements using the direct contact technique. Only three disks are needed for the complete characterization of a wood species.

The influence of the natural variability of specimens, due to the biological nature of wood, on velocity and attenuation may be studied by choosing the frequency of the source so that the acoustic wavelengths in the material lie in a range roughly between the maximum dimension of the anatomical elements and the minimum specimen dimension. Some comments on the accuracy of the results of velocity and attenuation measurements are appropriate here. Parameters such as the probe diameter, the maximum pulse width, the attenuation of ultrasonic waves in wood in the three principal anisotropic planes, or the separation time of quasilongitudinal and quasitransverse waves must be considered when determining sampling strategy. Deviation of the energy flux vector of the quasilongitudinal and quasitransverse waves should not be ignored when determining the size of the specimen to be tested. A second consequence of the energy flux deviation is that for certain directions out of the principal axes, the propagation of more than three modes may occur. This phenomenon must be accounted for to avoid misinterpretation of the readings.

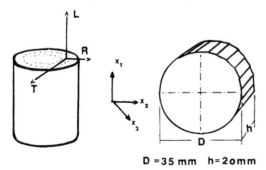

$D = 35\ mm \quad h = 20\ mm$

Figure 4.19. Multifaced disk-type specimen. (a) Natural wood symmetry and multifaced disk specimen; (b) velocities that can be measured. Note that one disk is cut parallel to an anisotropic plane. Three disks are needed for complete characterization of a wood species. (From Bucur and Perrin, 1987.)

In anisotropic and inhomogeneous materials such as wood, however, a mechanical transducer will simultaneously generate multiple modes. This phenomenon has special consequences for ultrasonic investigation of thin specimens in the RT plane because the repetition rate of the pulse is too great to allow the various modes to be separated in time. Thus, misinterpretation of travel time or attenuation measurements can occur.

A lower limit to the size of the specimen is also imposed by the requirement that waves should possess the character of plane waves in an infinite medium. The minimum size of the specimen ($<2\lambda$) must be established experimentally because no good theoretical criterion exists for this purpose.

The presence of inhomogeneities in a sample limits the achievable accuracy of the readings. For example, in a sample in which internal cracks or other discontinuities are comparable to the pulse wavelength, the pulse is attenuated by scattering at interfaces. The attenuation is frequency dependent. If the pulse shape is changed, errors in velocity and attenuation readings can occur. If the size of the discontinuities is much smaller than the wavelength, the change in pulse shape is smaller, allowing accurate data to be obtained. If inhomogeneity arises because of a gradual change in the measured property with position in a sample (caused perhaps by decay or other biological attack), then the wave velocity in a particular direction varies with the part of the specimen through which the ultrasonic pulse is traveling. Changes in the measurements of velocity or attenuation would indicate the degree to which the sample was inhomogeneous in different zones.

This is a good place to drawn attention to the fact that other factors such as internal and external conical refraction of the ultrasonic beams, or that pure mode axes do not occur along with the symmetry directions must be investigated further. However, the complexities discussed above need to be taken into account as regards the most frequently used configuration, which consists of sending an ultrasonic wave in the direction parallel to the L axis.

Finally, is worth remembering that the main advantage of using ultrasonic waves on wood specimens to measure velocity and attenuation is that the material under test is not affected by the propagation phenomena. The sample can be retested because no deformation or destruction occurs. The ultrasonic tests can be repeated on the same sample and show little variation. The results may be repeated for similar samples and experimental conditions.

4.3.2.2. Coupling Media

Commercially available ultrasonic transducers are constructed using a layer of material that covers and protects the piezoceramic element. Vincent (1987) and D'Souza et al. (1989) have dealt with the specific effects of coupling and matching layers on the efficiency of ultrasonic transducers. Several types of coupling from transducer to sample are possible, as are several arrangements of the transducers with respect to specimen (Truell et al., 1969). A single transducer can be both the source and the receiver for the echoes that result from a single pulse. Alternatively, two transducers on the opposite faces of the specimen can be used separately as emitter and receiver of an ultrasonic pulse. In the particular case of small, clear wood samples the most suitable scheme uses two transducers. A single transducer may be used only if the specimen does not exhibit too much attenuation. The limits depend on the thickness of the sample and its specific attenuation, which is important in the RT plane.

In this section attention is given first to the coupling media using wood specimens in the direct transmission technique with two transducers.

Coupling media are necessary to ensure bonding of the transducer to the wood specimen. This is accomplished using a variety of materials (solids or liquids), depending on the circumstances of use. Many bonding media have been used over a wide range of temperatures and moisture content. Silicone resins, wax, mineral greases, and glycerine are commonly used at room temperature. Alcohols can be used at lower temperatures. Methods of creating the bond vary with the testing conditions (nature of the method used for velocity or attenuation measurements, temperature, hardness of the wood sample, etc.). With hardwoods, when the specimen is large enough, the transducer may be screwed into the sample and no coupling medium is needed.

Because the aim of our analysis is to report the way in which accurate ultrasonic measurements may be performed, it is evident that coupling losses are to be made as small as possible at all frequencies. In any case the losses due to the transducer bond and to the system can be reasonably well measured by methods described in the previous sections of this chapter.

The effects of the thickness of the coupling medium can be neglected if a very thin coating couplant is used. For precise measurements a well-defined pressure applied to the specimen through the transducer

Table 4.11. Influence of coupling media on V_{TT} of propagation of longitudinal waves on *Pinus* spp. at 12% moisture content (Bucur, 1984).

Coupling media	Velocity (m/s)	Notes
No coupling	1000	Transducers are applied to specimen under small pressure
Cellophane sheets		Surface of the specimen is clean, the
0.03 mm	1029	reading easy to perform; during
0.02 mm	973	experiment, integrity of sheets must be verified
Mineral grease	1004	Grease could penetrate in specimen; easy handling
Medical gel	1040	No penetration into specimen; easy handling
SWC Panametrics	1050	Very good bond, very absorbent by wood

Note: A thin layer of SWC Panametrics resin on the surface of the transducer covered with a cellophane sheet could be useful for the protection of the surface of wood samples during measurements.

is required for reproducible measurements of attenuation. Force transducers or strain gauges can be installed on the transducer holders to measure the pressure applied to the specimen. Great care is required for low-density specimens (e.g., balsa) because the applied pressure can affect the measured velocity and attenuation, particularly for high frequency, or even break the fragile specimen. Table 4.11 reports observations concerning the influence of various types of coupling media on the ultrasonic velocity measurements.

In practice, the changes induced by the penetration of the coupling medium into the wood specimen can produce unexpected experimental errors (Table 4.12). The error initially encountered is the apparent unstable velocity and attenuation measurements. Our understanding of these results is that the wood specimen absorbs the coupling medium to a significant degree, the characteristic acoustic impedance is changed, and the transmit-receive response is modified. To avoid such situations, the thin layer of coupling medium on the transducer surface can be covered with a cellophane sheet (transparent wrapping material made from viscose). (Also, note that the acoustic impedance of cellophane matches well that of wood.) This protective layer keeps the surface of the wood sample clean and the results are reliable.

Progress in ultrasonic transducers used in the direct contact procedure was achieved recently by employing a dry coupling layer.

When the immersion technique is used, particular care is necessary to maintain the moisture content of the specimen. One of the best methods is to envelope the specimen in a surgical membrane, which protects the sample from the penetration of external humidity.

4.3.2.3. Specimens of Finite Dimensions

Specimens of finite length and width can affect the conditions of the propagation of ultrasonic waves in an infinite medium. It is usually desirable for the diameter of the sample to be several times greater than the diameter of the transducer (especially at low frequencies with highly divergent beams) in order

Table 4.12. Effect of penetration into wood by coupling materials on velocity measured in Sitka spruce*

Materials	Differences (%) in velocities in L, R, T directions					
	Coupling material effect			Penetration effect		
	L	R	T	L	R	T
Vaseline	3.08	3.98	3.96	3.24	0.54	1.41
Grease	5.99	4.21	4.79	1.49	2.49	1.05
Machine oil	4.95	4.94	4.10	2.62	1.59	1.69
Water	5.20	2.81	4.81	2.47*	13.66*	13.29*

Note: *, specimens immersed in water. The reference value is considered for measurements on specimens in direct contact with the transducer. (For small, clear specimens this reference must be considered with care.)

(From Kamioka, H. and Kataoka, A., *Mokuzai Gakkaishi*, 28(5), 274, 1982. With permission.)

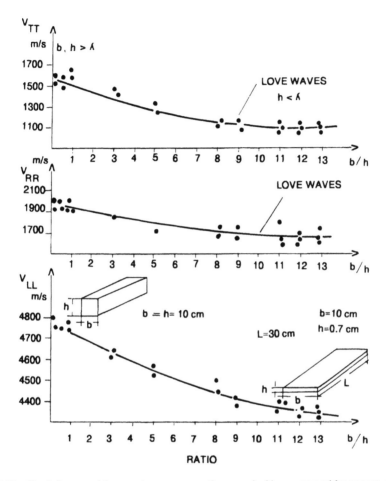

Figure 4.20. The influence of the specimen cross-section on velocities measured for spruce, when the length of the specimen is constant. Dimensions of the specimen — at the beginning of the experiment: L = 30 cm, b = 10 cm, h = 10 cm; at the end of the experiment: L = 30 cm, b = 10 cm, h = 0.7 cm. Corresponding anisotropic axis: L in the length of the specimen, R in b, T in h (Bucur, 1984).

to reduce side wall effects. However, larger-diameter transducers are more difficult to bond to the sample. If the transducer fully covers the end of the sample, the transmission phenomenon confines the wave to the specimen (cylindrical rod or slab, etc.), as in a wave guide.

Principally this section analyzes the influence of the finite size of specimens on ultrasonic longitudinal wave velocity measurements. To complete the discussion, two cases are presented: specimens of constant length and of variable cross-section, and specimens of constant cross-section, and variable length.

Spruce specimens having a constant length of 300 mm and an initial cross-section of 120 × 100 mm were repeatedly planed to modify the ratio of width to thickness from 1:14. For every value of this ratio the longitudinal wave velocities V_{LL}, V_{RR}, and V_{TT} were measured (Figure 4.20). The velocity V_{LL} is strongly and continuously affected by the ratio b:h. The maximum V_{LL} is obtained when the ratio b:h lies between 1 and 2 and the specimen is a rod and b and h are greater than the wavelength. (In this case the velocity of a longitudinal plane wave was probably measured.) The minimum V_{LL} was measured for the ratio b:h = 13 to 14, when the specimen was a plate and h was smaller than the wavelength. The measured velocity corresponds to the Love wave velocity. The values of the velocities V_{RR} and V_{TT}, corresponding to the measurements obtained from the transverse section of the specimen, are less affected by the modification of the section geometry. It seems that for the ratio b:h = 10 or higher, the cross-section does not influence V_{RR} and V_{TT}. We note that the dimensions of the specimen

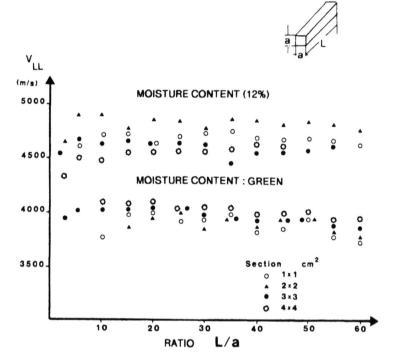

Figure 4.21. The influence of the length of the specimen on longitudinal V_{LL} when the cross-section is constant. Measurements on beech at 12% moisture content and under saturated conditions (Bucur, 1984).

corresponding to the propagation in R and T directions are greater than 2λ. For this reason the velocities V_{RR} and V_{TT} are not influenced by modifying the size of the specimen.

Figure 4.21 shows the measurements on beech specimens with constant cross-section and variable length. Specimens of 600 mm initial length and cross sections of 10×10, 20×20, 30×30, and 40×40 mm were shortened successively from the initial length to 25 mm. The longitudinal V_{LL} is nearly constant when the length: width ratio is varied from 20 to 40. Below this limit, V_{LL} diminishes. One may also deduce from Figure 4.21 that the length has no influence on V_{LL}. The velocity under dry conditions (12% moisture content) is always higher than the velocity under more humid conditions.

The influence of the path length on the V_{LL} when the specimen is simultaneously shortened and planed in cross-section is shown in Table 4.13. The V_{LL} diminishes about 12% between the shorter and the longer lengths. This fact is connected with the reduction of the dimensions of the transverse section and can be explained by mode conversion phenomena (from bulk longitudinal waves in an infinite solid to longitudinal waves in a rod).

Table 4.13. Influence of path length on ultrasonic V_{LL} measured on parallelepiped spruce specimens at 12% moisture content, 1 MHz, broadband transducers, and diameter $\phi8$ mm (Bucur, 1984).

Specimen	Dimension (mm)			Ratio, L:b	Velocity V_{LL} (m/s)
	L	b	h		
Infinite solid	180	60	60	3	5570
	120	40	40	3	5100
Rod	60	20	20	3	4900

Note: The specimen was simultaneously shortened and planed, and was oriented as L in the longitudinal direction, b in the radial direction, h in the tangential direction.

Table 4.14. Influence of path length on ultrasonic velocities (measurements on horse chestnut wood, at 12% moisture content, with tone burst technique, 1.5 MHz and direct contact) (Bucur and Perrin, 1988).

Specimen	L	R	T	Section	Ratio, R:T	Velocities (m/s)		
						V_{LL}	V_{RR}	V_{TT}
Cylinder	160	18	18	Circle	1	5298	1521	1269
Disk	70	70	22	Rectangle	3	4975	1532	1263
	35	35	22	Square	1.6	5013	1562	1261

Another interesting aspect is that of the influence of the geometrical shape of the specimen on velocity values. Table 4.14 gives the values for the longitudinal V_{LL} for horse chestnut and for transverse V_{LR} and V_{LT} on cylindrical specimens and on disks. From this it can be deduced that the velocity values are not affected by the geometry of the specimen.

The multiple modes propagating in wood can be seen in Figure 4.22, in which the typical behavior of cubic poplar wood specimens in time and frequency domains are presented. From the frequency spectrum of longitudinal waves we can see the lobe patterns corresponding to the geometry of the specimen as well as the peaks made by the vibration of fibers 2 to 3 mm long in the region corresponding to frequencies >1.6 MHz. From the frequency spectrum of shear waves the annual ring width and the proportion of latewood in the annual ring can be deduced. The complexity of modes propagating in wood can also be seen in Figure 4.23, when a disc-type specimen is excited at 45° with shear waves in four symmetric points. This figure generates two points:

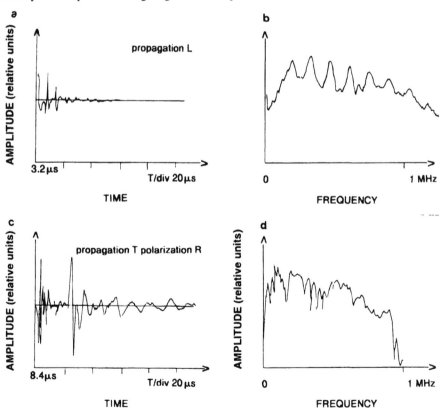

Figure 4.22. Pulse excitation with longitudinal (a, b) and transverse waves (c, d) on a cubic specimen of poplar, in time and frequency domain. Böhnke and Guyonnet (1991).

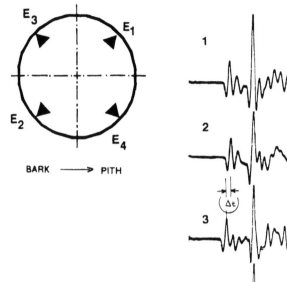

Figure 4.23. Mode conversion phenomena induced on a specimen excited at 45° with transverse waves. *Note:* 1, 2, 3 and 4 are signals corresponding to excitation points E_1, E_2, E_3, and E_4. The peak corresponding to the maximum of amplitude is related to the quasitransverse wave. The position of this peak is conserved for different points of excitation. The second peak in amplitude corresponds to the quasilongitudinal wave and is affected by the anatomical structure. This second peak shows displacement with Δt (Bucur and Perrin, 1988).

1. The main amplitude of the signal corresponds to the arrival of the quasitransverse wave. This component is conserved and arrives at the same time for all excitation positions. No interference occurs between the wavelength and the anatomic elements.

2. The smaller amplitude of the signal corresponds to the arrival of the quasilongitudinal wave. This component shows a displacement of Δt. The wavelength is comparable to the fiber length. We suppose that the time difference illustrates the misalignment of fibers.

In the future more elaborate techniques such as ultrasonic spectroscopy (Gerike, 1970; Fitting and Adler, 1981; Alippi, 1989, 1992), signal processing via high resolution spectral analysis (Chen, 1988), and evaluation in the time-frequency domain by means of the Wigner-Ville distribution (Flandrin, 1988) could improve our understanding of the phenomena related to the propagation of ultrasonic waves in wood microstructure.

4.3.2.4. Influence of the Physical Properties of Wood on Measurement of Ultrasonic Velocity

The physical properties focused upon in this section are density, a parameter which characterizes each species; moisture content, when wood is green or exposed to arid conditions; and details of the structure of the annual rings, with the corresponding proportions of early- and latewood.

The experimental relationship between density (ranging from 200 to 900 kg/m³) and sound velocity (2500 to 5800 m/s) on >40 species of soft- and hardwoods, delineated from the resonance frequency method by Barducci and Pasqualini (1948), is statistically significant at 5%. This means that the empirical relationship between these parameters is not very strong. Generally, small values of V_{LL} correspond to high densities. This seems natural if we consider the anisotropic nature of wood, as well as its structural organization. For this purpose Burmester (1965) produced a more refined analysis in which he plotted separately the ultrasonic velocity and density for two species, spruce and limba. Burmester found that spruce reacts with ultrasonic waves as a complicated natural composite material, whereas limba behaves more like an homogeneous orthotropic solid.

Let us now analyze the influence of earlywood and latewood on the ultrasonic velocity using the annual ring. The opinions of different authors are rather divergent. Burmester (1965) agrees that the velocity in isolated earlywood is slower than that measured in solid wood. Yiannos and Taylor (1967)

Table 4.15. Ultrasonic velocity on earlywood and latewood and density components in X-ray microdensitometric analysis on resonance spruce (Bucur, 1983).

Variables	Earlywood (74%)	Latewood (26%)	Solid wood (100%)
V_{LL} (m/s)	3226	3650	5500
V_{TT} (m/s)	1062	1468	1500
Density (X-ray) (kg/m³)	364	636	426
Correlation coefficient (V; ρ)	0.578**	0.613**	

Note: The measurements were made on very thin (3 × 3 mm²) radiographic specimens. This fact induces dispersion of ultrasonic waves and could explain the relatively small values measured on early- and latewood when compared to solid wood.

reported higher velocities in latewood than in earlywood in pine. Gerhards (1978) found that early- and latewood had no effect on stress wave velocity measurements parallel to the grain in Sitka spruce and southern pine. The data reported are strictly related to his experimental procedure, in which he used accelerometers clamped onto transverse sections of earlywood or latewood specimens. It is surmised that the probe was not small enough and the two zones overlapped.

Technological progress in ultrasonic transducers in the last 10 years allows fine measurements to be made on a contact surface of <1 mm². Measurements of spruce in the 1 MHz frequency range (Table 4.15) with longitudinal waves in L and T directions are related to the density determined by an X-ray technique. Ultrasonic measurements were performed on the specimens required for microdensitometric analysis (3 × 3mm² section). The values of longitudinal velocities in L and T directions, V_{LL} and V_{TT}, are different in earlywood and latewood and smaller than those measured in solid wood, probably because of the dispersion induced by the geometry of the thin specimen. The V_{LL} and V_{TT} in latewood are slightly higher than in earlywood.

Relationships statistically significant at 1% were established between velocities and the corresponding density of earlywood. Prior data are primarily interesting methodologically, nevertheless they can be used as references for producing composites for the musical instrument industry.

The influence of the curvature of rings is a pertinent question with respect to the correct measurement of V_{RR}, when bulk longitudinal waves propagate in the R direction perpendicular to the annual ring. As regards this case Woodhouse (1986) emphasized the sample-size contraint and points out that the continuum theory employed in ultrasonic characterization using an orthotropic model ignores this curvature. To clarify this point, several experiments were performed by Bucur and Perrin (1987). The effect of the curvature of the rings on V_{RR} was studied on two disks cut in RT plane. The first disk was 10 cm in diameter representing a cross-section of a young spruce tree on which the V_{RR} was 1574 m/ s. The second disk was cut from the first and was 4 cm in diameter. The relative curvature of the second disk was greater but the velocity was the same, V_{RR} = 1579 m/s. Similarity of the values enables us to conclude that for the direct transmission technique, when longitudinal bulk waves of 1 MHz are used, ring curvature has no influence on V_{RR}. The situation may be different for transverse waves (Figure 4.24).

Using the acousto-ultrasonic technique, Lemaster and Quarles (1990) analyzed the influence of the layered structure on a parameter related to the amplitude and energy of the ultrasonic signal (measured in rms voltage) induced by both the alteration of early- and latewood and that of the radius of the curvature of the annual ring. When a direct transmission technique is employed, which probably leads to the propagation of a bulk longitudinal wave in the radial direction of the specimen, the curvature of the rings has no effect on the rms voltage. When "side excitation" was used, the measured rms signal (probably corresponding to a surface wave propagating in LT plane) for the concave interface of pine was 1.2 V and was greater than that for the convex interface (0.75 V). The high amplitude of the signal observed can be explained by the fact that the signal travels from earlywood to latewood and the acoustic impedance of layers increases gradually in the concave interface. A similar situation was observed for shear waves when V_{TR} was measured.

It is well known that to define the C_{44} term of the stiffness matrix, the shear V_{TR} or V_{RT} (or both) are measured. (Note that the first index is related to the propagation direction and the second is related to the polarization direction.) Modulation of the transverse wave by the structure is strongly related to

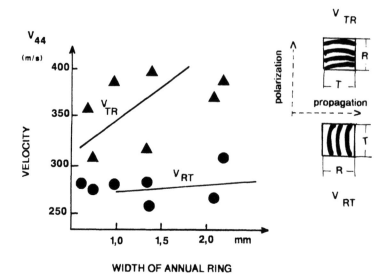

WIDTH OF ANNUAL RING

Figure 4.24. Discrepancies in transverse velocities V_{RT} and V_{TR} due to a waveguide effect in spruce. Note that the influence of the width of the annual ring is evident in V_{TR}, where the propagation vector is parallel to the layering (T). No influence on V_{RT} where the propagation vector is parallel to R (Bucur and Perrin, 1988).

the propagation and polarization directions. Data are given in Figure 4.24. It can be seen that V_{RT} is not affected by the ring curvature because the direction of propagation is R.

The opposite effect is observed on V_{TR} in species with a pronounced difference between latewood and earlywood. This velocity is probably related to the waveguide effect induced by latewood with thick walls and high density. For softwoods (spruce, Douglas fir) the difference between V_{TR} and V_{RT} is on the order of 10 to 15%, and the figure for ring porous oak is 17%. The discrepancy between the two values in diffuse, porous hardwoods is only 5%. Further investigation of these effects requires that the waveguide effects and dispersive propagation due to the interaction of ultrasonic waves with the microstructure be related to the frequency and the wavelength (Figure 4.25).

The influence of frequency on ultrasonic velocities was studied using the sinusoidal burst direct transmission technique, as described by Bucur and Feeney (1992). The frequencies studied were 100, 250, 500, 1000, and 1500 kHz. Both longitudinal and transverse waves were used. A pulse length of four cycles was employed, which produced a narrow band ultrasonic wave at driven frequency. The following observations can be drawn from Figure 4.25a. V_{11} is strongly influenced by frequency, with a large increase in velocity from 100 to 250 kHz and a steady but smaller increase in velocity from 250 kHz to 1.5 MHz. The relatively small value of velocity measured at 100 kHz was probably induced by geometric dispersion. V_{22}, V_{33}, V_{44}, V_{55}, and V_{66} are insensitive to the frequency variation for frequencies >250 kHz.

Up to this point it has been assumed that the wavelength is much longer than the structural dimensions of the material. However, as soon as the wavelength matches the layer or cell dimensions, the velocity becomes dependent on frequency. The choice of the most interesting frequency field of investigation must be related to a wavelength that is comparable to the dimensions of the anatomical elements, which vibrate as elementary resonators. Only the frequency component that matches the natural frequency of these resonators (Figure 4.25b: the fibers vibrate in the range of 1 to 1.5 MHz) can give a detailed answer to ultrasonic wave-wood structure interaction, and can at the same time explain the overall wood acoustic properties.

The interaction of microstructure and wavelength, for the solid wood that behaves like a filter with alternating pass bands and stop bands, using ultrasonic spectroscopy is an interesting research aspect for the future.

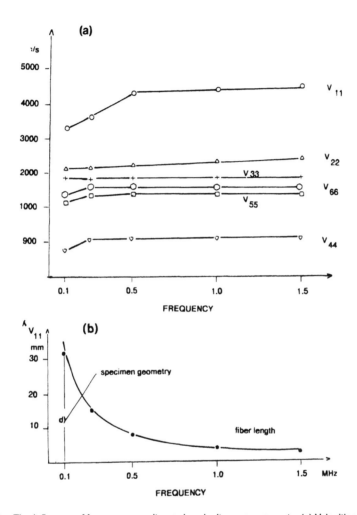

Figure 4.25. The influence of frequency on ultrasonic velocity measurements. (a) Velocities vs. frequencies from 0.1 to 1.5 MHz; (b) wavelength corresponding to V_{LL} vs. frequency. Specimen geometry is seen at 0.1 MHz and fiber length at 1-1.5 MHz.

4.4. ATTENUATION OF ULTRASONIC WAVES

The basis of the ultrasonic evaluation of the viscoelastic behavior of wood is associated with the measurement of attenuation coefficients. Pioneering work on polycrystalline media (Papadakis, 1965, 1967, 1968, 1990) and on biological tissues, polymers, and inhomogeneous media in general (Chivers, 1973, 1991) has shown that attenuation is a valuable parameter which may give information about the structure and the environmentally influenced conditions of the polycrystalline and biological materials through which the ultrasonic waves are propagated. The parameters of ultrasonic waves propagating in a solid structure can be influenced by a broad range of factors: the physical properties of the substrate, the geometric characteristics of the specimen under test (macro- and microstructural features), the environmental conditions (temperature, moisture content, mechanical loading), and the measurement conditions (sensitivity and frequency response of the transducers, their size and location, coupling medium, dynamic characteristics of the electronic equipment).

The numerical significance of attenuation depends on the specific measurement conditions (Bucur and Feeney, 1992). Attenuation coefficient values were reported by Okyere and Cousin (1980) for wood material [$\alpha_L = 5$ dB/10^{-2} m; $\alpha_R = 22$ dB/10^{-2} m for red pine at 1 MHz and 48% moisture content], and more recently by Böhnke (1992) (for dry sugi wood and longitudinal waves: $\alpha_L = 2.1$ dB/10^{-2} m and $\alpha_R = 4.7$ dB/10^{-2} m; for shear waves: $\alpha_{LR} = 3.4$ dB/10^{-2} m and $\alpha_{RL} = 14.3$ dB/10^{-2} m).

4.4.1. THEORETICAL CONSIDERATIONS

In positing these considerations we assume that wood is a viscoelastic linear solid, having an orthotropic symmetry. Therefore, we will consider the effective properties of an equivalent solid medium.

The scattering of an ultrasonic wave in solid medium results in frequency-dependent wave velocity and attenuation. The dispersion equation (Christensen 1971, 1979; Hosten et al., 1987; Chevalier, 1988, 1989; Hosten, 1991) relating all the parameters of propagation phenomena in anisotropic solids is

$$[\Gamma^*_{ij}(\omega) - \Lambda^*\delta_{ij}] = 0 \tag{4.63}$$

In axis propagation, the dispersion equation takes the following form:

$$\begin{bmatrix} \Gamma^*_{11} - \Lambda^* & 0 & 0 \\ 0 & \Gamma^*_{22} - \Lambda^* & 0 \\ 0 & 0 & \Gamma^*_{33} - \Lambda^* \end{bmatrix} = 0 \tag{4.64}$$

The eigenvalue Λ^* of the dispersion equation (Chevalier, 1989) is

$$\Lambda^{\wedge*} = \rho v^2_{ij}/(1 - i\,\alpha_{ij}v_{ij}/\omega)^2 \tag{4.65}$$

where ρv^2_{ij} is the real part of the diagonal stiffnesses and $i\,\alpha_{ij}\,v_{ij}/\omega$ is the ratio imaginary: real part of the stiffnesses ($i = \sqrt{-1}$).

4.4.2. MEASUREMENT TECHNIQUE

Measurement of attenuation can be performed either with the broad band pulses (containing a wide range of frequencies) of ultrasonic spectroscopy, or with narrow band pulses of burst excitation at a fixed frequency (Bucur and Böhnke, 1994). Both cases are analyzed below.

The calculation relationship adopted for attenuation coefficients is

$$\alpha_{ij} = -1/d \ln A/Ao \tag{4.66}$$

4.4.3. FACTORS AFFECTING ATTENUATION MEASUREMENTS

The propagation of ultrasonic waves in wood may be attenuated by three main factors: geometry of the radiation field, scattering, and absorption. The first factor is concerned with both the properties of the radiation field of the transducer used for measurements (beam divergence and diffraction) and the wave reflection and refraction occurring at macroscopic boundaries of the medium. *These factors are related to the geometry of the specimen.* Scattering and absorption are phenomena related to characteristics of the material.

4.4.3.1. Geometry of the Specimen

To study the influence of the geometry of specimens on ultrasonic attenuation, cylindrical samples of the same diameter but of different length were selected. Figure 4.26 shows attenuations expressed as amplitudes for cylindrical specimens of beech (diameter: 20×10^{-3} m, length: 50, 100, 135, 200×10^{-3} m). For small specimens, attenuation decreased linearly with the frequency over the range 1 to 2 MHz. For more lengthy specimens, the central frequency moved to the lower frequency and no linearity was observed.

Both cutoff frequency and frequency for which an amplitude maximum exists depend on the insonified volume of the specimen, which increases with the length of the sample. The waveguide effect is more evident in a small diameter specimen. For a 50-mm long specimen excited in the L direction, the cutoff frequency is about 2 MHz, for 135-mm long specimens the cutoff frequency is about 1.35 MHz. For long specimens the frequencies are shifted to the lower range, which means that attenuation is greater in long than in short specimens. Thus, the proposed linear viscoelastic model seems to be quite satisfactory. It is possible then that a recommendation could be made that relatively small specimens (20 to 50 mm long) be

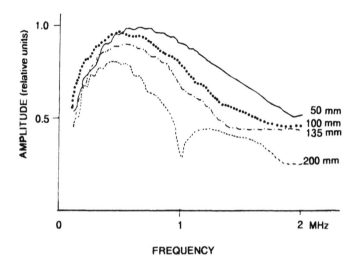

Figure 4.26. Attenuation (expressed as amplitude) vs. frequency for beech cylindrical specimens. (From Bucur, V. and Böhnke, I., *Ultrasonics,* 32(5), 385, 1994. With permission.)

Table 4.16. Length of the radiation field, wavelength, and attenuation measurements of beech, expressed by signal amplitude at a given frequency (Böhnke, 1992).

Anisotropic direction	Velocity (m/s)	Attenuation (dB)	Frequency (kHz)	Wavelength (mm)	$\phi^2/4\lambda$ (mm)
Cylinder, 20 mm Length; Transducers, 1 MHz; ϕ 14 mm					
L	4887	17	151	32	1.5
R	2492	12	79	32	1.5
T	1735	14	75	23	2.0
Cube, 20 mm; Transducers, 1 MHz; ϕ 14 mm					
L	5609	24	705	9	5.5
R	2648	35	635	4	11.0
T	1772	48	584	3	16.3
Cylinder, 20 mm Length; Transducers, 5 MHz; ϕ 14 mm					
L	4734	16	195	24	0.5
R	2151	13	254	9	1.5
T	1576	13	159	8	1.5
Cube, 20 mm; Transducers, 5 MHz; ϕ 7 mm					
L	4305	27	608	7	1.8
R	2196	40	591	4	3.3
T	1609	57	217	7	4.5

Note: $\phi^2/4\lambda$ is the length of the nearfield of radiation and λ is the wavelength.

selected for laboratory measurement. In such a case it is vital to verify that the propagation phenomena take place in the far field and that no resonance occurs. For this reason the wavelength λ and the nearfield length were calculated (Table 4.16) when measurements were performed on short specimens, cubes, and cylinders 20×10^{-3} m long. The reader should note that the dimensions of the cubic specimens are greater than the wavelength in all anisotropic directions. The nearfield of radiation in the longitudinal direction is between 1.8 and 5.5×10^{-3} m; between 3.3 and 11×10^{-3} m in the radial direction; and between 4.5 and 16.3×10^{-3} m in the tangential direction. Short, cylindrical specimens must therefore be avoided.

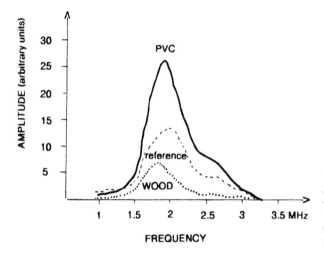

Figure 4.27. Amplitude spectra on spruce and polyvinylchloride with broad band transducers. (From Bucur, V. and Böhnke, I., *Ultrasonics*, 32(5), 385, 1994. With permission.)

As the microstructure of a solid governs its mechanical properties, it seems natural to consider that the mechanisms of wave attenuation should be related to the characteristics of the individual grain of the microstructure (the cell, in the case of wood). Moreover, it is interesting to compare the wavelength with some data related to the dimensions of the anatomic elements of wood. The length of wood cells is well known (Panshin and De Zeeuw, 1980): tracheids in conifers are about 3 to 4 mm and the fiber length of broadleaf species is about 1 mm. The dimension of spruce cells in the T direction is about 30 μm; in the radial direction this dimension may be 50 μm. The proportion of medullary rays is about 7% of the total volume.

Broadleaved species are very different and it is difficult to provide figures for general characteristics. Thus, maple (*Acer* spp.) was selected as an example. The percentages of the different elements comprising the total volume of *Acer* wood are 18% vessels, 65% fibers, and 15% rays. The fiber length is about 1 mm and the vessel length is about 0.5 mm. The diameter of the vessels is about 300 μm. The fiber diameter is approximately 30 μm.

Returning to the connections between wavelength and dimensions of the anatomical elements, we note that in cubic specimens that are insonified with longitudinal waves in the longitudinal direction, the wavelength and cell length are measured in millimeters. Propagation probably takes place in the stochastic scattering regime. In radial and tangential directions the wavelengths significantly exceed the mean cell dimensions (i.e., in the R direction $\lambda = 4$ mm and the dimension of the cell may be 50 μm; in the T direction $\lambda = 3$ mm and the dimension of the cell may be 30 μm). Here, propagation probably takes place within the Rayleigh scattering regime.

4.4.3.2. Characteristics of the Materials

The influence of the material characteristics on scattering is analyzed in three typical situations: when wood is compared to an isotropic solid, when different anisotropic axes are compared in the same species, and when the same anisotropic axis is considered in different species. When the wood material is compared to an isotropic solid we must analyze the amplitude spectrum (Figure 4.27), in which we note for wood the shift of central frequency with respect to the low frequency region.

The influence of the different material anisotropic directions on attenuation is shown in Figure 4.28a, in which the signal in the time domain is shown for the L, R, and T directions. It should be noted that for the same cubic specimen of beech, the more dispersive the direction of propagation, the more the signal loses high frequency components and consequently becomes wider. In the frequency domain (Figure 4.28b) we analyze the amplitude spectra corresponding to the L and T directions, insonified with longitudinal waves. In the spectrum corresponding to the L directions the structural vibration is between 600 and 700 kHz. From this spectrum, if the value of phase velocity is known (e.g., 4000 m/s), it is possible to determine the fiber length (approximately 5 mm), but for this purpose the appropriate methodology for phase velocity measurements must first be developed.

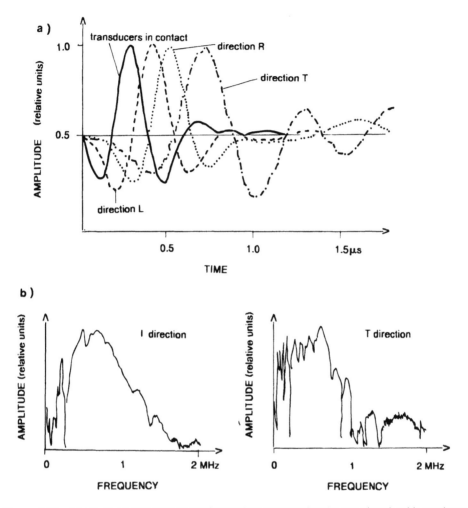

Figure 4.28. Amplitude on different anisotropic directions measured on the same beech cubic specimen (20 mm). (a) In the time domain for propagation in *L, R,* and *T* directions, using broadband transducers of 1 MHz central frequency and longitudinal waves. (b) In the frequency domain for propagation in *L* and *T* directions, using broadband transducers of 5 MHz central frequency and longitudinal waves (Böhnke, 1993).

In the *T* direction the spectrum is very complex and is similar to the spectrum obtained for periodic layered composites (Scott and Gordon, 1977). The periodicity observed in the specimen is induced by the annual rings, and coincides with the periodicity of the highest amplitude peaks in the spectrum.

When we compare the same anisotropic direction *L* (Figure 4.29) in different species (Douglas fir and poplar), we note that the broader signal is observed in the most heterogeneous species (Douglas fir). This behavior could be associated with the anatomical scatter in the structure. The richer the structure is in scatterers of various densities, the more important are the dispersion phenomena, as seen in Douglas fir.

Table 4.17 shows the attenuation coefficients for all the anisotropic axes and planes calculated for excitation by both longitudinal and shear waves. It is seen that:

- The attenuation coefficients increase with frequency for both longitudinal and transverse waves.
- For longitudinal waves, the *T* direction exhibits the highest attenuation.
- For transverse waves, generally no significant differences occurred between the attenuation coefficients of waves propagating in different directions.

78

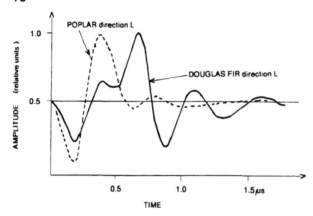

Figure 4.29. Propagation of ultrasonic waves in the longitudinal direction in a cubic specimen on poplar and Douglas fir. (From Bucur, V. and Böhnke, I., *Ultrasonics*, 32(5), 385, 1994. With permission.)

Table 4.17. Attenuation coefficients ($\alpha_{ij} = -1/d \ln P/P_o$; nepers/cm) and velocities (m/s) on curly maple and common spruce, with sinusoidal burst transmission technique and longitudinal and transverse waves

Coefficients	Frequencies (MHz)				
	0.10	0.25	0.50	1.0	1.5
Curly Maple					
Longitudinal Waves					
α_{11}	1.55	1.62	1.62	1.75	1.90
V_{11}	4332	4409	4540	4706	4559
α_{22}	2.30	2.25	2.29	2.47	2.63
V_{22}	2285	2270	2279	2325	2340
α_{33}	2.82	2.82	3.03	3.22	3.22
V_{33}	1254	1291	1321	1316	1345
Transverse Waves					
α_{44}	1.64	1.85	1.85	2.32	2.47
V_{44}	869	966	918	918	—
α_{55}	1.68	1.77	2.10	2.47	2.39
V_{55}	1214	1350	1394	1428	1399
α_{66}	1.77	1.94	2.05	2.27	2.34
V_{66}	1342	1552	1566	1602	1580
Common Spruce					
Longitudinal Waves					
α_{11}	2.17	2.07	2.10	2.29	2.47
V_{11}	4458	4847	5343	5327	5401
α_{22}	2.83	3.02	3.22	3.22	3.29
V_{22}	1612	1741	1832	1832	—
α_{33}	2.71	3.02	3.03	3.02	3.22
V_{33}	1283	1400	1321	1325	1346
Transverse Waves					
α_{44}	Impossible to analyze signal				
V_{44}					
α_{55}	1.62	1.62	1.85	2.06	2.17
V_{55}	1310	1320	1356	1383	1372
α_{66}	1.64	1.71	1.71	2.17	2.36
V_{66}	1250	1372	1383	1822	1839

Note: 1 dB/10^{-2} m = 8.69 neper/10^{-2} m.

From Bucur, V. and Böhnke, I., 32(5), 385, 1994.

Table 4.18. Statistical analysis of the influence of the natural variability of spruce wood on attenuation expressed by the measured amplitude values (Böhnke, 1992)

Statistical parameter	Density (kg/m³)	Velocity (m/s)	Amplitude (dB)	Frequency (kHz)
Minimum	345	5512	22.5	628
Maximum	600	6694	69.7	980
Average	493	6209	47.0	854
Coefficient of variation (%)	11	4	19	8

- The attenuation coefficients of a longitudinal wave in L and R directions are higher in spruce than in maple. This is probably due to important density differences in annual ring zones (e.g., in earlywood spruce the density is 300 kg/m³ and in latewood the density is 900 kg/m³). The proportion of latewood spruce is 15 to 20% of the annual ring width. In maple the latewood zone is very narrow—about 5%.

The propagation phenomena of ultrasonic waves in wood on the scale of the structure can be understood using a simplified acoustical model (Bucur, 1980). Wood cells may be treated as "tubes" of cellulosic crystalline substance embedded in an amorphous matrix, lignin. Thus, solid wood is a rectangular array of tubes embedded in a matrix. The longitudinal orientation of the tubes is slightly disturbed by horizontal elements, the medullary rays. The dissipation of acoustic energy takes place at the edges of the tubes in the longitudinal direction. Accordingly, the longitudinal axes, which are constructed from long elements, provide high velocity values and relatively small attenuation values. The highest attenuation is expected in the T direction in which no continuous structural elements exist.

Statistical analysis of the influence of the natural variability of wood material on attenuation expressed by measured amplitude is given in Table 4.18. The coefficient of variation in the longitudinal direction was 19%, which is in the same range as that for other mechanical properties of wood.

This section obviously does not cover all aspects of attenuation measurements in wood, but it may be noted that the main factors affecting ultrasonic attenuation measurements in wood are related to the geometry of the radiation field, to wave reflection and refraction occurring at the macroscopic boundary of the medium (edge of the specimen), and to the scattering phenomena.

Cubic specimens are appropriate for attenuation measurements. In cubes insonified using longitudinal waves, the wavelength and the wood cell length are measured in millimeters. In the longitudinal direction propagation takes place in the stochastic scattering regime, whereas in the radial and tangential directions it takes place in the Rayleigh scattering regime because the wavelengths exceed the dimensions of the cell.

Attenuation coefficients increase with the frequency of both longitudinal and shear waves. Attenuation is lowest in the longitudinal direction and highest in the tangential direction. The richer the wood structure is in scatterers of various densities, the more important are the dispersion phenomena.

4.5. INTERNAL FRICTION IN WOOD IN THE AUDIBLE FREQUENCY RANGE

In the audible frequency range the viscoelastic behavior of wood is associated with the magnitude of the damping coefficients. Theoretical bases for the estimation of damping coefficients in solids are given by Main (1985), Beltzer (1988), Read and Dean (1978), and Cremer and Heckl (1973). Several parameters are used to describe the internal friction or the absorption of mechanical energy by solids in the audible frequency range. The most common are

Mechanical damping, defined as the "logarithmic decrement", e.g., the logarithm of the ratio of two subsequent amplitudes of free vibrations calculated as $2\pi \tan \delta$, where δ is the phase angle.

Quality factor, Q, for steady-state forced vibrations, defined by analogy with the theory of electric circuits as the ratio of the width (Δf) of the resonance curve at half maximum amplitude or at half-power level (or at -3 dB) to the resonance frequency f,

$$Q = \Delta f / f, \qquad (4.67)$$

Very often the experimental results are expressed as Q^{-1}.

Table 4.19. Internal friction (Q^{-1}) in several species, determined on L, R, and T directions, with longitudinal vibrations

Direction	Density (kg/m³)	Q^{-1} (10⁻³)	Anisotropy
	Sitka Spruce (*Picea sitchensis*)		
L	460	11.2	1
R	449	23.2	2.10
T	454	24.4	2.20
	Lauan (*Pentacme contorta*)		
L	481	7.6	1
R	489	19.7	2.59
T	478	20.2	2.63
	Makore (*Tieghemella heckelii*)		
L	669	9.5	1
R	673	28.0	2.95
T	670	33.0	3.47
	Matoa (*Pometia pinnata*)		
L	795	9.4	1
R	700	27.3	2.91
T	674	27.5	2.93
	Mizunara (*Quercus mongolica*)		
L	630	10.7	1
R	654	25.8	2.41
T	620	28.9	2.70
	Yachidamo (*Fraxinus mandshurica*)		
L	570	8.8	1
R	548	25.1	2.82
T	517	26.9	3.03

From Ono, T. and Norimoto, M., *Jpn. J. Appl. Phys.*, 24(8) 960, 1985. With permission.

The different parameters relevant to the internal friction phenomena in solids in the audible frequency and ultrasonic ranges are:

$$2\pi \tan \delta = 2\pi \, \Delta f/f_r = 2\pi \, Q^{-1} = 2\alpha V/f \tag{4.68}$$

where V is the ultrasonic velocity and α is the ultrasonic attenuation. This relationship is valid for materials having $\tan \delta \leq 0.2$ (Read and Dean, 1978). Measurements of the internal friction parameters and of the ultrasonic attenuation coefficient in various kinds of solids (alloys, glass, ceramics) were reported by Smith (1980).

4.5.1. TYPICAL DAMPING COEFFICIENT VALUES

The most complete list of values of logarithmic decrement in the L and R directions for a number of species was compiled by Haines (1979). Measurements in three anisotropic directions for different modes are scarce, however, because of the difficulty in obtaining corresponding specimens for T direction tests. The measurements using the longitudinal mode reported by Ono and Norimoto (1985) are reproduced in Table 4.19.

It is worth noting that $Q_L^{-1} << Q_R^{-1} < Q_T^{-1}$ and the anisotropy ratio deduced as Q_R^{-1}/Q_L^{-1} is typically between 2.1 and 3.0 and Q_T^{-1}/Q_L^{-1} is between 2.2 and 3.5. In addition, it may be said that internal friction is lower in the fiber direction and higher in the T direction. The damping mechanism in solid wood is caused by the lignin regions of the cellular wall. Cellulosic microfibrils are highly crystalline and they consequently have low damping.

Table 4.20. Internal friction (Q^{-1}) as a function of grain angle on spruce in *LR* plane

Angle	0°	15°	45°	65°	75°	90°
Density (kg/m³)	397	431	426	420	421	423
Q^{-1}	6.6	9.81	15.0	17.7	18.3	18.7

From Ono, T., *J. Soc. Mater. Sci. Jpn.*, 32(352) 108, 1983. With permission.

The variation of damping using the grain angle was reported by Ono (1983), and some data are reproduced in Table 4.20. Ono's general equation for predicting the internal friction using the grain angle is

$$Q_\alpha^{-1} = Q_{0°}^{-1} + Q_{90°}^{-1} - [Q_{0°}^{-1} Q_{90°}^{-1}][Q_{0°}^{-1} \cos^2 \alpha + Q_{90°}^{-1} \sin^2 \alpha] \tag{4.69}$$

where α is the grain angle.

4.5.2. DAMPING COEFFICIENTS AS INDICATORS OF MICROSTRUCTURAL MODIFICATIONS INDUCED BY DIFFERENT FACTORS
4.5.2.1. Temperature and Moisture Content
We chose to indicate the way in which the parameter Q^{-1} is able to reflect fine structural modifications in wood induced by an increase of temperature. For this purpose the internal friction spectra of beech as a function of temperature are given in Figures 4.30 and 4.31. In Figure 4.30 both anisotropic directions (*L* and *R*) two peaks of Q^{-1} were observed to be associated with corresponding changes of frequencies. At low temperatures (e.g., <170 K) the molecular motion in amorphous lignin is probably frozen, and therefore the frequency variations are very small. The increase of temperature permits dislocation and rearrangement of the structure, probably at the hemicellulose level. This begins at 210 K for the *L* direction and at 230 K for specimens in the *R* direction. A larger damping was observed for acoustic waves which propagate across the fibers (*R* direction). The increase of temperature softens the amorphous regions. This is followed by a second transition, occurring in a range of 260 to 300 K, and characterized by a broadened peak in *L* type specimens. The softening of the material has led to a large decrease in corresponding resonance frequency. With further increase of the temperature (390 K), a third transition is observed, probably corresponding to the glass transition of wood.

The effect of heat treatment on the internal friction in wood used for musical instruments was also reported by Nakao et al. (1983) and Yano and Minato (1992).

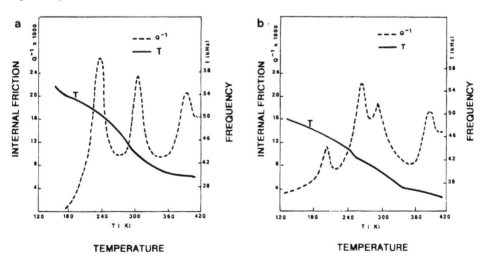

Figure 4.30. Internal friction (Q^{-1}) and resonance frequency in beech as a function of temperature and moisture content (a) in the longitudinal direction, (b) in the radial direction (Khafagy et al., 1984).

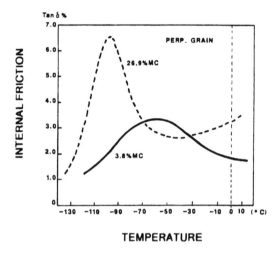

Figure 4.31. Internal friction in the low temperature range for beech with different moisture contents. (From Sellevold et al., *Wood Fiber*, 7(3)162. 1975. With permission.)

The influence of moisture content on damping at room temperature was reported by Suzuki (1980) and Sasaki et al. (1988). It was noted for hinoki (*Chamaecyparis obtusa*) that tan δ increased from 5.5 to 18.5×10^{-3} when moisture content increased from 5 to 35%. In the low temperature range (Figure 4.31) of beech conditioned at varying equilibrium moisture content (4 or 27%) a phase transition was observed, similar to those reported in other porous water-absorbing materials. Near the fiber saturation point, or more exactly at 27% moisture content, a phase transition of the same nature as "the glass transition of the absorbed water" was observed at −100°C. Decreasing the moisture content of specimens at 4% shifted the peak of internal friction to −60°C, probably because of the presence of water mainly in cell walls.

4.5.2.2. Chemical Treatment

Chemical modification of the structure of *Tilia japonica* Simk, induced by a half-esterified treatment with zinc to obtain a succinylated wood ionomer, can be observed by measuring tan δ over a wide range of temperatures (−120° to 230°C) (Figure 4.32). Two peaks are observed on this graph, 180°C and −60°C, probably corresponding to the restricted motion of the main cellulosic chain and the side chain motion, respectively (Nakao et al., 1990). In the region between −60° and 20°C, corresponding to the micro-Brownian motion of half-esterified cellulose in the amorphous region and to the local vibration of carboxyl groups related to water molecules, tan δ decreases dramatically. At 20°C its value is at a minimum. The increase in tan δ as the temperature rises to 60°C indicates an increase in internal

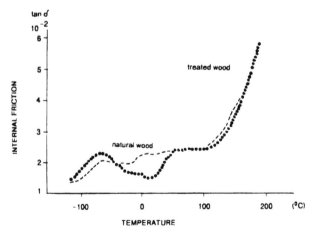

Figure 4.32. Influence of zinc content on the dynamic behavior of wood ionomer-succinylated wood containing zinc compared with natural, untreated wood of *Tilia japonica*. (From Nakano et al., *Mokuzai Gakkaishi*, 29(10), 657, 1990. With permission.)

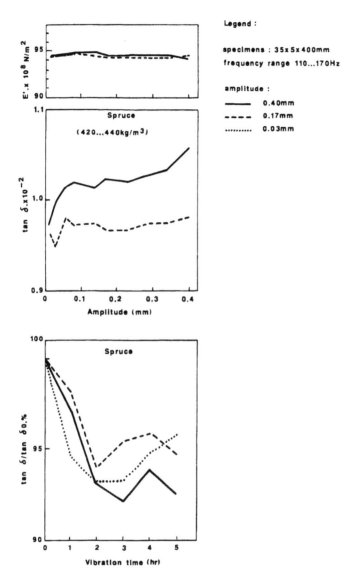

Figure 4.33. Internal friction (tan δ_0), the ratio of tan δ: tan δ_0, and the continuous flexural vibration of small amplitude (solid line, 0.40 mm; broken line, 0.17 mm; dotted line, 0.03 mm) in spruce, measured on specimens of $35 \times 5 \times 400$ mm, at a frequency range of 110 to 170 Hz. (From Sobue, N. and Okayasu, S., *J. Soc. Mater. Sci. Jpn.*, 1992. With permission.)

friction. Zinc ions form intermolecular crosslinks between cellulose chains. This structural modification of solid wood was also observed on tan δ measurements.

The presence of metallic ions in the wood ionomer retards the thermal degradation of this material above 100°C, as seen in the plateau of the graph in Figure 4.32.

4.5.2.3. Dynamic Loading

Sobue and Okayasu (1992) showed that internal friction parameters measured on wood could be used for the estimation of very fine modifications of cellular wall structure induced by the continuous vibration of specimens, even with relatively small amplitudes. From Figure 4.33 we observe that the value of

E_L is not affected by the amplitude of 5-h bending vibration, although the ratio tan δ:tan δ_0 (tan δ_0 corresponding to the initial state) decreased at a rate dependent on the vibration time and amplitude (e.g., 8% for 3 h at 0.40 mm amplitude). These results were related to the chemical hydrogen bonds between the chains that connect the microfibrils and the lignin, and were explained via the rate theory. Detailed calculation using the rate theory in wood can also be seen in Cousins (1974) and Mark (1967).

The results reported by Sobue and Okayasu (1992) can be combined with earlier experiments by Murphy (1963) using static tests at low stress levels. Murphy found that the crystallinity of wood, measured by X-ray diffraction, increases with the rate of applied load. This behavior was related to the moderate strength hydrogen bonds. Murphy proposed that when the matrix of the material was squeezed under stress, the connections were readjusted to provide new bond positions. The intensity of the stress then produced either an elastic or a plastic response of the specimen.

REFERENCES

Achenbach, J. D. (1973). *Wave Propagation in Elastic Solids*. North Holland, Amsterdam.

Akitsu, H., Norimoto, M., and Morooka, T. (1991). Vibrational properties of chemically modified wood. *Mokuzai Gakkaishi*. 37(7), 590–597.

Alippi, A. and Mayer, W. G. (Eds.) (1987). *Ultrasonic Methods in the Evaluation of Inhomogeneous Materials*. NATO ASI Ser. E: Applied Science no. 126. Martinus Nijhoff, Dordrecht.

Alippi, A. (Ed.) (1989). *Ultrasonic Signal Processing*. World Scientific, Singapore.

Alippi, A. (Ed.) (1992). *Acoustic Sensing and Probing*. World Scientific, Singapore.

Archer, R. R. (1987). *Growth Stresses and Strains in Trees*. Springer-Verlag, Berlin.

Arima, T., Nakamura, N., Maruyama, N., and Hayamura, S. (1991). Classification of logs based on sound analysis and its application in product processing. *Proc. Int. Timber Eng. Conf.*, London, 2.266–2.273.

Auld, B. A. (1973). *Acoustic Fields and Waves in Solids*. Vol. 1. Wiley Interscience, New York.

Barducci, I. and Pasqualini, G. (1948). Misura dell'attrito interno delle costanti elastiche del legno. *Nuovo Cimento*, 5 (Octobre), 416–502.

Baum, G. A. and Bornhoeft, L. R. (1979). Estimating Poisson ratios in paper using ultrasonic technique. *Tappi J*. 63(5), 87–90.

Becker, H. and Noack, D. (1968). Studies on the dynamic torsional viscoelasticity of wood. *Wood Sci. Tech*. 2, 213–230.

Beltzer, A. (1988). *Acoustics of Solids*. Springer-Verlag, Berlin.

Bodig, J. and Jayne, B. A. (1982). *Mechanics of Wood and Wood Composites*. Van Nostrand Reinhold, New York.

Bodig, J. and Goodmann, J. R. (1973). Prediction of elastic parameters for wood. *Wood Sci*. 5(4), 249–264.

Böhnke, I. (1993). Etude Experimentale et Theorique des Traitements Thermiques du Bois. Ph.D. thesis no. 81ED, Ecole de Mines, St. Etienne, France.

Böhnke, I. and Guyonnet, R. (1991). Spectral analysis of ultrasonic waves for the characterization of thermally treated wood. in *Proc. Int. Ultrasonic Conf.* July 1 to 4, Le Touquet, France, 499–502.

Brenndörfer, D. (1972). Nondestructive testing of wood with acoustic vibrations. *Ind. Lemnului*. 6,234–240.

Bucur, V. (1980). Anatomical structure and some acoustical properties of resonance wood. *Catgut Acoust. Soc. Newslett*. no. 33, 24–29.

Bucur, V. (1983). X-ray microdensitometric analysis and ultrasonic velocities on resonance spruce. Unpublished data.

Bucur, V. (1984a). Ondes Ultrasonores dans le Bois. Caracterisation Mécanique et Qualité de Certaines Essences de Bois. Ph.D. thesis, Institut Superieur des Matériaux, Paris.

Bucur, V. (1984b). Relationship between grain angle of wood specimens and ultrasonic velocity. *Catgut Acoust. Soc. Newslett*. no. 41, 30–35.

Bucur, V. (1987). Wood characterization through ultrasonic waves. in *Ultrasonic Methods in the Evaluation of Inhomogeneous Materials*. NATO ASI Ser. E: Applied Science no. 126, Alippi, A. and Mayer, W. G. (Eds.), Martinus Nijhoff, Dordrecht, 323–342.

Bucur, V. and Böhnke, I. (1994). Factors affecting ultrasonic attenuation measurements in solid wood. *Ultrasonics*. 32(5), 385–390.

Bucur, V. and Perrin, J. R. (1988). Ultrasonic waves-wood structure interaction. in *Proc. Inst. Acoust. Acoustics '88*, Vol. 10, Part 2, University of Cambridge, U.K., April 5 to 8 199–206.

Bucur, V. and Archer, R. R. (1984). Elastic constants for wood by an ultrasonic method. *Wood Sci. Tech.* 18(4), 255–265.

Bucur, V., Janin, G., Herbe, C. and Ory, J. M. (1991). Ultrasonic Detection of Reaction Wood in European Species. Voluntary paper at 10th World Forestry, Paris. September 16 to 27.

Bucur, V. and Feeney, F. (1992). Attenuation of ultrasound on solid wood. *Ultrasonics.* 30(2), 76–81.

Bucur, V. and Perrin, J. R. (1988). Influence of experimental conditions on ultrasonic velocities measurements on specimens of different type. Unpublished data.

Burmester, A. (1965). Relationship between sound velocity and the morphological, physical and mechanical properties of wood. *Holz Roh Werkst.* 23(6), 227–236.

Caldersmith, G. W. (1984). Vibrations of orthotropic rectangular plates. *Acustica.* (56), 144–152.

Caldersmith, G. W. and Rossing, T. D. (1983). Ring modes, X-modes and Poisson coupling. *Catgut Acoust. Soc. Newslett.* no. 39, May, 12–14.

Chen, C. H. (1988). High resolution spectral analysis NDT techniques for flaw characterization, prediction and discrimination. in *Signal Processing and Pattern Recognition in Nondestructive Evaluation of Materials.* NATO ASI Ser. F, Vol. 44, Chen, C. H. (Ed.) Springer-Verlag, Berlin, 155–174.

Chevalier, Y. (1988). Comportement élastique et viscoélastique des composites. Techniques de l'Ingenieur, Paris. précis A7750; A7751, traité A7 Matériaux Industriels.

Chevalier, Y. (1989). Testing composite materials by ultrasound. in *Caractérisation Mécanique des Composites.* Vautrin, A. (ed.), Pluralis, Paris.

Chivers, R. C. (1973). The scattering of ultrasound by human tissues. Some theoretical models. *Ultrasound Med. Biol.* 3, 1–13.

Chivers, R. C. (1991). Measurement of ultrasonic attenuation in inhomogeneous media. *Acustica.* 74, 8–15.

Christensen, R. M. (1971). *Theory of Viscoelasticity, an Introduction.* Academic Press, New York.

Christoffel, E. B. (1877). Über die Fortpflanzung von Stössen durch elastische feste Körper. *Ann. Matematica,* serie II, tomo VIII, 193–243, Milano, G. Bernardoni (Ed.)

Chui, Y. H. (1991). Simultaneous evaluation of bending and shear moduli of wood and the influence of knots on these parameters. *Wood Sci. Tech.* 25, 125–134.

Cremer, L. and Heckl, M. (1973). *Structure-Borne Sound. Structural Vibrations and Sound Radiation in Audio Frequencies.* Springer-Verlag, Berlin.

Cousins, W. J. (1974). Effects of strain-rate on the transverse strength of *Pinus radiata* wood. *Wood Sci. Tech.* 8, 307–321.

D'Souza, D. P., Anson, L. W., and Chivers, R. C. (1989). Effects of matching layers on measurements of ultrasonic attenuation. *Acustica.* 69(2), 88–92.

Dieulesaint, E. and Royer, D. (1974). *Ondes élastiques dans les Solides.* Masson et Cie, Paris.

Dinwoodie, J. M. (1981). *Timber, Its Nature and Behaviour.* Van Nostrand Reinhold, New York.

Dunlop, J. I. (1980). Testing of particleboard by acoustic techniques. *Wood Sci. Tech.* 14, 69–78.

Dunlop, J. I. (1978). Damping loss in wood at mid kilohertz frequencies. *Wood Sci. Tech.* 12, 49–62.

Edmonds, P. D. (Ed.) (1981). *Ultrasonics.* Academic Press, New York.

Facaoaru, I. and Jones, R. (1971). *Nondestructive Testing of Concrete.* Ed. Technica, Bucharest (in Romanian).

Flandrin, P. (1988). Nondestructive evaluation in the time-frequency domaine by means of the Wigner-Ville distribution. in *Signal Processing and Pattern Recognition in the Nondestructive Evaluation of Materials.* NATO ASI Ser. F, Vol. 44, Chen, C. H. (Ed.). Springer-Verlag, Berlin, 109–118.

Fidler, B. (1983). *Mesures des Constantes Élastiques du Bois. Mémoire de Fin d'Études.* C.N.A.M., Paris.

Fitting, D. W. and Adler, L. (1981). *Ultrasonic Spectral Analysis for Nondestructive Evaluation.* Plenum Press, New York.

Fujisawa, Y., Ohta, S., Nishimura, K., and Tajima, M. (1992). Wood characteristics and genetic variations in sugi. Clonal differences and correlations between location of dynamic moduli of elasticity and diameter growth in plus-tree clones. *Mokuzai Gakkaishi.* 38(7), 638–644.

Gerhards, C. C. (1982). Longitudinal stress waves for lumber stress grading: factors affecting applications. State of art. *For. Prod. J.* 32(2), 20–35.

Gerhards, C. C. (1978). Effect of earlywood and latewood on stress wave measurements parallel to the grain. *Wood Sci.* 11(2), 69–72.

Gerhards, C. C. (1987). The equivalent orthotropic elastic properties of plywood. *Wood Sci. Tech.* 21, 335–348.

Gerike, O. R. (1970). Ultrasonic spectroscopy. in *Research Technique in Non-Destructive Testing.* Sharpe, R. S. (Ed.). Academic Press, London, 31–61.

Green, A. E., Zerna, W. (1968). *Theoretical Elasticity.* Oxford University Press, Oxford.

Green, R. E., Jr. (1973). *Ultrasonic Investigation of Mechanical Properties.* Academic Press, New York.

Guilbot, J. (1992). Célérités de phase et de groupe: relations explicites entre les lois de dispersion. *J. Acoust.* 5, 635–638.

Guitard, D. (1987). *Mécanique du Matériau Bois et Composites.* Cepadues Editions, Toulouse, France.

Haines, D. (1979). On musical instrument wood. *Catgut Acoust. Soc. Newslett.* no. 31, 23–32.

Harris, C. M. and Crede, C. E. (1961). *Shock and Vibration Handbook. Vol. 1. Basic Theory and Measurements.* McGraw-Hill, New York.

Hearmon, R. F. S. (1961). *An Introduction to Applied Anisotropic Elasticity.* Oxford University Press, Oxford.

Hearmon, R. F. S. (1948). The Elasticity of Wood and Plywood. Dept. Sci. Ind. Res. For. Prod. Res. Spec. Rep. No. 7. His Majesty's Stationery Office; London.

Hosten, B. (1991). Reflection and transmission of acoustic plane waves on an immersed orthotropic and viscoelastic solid layer. *J. Acoust. Soc. Am.* 89(6), 2745–2752.

Hosten, B. Deschamps, M. and Tittmann, B. R. (1987). Inhomogeneous wave generation and propagation in lossy anisotropic solids. Application to the characterization of viscoelastic composite materials. *J. Acoust. Soc. Am.* 82(5), 1763–1770.

Hutchins, D. A. and Hayward, G. (1990). Radiated fields of ultrasonic transducers. in *Physical Acoustics, Ultrasonic Measurement Methods,* Vol. 19, Thurston, R. N. and Pierce, A. D. (Eds.). Academic Press, New York, 1–80.

James, B. A. (1986). Effect of Transverse Moisture Content Gradients on the Longitudinal Propagation of Sound in Wood. For. Prod. Lab. Res. Pap. no. 466. USDA, Washington, D.C.

Jayne, B. A. (1972). *Theory and Design of Wood and Fiber Composite Materials.* Syracuse University Press, Syracuse, NY.

Jones, R. M. (1975). *Mechanics of Composite Materials.* McGraw-Hill, New York.

Kamioka, H. and Kataoka, A. (1982). The measurement error factor of velocity in wood. *Mokuzai Gakkaishi.* 28(5), 274–283.

Kataoka, A. and Ono, T. (1975/1976). The relations of experimental factors to the measured values of dynamic mechanical properties of wood. *Mokuzai Gakkaishi.* Part 1, 21(10), 543–550; Part 2, 22(1), 1–7.

Khafagy, A. H., Britton, W. G. B., and Stephens, R. W. B. (1984). Acoustic emission and internal friction in wood. in Progress in Acoustic Emission II. Proc. 7th Int. Symp. in Acoustic Emission, Zao, Japan, October 23 to 26, 525–532.

Kollmann, F. F. P. and Krech, M. (1960). Dynamische Messung der elastischen Holzeigenschaften und der Dämpfung. *Holz Roh Werkst.* 18(2), 41–54.

Krautkrämer, J. and Krautkrämer, H. (1969). *Ultrasonic Testing of Materials.* Springer-Verlag, Berlin.

Kriz, R. D. and Ledbetter, H. M. (1986). Elastic-wave surfaces in anisotropic media. in *Rheology of Anisotropic Materials.* Huet, C., Bourgoin, D., and Richemond, S. (Eds.). Editions Cepadues, Toulouse, France.

Kriz, R. D. and Stinchcomb, W. W. (1979). Elastic moduli of transversely isotropic graphite fibers and their composites. *Exp. Mech.* 19(2), 41–49.

Leeman, S. (1990). Personal communication related to attenuation measurements in solids. King's College, London (cited in Bucur and Böhnke, (1994)).

Lemaster, R. L. and Quarles, S. L. (1990). The effect of same-side and through-thickness transmission modes on signal propagation in wood. *J. Acoust. Emiss.* 9(1), 17–24.

Lipsett, A. W. and Beltzer, A. I. (1988). Reexamination of dynamic problems of elasticity for negative Poisson's ratio. *J. Acoust. Soc. Am.* 86(6), 2179–2186.

Love, A. E. H. (1944). A treatise on the mathematical theory of elasticity. Dover Publ. Inc. New York.

Lynnworth, L. (1989). *Ultrasonic Measurements for Process Control.* Academic Press, New York.

Main, I. G. (1985). *Vibrations and Waves in Physics.* Cambridge University Press, Cambridge.

Mann, R. W., Baum, G. A., and Haberger, C. C. (1980). Determination of all nine orthotropic elastic constants for machine-made paper. *Tappi J.* 63(2), 163–166.

Mark, R. E. (1967). *Cell Wall Mechanics of Tracheids*. Yale University Press, New Haven, CT.

McDonald, K. A. (1978). Lumber Defect Detection by Ultrasonics. For. Prod. Lab. Res. Pap. No. 311, USDA, Washington, D.C.

McIntyre, M. E.; Woodhouse, J. (1978). The influence of geometry on linear damping. *Acustica*. 39(4), 209–224.

McIntyre, M. E. and Woodhouse, J. (1985–1986). On measuring wood properties. *J. Catgut Acoust. Soc*. Part 1, no. 42, 11–15; Part 2, no. 43, 18–24; Part 4, no. 45, 14–23.

McIntyre, M. E. and Woodhouse, J. (1988). On measuring the elastic and damping constants of orthotropic sheet materials. *Acta Metall*. 36(6), 1397–1416.

McSkimin, H. J. (1964). Ultrasonic methods for measuring the mechanical properties of liquids and solids. in *Physical Acoustics*. Vol. 1, Part A, Mason, W. P. (Ed.). Academic Press, New York.

Molin, N. E. and Jansson, E. V. (1989). Transient wave propagation in wooden plates for musical instruments. *J. Acoust. Soc. Am*. 85(5), 2179–2184.

Moore, G. R., Kline, D. E., and Blankenhorn, P. R. (1983). Dynamic mechanical properties of epoxy poplar composites materials. *Wood Fiber Sci*. 15(4), 358–375.

Morizumi, S., Fushitani, M., Kaburagi, J. (1971–1973). Viscoelasticity and structure of wood. *Mokuzai Gakkaishi*. Part 1, 17(12), 431–436; Part 2, 19(2), 81–87; Part 3, 19(3), 109–115.

Morze, Z., Olszewski, J., and Paprzycki, O. (1979). Untersuchungen der Elemente des dynamischen Schubmoduls des Holzes. *Holztechnologie*. 20(3), 179–183.

Murakami, K. and Matsuda, H. (1990). Dynamic mechanical properties of oligoesterified woods. *Mokuzai Gakkaishi* 36(1), 49–56.

Murphy, W. K. (1963). Cell-wall crystallinity as a function of tensile strain. *For. Prod. J*. 13(4), 151–155.

Musgrave, M. J. P. (1970). *Crystal Acoustics*. Holden-Day, San Francisco.

Nakano, T., Honma, S., and Matsumoto, A. (1990). Physical properties of chemically modified wood containing metal. *Mokuzai Gakkaishi*. 36(12), 1063–1068.

Nakao, T., Okano, T., and Asano, I. (1983). Effect of heat treatment on the loss tangent of wood. *Mokuzai Gakkaishi*. 29(10), 657–662.

Norimoto, M., Gril, J., Sasaki, T., and Rowell, R. M. (1988). Improvement of acoustical properties of wood through chemical modifications. in *Actes Colloque Comportement Mecanique du Boise*, Bordeaux, June 8 to 9, 37–44.

O'Donnell, M., Busse, L. J., and Miller, J. G. (1981). Piezoelectric transducers. in *Methods of Experimental Physics, Ultrasonics*, Vol. 19, Edmonds, P. D. (Ed.). Academic Press, New York, 29–65.

Okyere, J. G. and Cousin, A. J. (1980). On flaw detection in live wood. *Mater. Eval*. 38(3), 43–47.

Olszewski, J. and Struk, K. (1983). Effect of moisture on the orientation of dynamic shear moduli of a particleboard. *Holztechnologie*, 24(3), 165–168.

Ono, T. (1983). Effect of grain angle on dynamic mechanical properties of wood. *J. Soc. Mater. Sci. Jpn*. 32(352), 108–113.

Ono, T. and Norimoto, M. (1983). Study on Young's modulus and internal friction of wood in relation to the evaluation of wood for musical instruments. *Jpn. J. Appl. Phys*. 22(4), 611–614.

Ono, T. and Norimoto, M. (1985). Anisotropy of dynamic Young's modulus and internal friction in wood. *Jpn. J. Appl. Phys*. 24(8), 960–964.

Panametrics, Inc. (1988). *Ultrasonic Products for Non-Destructive Testing*.

Panametrics, Inc. (1992). *Transducer Certification for V153*.

Panshin, A. J. and De Zeeuw, C. (1980). *Textbook of Wood Technology*. Vol. 1, McGraw-Hill, New York.

Papadakis, E. P. (1965). Ultrasonic attenuation caused by scattering in polycrystalline metals. *J. Acoust. Soc. Am*. 37(4), 711–717.

Papadakis, E. P. (1967). Ultrasonic velocity and attenuation. Measurement methods with specific and industrial applications. in *Physical Acoustics*. Vol. 12, Mason, W. P. (Ed.). Academic Press, New York, 277–375.

Papadakis, E. P. (1968). Ultrasonic attenuation caused by scattering in polycrystalline media. in *Physical Acoustics*. Vol. 4B, Mason, W. P. (Ed.). Academic Press, New York, 269–328.

Papadakis, E. P. (1990). The measurement of ultrasonic attenuation. in "*Physical Acoustics*. Vol. 19, Thurston, R. N. and Pierce, A. D. (Eds.). Academic Press, New York, 108–155.

Preziosa, C., Mudry, M., Launay, J., and Gilletta, F. (1981). Détermination des constantes élastiques du bois par une méthode acoustique goniométrique. *C.R. Acad. Sci. Paris*, 293, (II) 91–94.

Radu, A. and Brenndörfer, D. (1976). Zur zerstörungsfreien Prüfung des Holzes durch Schwingungsversuche. *Holz Roh Werkstoff.* (34)219–222.

Read, B. E. and Dean, G. D. (1978). *The Determination of Dynamic Properties of Polymers and Composites.* Adam Hilger Ltd., Bristol, U.K.

Rebic, M. and Srepcic, J. (1988). Correlation between static and dynamic modulus of elasticity at various moisture conditions. in *Actes Colloque Comportement Mecanique du Bois,* Bordeaux, June 8 to 9, 1988, 81–90.

Sachse, W. and Hsu, N. N. (1979). Characterization of ultrasonic transducers used in material testing. in *Physical Acoustics.* Vol. 14, Mason, W. P. and Thurston, R. N. (Eds.). Academic Press, New York.

Sachse, W. and Pao, Y. H. (1978). On the determination of phase and group velocities of dispersive waves in solids. *J. Appl. Phys.* 49(8), 4320–4327.

Sahay, S. K., Kline, R. A., and Mignona, R. (1992). Phase and group velocity considerations for dynamic modulus measurement in anisotropic media. *Ultrasonics.* 30(6), 373–382.

Sasaki, T., Norimoto, M., Yamada, T., and Rowell, R. M. (1988). Effect of moisture content on the acoustic properties of wood. *Mokuzai Gakkaishi.* 34(10), 794–803.

Schumacher, R. T. (1988). Compliances of wood for violin top plates. *J. Acoust. Soc. Am.* 84(4), 1233–1235.

Scott, W. R. and Gordon, P. F. (1977). Ultrasonic spectrum analysis for nondestructive testing of layered composite materials. *J. Acoust. Soc. Am.* 62(1), 108–116.

Sellevold, E. J., Radjy, F., Hoffmeyer, P., and Bach, L. (1975). Low temperature internal friction and dynamic modulus for beech wood. *Wood Fiber.* 7(3), 162–169.

Smith, C. C. (Ed.) (1980). Internal friction and ultrasonic attenuation in solids. in *Proc. 3rd Conf. University of Manchester, England, 18–20 July.* Pergamon Press, Oxford.

Snowdon, J. C. (1968). *Vibration and Shock in Damped Mechanical Systems.* John Wiley & Sons, New York.

Sobue, N. (1983). Prediction of dynamic Young's modulus and loss tangent of plywood in bending. *Mokuzai Gakkaishi.* 29(1), 14–19.

Sobue, N. (1986a). Measurement of Young's modulus by the transient longitudinal vibration of wooden beams using a FFT spectrum analyser. *Mokuzai Gakkaishi.* 32(9), 744–747.

Sobue, N. (1986b). Instantaneous measurement of elastic constants by analysis of the tap tone of wood. Application to flexural vibration of beams. *Mokuzai Gakkaishi.* 32(4), 274–279.

Sobue, N. (1987). Simultaneous determination of Young's modulus and of shear modulus of commercial lumber by complex vibration of bending and twisting. C. R. Coll. Ann. *Mokuzai Gakkaishi,* 189.

Sobue, N. (1988). Simultaneous determination of Young's modulus and shear modulus of structural lumber of complex vibrations of bending and twisting. *Mokuzai Gakkaishi.* 34(8), 652–657.

Sobue, N. (1990). Correction factors of the resonance frequency for tapering and shear deformation of a log in flexural vibration. *Mokuzai Gakkaishi.* 36(9), 760–764.

Sobue, N. and Asano, I. (1976). Studies on the fine structure and mechanical properties of wood. On the longitudinal Young's modulus and shear modulus of the cell wall. *Mokuzai Gakkaishi.* 22(4), 211–216.

Sobue, N. and Katoh, A. (1990). Simultaneous measurements of anisotropic elastic constants of standard full-size plywood by vibration technique. *Proc. Int. Timber Eng. Conf.,* October 23 to 25, Tokyo.

Sobue, N. and Kitazumi, M. (1991). Identification of power spectrum peaks of vibrating completely free wood plates and moduli of elasticity measurement. *Mokuzai Gakkaishi.* 37(1), 9–15.

Sobue, N. and Okayasu, S. (1992). Effects of continuous vibration on dynamic viscoelasticity of wood. *J. Soc. Mater. Sci. Jpn.* 41(461), 164–169.

Sobue, N. and Yvasaki, Y. (1981a). Effect of veneer construction on anisotropy of dynamic viscoelasticity of plywood. *Mokuzai Gakkaishi.* 27(6), 457–462.

Sobue, N. and Yvasaki, Y. (1981b). Effect of adhesive on dynamic viscoelasticity of laminated lumber of two-ply sawn laminae. *Mokuzai Gakkaishi.* 27(7), 597–601.

Suzuki, M. (1980). Relationship between specific gravity and dynamic decrement with water contents. *Mokuzai Gakkaishi.* 26(5), 299–304.

Takada, K., Koizumi, A., and Ueda, K. (1989). Growth and modulus of elasticity of plus-tree clones. *Res. Bull. Hokkaido Univ.* 46(4), 989–1001.

Tang, R. C. and Hsu, N. N. (1972). Dynamic Young's moduli of wood related to moisture content. *Wood Sci.* 5(1), 7–14.

Tonosaki, M. and Okano, T. (1985). Evaluation of acoustical properties of wood by plate vibration tests. *Mokuzai Gakkaishi.* 31(8), 627–632.

Tonosaki, M., Okano, T. and Asano, T. (1983). Vibrational properties of Sitka spruce with longitudinal and flexural vibration. *Mokuzai Gakkaishi.* 29(9), 547–552.

Truell, R., Elbaum, C., and Chick, B. B. (1969). *Ultrasonic Methods in Solid State Physics.* Academic Press, New York.

Vincent, A. (1987). Influence of wearplate and coupling layer thickness on ultrasonic velocity measurement. *Ultrasonics.* 25(July), 237–243.

Vinh, T. (1982). Etude critique des méthodes ultrasonores et vibratoires pour caractériser les matériaux composites. in *Proc. Euromech* 154. Bordeaux, April 27 to 29.

Woodhouse, J. (1986). Spruce for soundboards: elastic constants and microstructure. in *Proc. Inst. Acoust.* Vol. 18, Part 1, Edinburgh, U.K., 99–106.

Yano, H. Minato, K. (1992). Improvement of the acoustic and hygroscopic properties of wood by chemical treatment and application to violin parts. *J. Acoust. Soc. Am.* 92(3), 1222–1227.

Yiannos, P. N. and Taylor, D. L. (1967). Dynamic modulus of thin wood sections. *Tappi J.* 50(1), 40–47.

Chapter 5

Elastic Constants of Wood Material

5.1. GLOBAL CHARACTERIZATION

Global characterization of wood is based on the assumption that its properties can be represented by an equivalent homogeneous anisotropic continuum. The orthotropic symmetry closely approximates the actual wood structure. Applying the principles of crystal physics and solid mechanics to obtain precise estimates of the mechanical properties of wood leads to the development of an ultrasonic technique for measuring elastic constants. This approach was nurtured by the development of a specific procedure for measuring the nondiagonal terms of the stiffness matrix. It is useful to mention here that the properties of artifical composite materials of orthotropic or transverse (also called plane) symmetry are strongly dependent on the orientation of the reference coordinates. Optimization procedures to determine elastic constants were developed by Roklin and Wang (1989a, b); Wu et al. (1973); Kriz and Stinchcomb (1979); Every and Sachse (1990); Castagnede et al. (1990); and Every et al. (1991).

5.1.1. OPTIMIZATION CRITERIA FOR OFF-DIAGONAL TERMS OF THE STIFFNESS MATRIX DETERMINED USING BULK WAVES

Chapter 4 explained that the ultrasonic determination of elastic constants is based on wave speed measurements. Measuring the six diagonal terms of the stiffness matrix from the velocities of longitudinal and transverse (or shear) waves that propagate along the principal symmetry axes is fairly straightforward and well established. Linear equations must be solved in density and squared velocities as $C_{ii} = \rho V^2$. For the three off-diagonal terms of the stiffness matrix $[C_{ij}]$ the velocities measured out of principal symmetry directions must be considered. More velocity measurements than independent elastic constants are to be determined; e.g., on disk-type specimens with 24 facets it is possible to measure 48 velocities for only one off-diagonal term of the stiffness matrix. The directional dependency of wood constants renders conventional averaging techniques inapplicable on redundant measurements. To solve this problem, optimization procedures are used. This section provides several procedures for data averaging the directionally dependent ultrasonic measurements of coniferous and broadleaves species.

Chapter 4 described that the off-diagonal terms of the stiffness matrix of orthotropic solids are related to the velocities and to the diagonal terms by Equation (4.45). More explicitly, the term C_{12} can be calculated as:

$$C_{12} = (n_1 \cdot n_2)^{-1}[(C_{11}n_1^2 + C_{66}n_2^2 - \rho V^2)(C_{66}n_1^2 + C_{22}n_2^2 - \rho V^2)]^{1/2} - C_{66} \tag{5.1}$$

where $n_1 = \cos \alpha$, and $n_2 = \sin \alpha$ (α is the propagation angle).
Via permutations of the indices we obtain the corresponding expressions of C_{13} and C_{23}:

$$C_{13} = (n_1 \cdot n_3)^{-1}[(C_{11}n_1^2 + C_{55}n_3^2 - \rho V^2)(C_{55}n_1^2 + C_{33}n_3^2 - \rho V^2)]^{1/2} - C_{55} \tag{5.1'}$$

$$C_{23} = (n_2 \cdot n_3)^{-1}[(C_{22}n_2^2 + C_{44}n_3^2 - \rho V^2)(C_{44}n_2^2 + C_{33}n_3^2 - \rho V^2)]^{1/2} - C_{44} \tag{5.1''}$$

In the interest of clarity, note that for the calculation of the nondiagonal terms of the stiffness matrix the value of velocity V of either a quasilongitudinal or quasitransverse wave, or both is necessary. Also, these values are dependent on the propagation vector and consequently on the orientation of the specimen and on angle α (Table 5.1). The quasilongitudinal or quasitransverse velocities can be measured on specimens cut at convenient angles to the principal direction when the ultrasonic transmission technique is used.

One of the most common methods used to check the validity of these procedures is to calculate slowness curves fitting the optimized values for C_{12}, C_{13}, or C_{23} in Christoffel's equation and to compare those graphs with experimental measurements. Good agreement between theoretical and experimental values may be considered a test of the validity of the proposed optimization procedure. In developing optimization procedures for wood one must observe that when testing this material variations in the

Table 5.1. Velocities (m/s) out of principal axes of propagation, measured at 1 MHz, in curly maple (Bucur, 1987)

	Type of waves used for measurements in all anisotropic planes					
	Quasilongitudinal			Quasitransverse		
Angle	LR	RT	LT	LR	LT	RT
	Velocities (m/s)					
15°	3875	2348	3745	1397	1703	1520
30°	3620	2289	3303	1297	1621	1516
45°	3075	2104	2419	1064	1577	1027
60°	2815	1933	1820	1536	1482	818
75°	2658	1790	2282	1679	1443	941

experimental results are attributable to the inherent variability of the material and to the accuracy of the measurement techniques. These sources of scatter are accounted for in beech and Douglas fir by Table 5.2, when diagonal stiffness terms are considered.

Measurement errors were estimated using the most conservative error prediction model, which employs the formulas of standard error calculation with partial derivatives. For ultrasonic velocities the error is <1%, and for diagonal terms of the stiffness matrix the error is <4%. However, this experimental measurement error is small compared to the inherent sample to sample material variability expressed in Table 5.2 by the coefficient of variation (calculated as the ratio between standard deviation and the average value of the tested samples).

Figure 5.1 gives a typical variation in the measurement error on nondiagonal term C_{13} as a function of angle with respect to the principal direction. In this case an accumulation of measurement error on different velocities and diagonal stiffness terms can reach an error of 11%. More detail on this aspect is given by Bucur and Archer (1984). Error accumulation greatly affects the off-diagonal terms, and consequently, the error can be >100% (e.g., for an angle error of 3° when the propagation angle is 45°).

Table 5.2. Velocities, stiffnesses, and errors for beech and Douglas fir (Bucur, 1985)

| | | Velocities | | Species variability | |
| | | Measured error | | | |
	Value (m/s)	m/s	%	Range (m/s)	Coefficient of variation (%)
Beech					
V_{LL}	5000	46	0.9	4540–5345	3.69
V_{TT}	1524	10	0.7	1333–1708	5.39
V_{LT}	1131	8	0.7	1028–1151	2.81
Douglas Fir					
V_{LL}	5161	49	0.9	4815–5352	6.72
V_{TT}	1584	11	0.8	1275–1844	7.51
V_{LT}	1306	9	0.7	1147–1372	3.00
		Stiffness Terms			
	10^8 N/m²	10^8 N/m²	%	10^8 N/m²	%
Beech					
C_{LL}	168.4	6.37	3.78	122.57–220.37	14.95
C_{TT}	15.6	0.52	3.33	11.55–19.70	15.34
G_{LT}	8.6	0.28	3.25	6.05–10.81	134.55
Douglas Fir					
C_{LL}	116.66	44.63	3.83	111.00–137.51	20.12
C_{TT}	10.98	0.37	3.37	7.76–15.94	18.22
G_{LT}	7.47	0.25	3.35	5.12–7.43	11.28

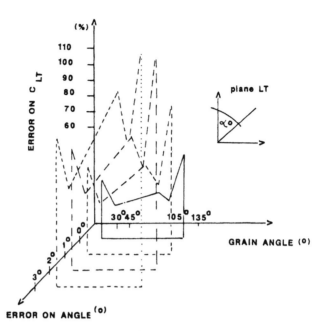

Figure 5.1. Beech. Error on the nondiagonal term C_{LT} as a function of the propagation angle. (From Bucur, V., *J. Catgut Acoust. Soc.*, Series I, (No. 44), 23, 1985.)

In order to achieve proper weighting in averaging, we need to discuss further the development of comparative procedures for optimizing redundant experimental measurements following four hypotheses:

1. The influence of shear moduli on the off-diagonal terms of the stiffness matrix
2. Computation of the minimal partial derivatives of the off-diagonal terms vs. each diagonal term
3. Computation of the most conservative error prediction for all terms of the stiffness matrix
4. The redundant experimental values of both quasilongitudinal and quasitransverse velocities, considered in mean time

Figure 5.2 (first hypothesis) shows the influence of shear diagonal term C_{55} on the off-diagonal term C_{13} for different angles of propagation when a simulation procedure is used. A variation of 20% around the measured value of the shear term was considered ($0.90\ C_{55} < C_{55} < 1.10\ C_{55}$). The optimum value for the off-diagonal term was chosen as the value exhibiting less dispersion for a definite angle of propagation (e.g., 120°). The corresponding elastic constants for this hypothesis were calculated and are presented in Tables 5.3 and 5.4. This procedure appears very simple, but it quite limited due to the large number of specimens necessary for measurements corresponding to different angles.

The second hypothesis (Figure 5.3) relates the influence of diagonal terms on off-diagonal terms of the stiffness matrix, when the propagation angle was considered. For example in the 1,2 plane we consider the influence of terms C_{22}, C_{11}, and C_{66} on C_{12} when the variation of the values of partial derivatives $\partial C_{12}/\partial C_{66}$, $\partial C_{12}/\partial C_{22}$, $\partial C_{12}/\partial C_{11}$ (Equations (5.2 to 5.6)) were calculated.

$$\partial C_{12}/\partial C_{11} = [C_{22}n_2^2 + C_{66}n_1^2 - \rho V^2]/[2 \cdot a_{12}^{1/2} \cdot n_2/n_1] \tag{5.2}$$

$$\partial C_{12}/\partial C_{22} = [C_{11}n_1^2 + C_{66}n_2^2 - \rho V^2]/[2 \cdot a_{12}^{1/2} \cdot n_1/n_2] \tag{5.3}$$

$$\partial C_{12}/\partial C_{11} = [[2C_{66} + C_{11}(n_1^2/n_2^2) + C_{22}(n_2^2/n_1^2) - \rho V^2(n_1^2 + n_2^2)/n_1^2 n_2^2]/[2 \cdot a_{12}^{1/2}/n_2 n_1]] - 1 \tag{5.4}$$

$$a_{12} = (C_{11}n_1^2 + C_{66}n_2^2 - \rho V^2)(C_{66}n_1^2 + C_{22}n_2^2 - \rho V^2) \tag{5.5}$$

$$n_1 = \cos \alpha \qquad n_2 = \sin \alpha, \text{ when } \alpha \text{ is taken relative to axis 1} \tag{5.6}$$

From Figure 5.3, we see that the diagonal term C_{11} exhibits less influence on C_{12}. The influence of α on C_{22} and C_{66} is more important. Both terms were optimized by simulation, following the previous

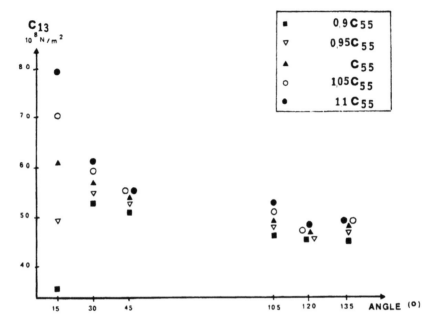

Figure 5.2. The influence of the shear term on the off-diagonal terms of the stiffness matrix. First hypothesis. (From Bucur, V., *Holzforschung*, 40(5), 315, 1986. With permission.)

hypothesis. The same argument was used for all constants related to all symmetry planes. The corresponding engineering constants are reported in Tables 5.3 and 5.4. Measurements were performed by using the direct ultrasonic transmission technique on cubic-type specimens.

An averaging procedure based on partial derivatives was also presented by Hosten and Castagnede (1983). The results of the immersion technique on plates (50 × 50 × 10 mm) are related to the tropical species, endranendrana, which has a density of 1280 kg/m³. The agreement between theoretical curves and experimental points is very good for quasilongitudinal waves in all anisotropic planes. For shear waves, the most unfavorable case was reported for the plane *RT*.

The third hypothesis is based on the most conservative error prediction, using the formulas of the standard error calculation with partial derivatives (note that V is the corresponding velocity at angle α):

- For the diagonal term:

$$C_{11} = \rho V^2 \tag{5.7}$$

$$\Delta C_{11} = \Delta \rho V^2 + \rho 2 V \Delta V \tag{5.7'}$$

$$\text{The error (\%)} = \Delta C_{11}/C_{11} \tag{5.7''}$$

- For the off-diagonal term:

$$C_{12} = \Gamma_{12}/n_1 n_2 - C_{66} \tag{5.8}$$

$$\text{The error (\%)} = \Delta \Gamma_{12}/\Gamma_{12} \tag{5.8'}$$

Table 5.3. Engineering constants (10^8 N/m^2) based on several hypotheses

Species	MHz	Young's moduli			Shear moduli		
		E_L	E_R	E_T	G_{RT}	G_{LT}	G_{LR}
			Hypothesis No. 1				
Tulip tree	0.5	80.26	12.73	8.78	2.90	7.45	10.56
	1.0	73.16	13.74	9.63	2.81	7.47	10.21
Oak	0.5	51.29	11.82	11.46	3.03	7.58	8.91
	1.0	45.43	13.52	11.22	2.92	7.58	8.45
Beech	0.5	91.32	15.02	8.27	3.53	8.62	12.43
	1.0	89.65	18.01	6.41	3.56	9.78	13.96
Douglas fir	0.5	47.70	2.98	1.44	0.57	7.47	9.66
			Hypothesis No. 2				
Tulip tree	0.5	87.37	15.23	10.14	2.90	8.29	11.62
	1.0	78.87	16.71	10.98	2.81	8.76	11.23
Oak	0.5	54.83	14.39	13.23	2.72	9.78	9.36
	1.0	48.00	16.01	13.08	2.92	8.09	9.33
Beech	0.5	96.19	18.43	9.88	3.53	9.58	13.68
	1.0	95.84	21.52	11.93	3.56	10.88	15.44
Douglas fir	1.0	52.56	7.09	3.31	0.57	9.62	10.91
			Hypothesis No. 3				
Tulip tree	0.5	88.31	11.74	8.08	3.85	7.69	10.91
	1.0	84.54	9.77	6.82	3.72	7.74	10.58
Oak	0.5	52.57	14.81	9.51	3.10	9.48	9.20
	1.0	47.45	13.99	11.22	3.86	7.83	8.73
Beech	0.5	95.65	14.86	8.82	3.64	8.90	12.84
	1.0	91.60	18.51	10.37	3.67	10.10	14.42
Douglas fir	1.0	84.75	4.49	1.87	0.69	7.72	9.97
Values Deduced from Ultrasonic Technique Reported by Preziosa et al. (1981) and Fidler (1983)							
Oak	0.5	94.00	22.20	13.20	6.07	11.05	13.19
Douglas fir	1.0	132.90	22.80	15.70	0.25	11.00	13.50
Values Deduced from Static Measurements Reported by Keylwerth (1951)							
Beech	—	140.00	22.80	11.60	4.70	10.80	16.40

From Bucur, V., *Holzforschung*, 40(5), 315, 1986. With permission.

where:

$$\Gamma_{11} = C_{11}n_1^2 + C_{66}n_2^2 \tag{5.9}$$

$$\Delta\Gamma_{11} = 2\cos\alpha\,|(-\sin\alpha)|\Delta\alpha C_{11} + \cos^2\alpha\Delta C_{11} + 2\sin\alpha\cos\alpha\Delta\alpha C_{66} + \sin^2\alpha\Delta C_{66} \tag{5.9'}$$

$$\Gamma_{12}^2 = \Gamma_{11}\Gamma_{22} - \rho V^2(\Gamma_{11} + \Gamma_{22}) + (\rho V^2)^2 \tag{5.10}$$

$$2\Gamma_{12}\Delta\Gamma_{12} = \Gamma_{11}\Delta\Gamma_{22} + \Gamma_{22}\Delta\Gamma_{11} - \Delta\rho V^2(\Gamma_{11} + \Gamma_{22}) - \rho 2V\Delta V(\Gamma_{11} + \Gamma_{22}) - \rho V^2(\Delta\Gamma_{11} + \Delta\Gamma_{22})$$
$$+ 2\rho\Delta\rho V^4 + \rho^2 4V^3\Delta V \tag{5.10'}$$

A typical variation of the nondiagonal terms C_{ij} is shown in Table 5.5 as a function of the angle of propagation. In each case the percentage error was considered. From this table the choice of C_{ij} for subsequent calculations was made by rejecting values with extreme error estimates which were obviously inaccurate, and by taking the midpoint of the range of the remaining values, as well as the highest and lowest values. The procedure was followed for all off-diagonal terms of the stiffness matrix. The entire set of midpoint values of C_{ij} were compiled (Table 5.6). Having calculated the full stiffness matrix, it is possible to predict the propagation velocity for all directions in the principal planes. Figure 5.4 shows

Table 5.4. Dynamic Poisson's ratio calculated using various hypotheses

Species	MHz	νLT	νTL	νLR	νRL	νTR	νRT
			Hypothesis No. 1				
Tulip tree	0.5	1.250	0.136	1.494	0.237	0.070	0.103
	1.0	1.080	0.142	1.261	0.236	0.067	0.096
Oak	0.5	0.411	0.092	1.217	0.280	0.291	0.300
	1.0	0.735	0.181	0.871	0.259	0.228	0.274
Beech	0.5	0.823	0.074	1.189	0.195	0.290	0.527
	1.0	0.863	0.100	1.281	0.257	0.147	0.255
Douglas fir	0.5	−0.095	−0.003	1.664	0.103	0.595	1.232
			Hypothesis No. 2				
Tulip tree	0.5	1.204	0.139	1.372	0.239	0.059	0.089
	1.0	1.043	0.145	1.126	0.238	0.055	0.085
Oak	0.5	0.107	0.444	1.094	0.287	0.245	0.267
	1.0	0.705	0.192	0.804	0.268	0.195	0.239
Beech	0.5	0.843	0.087	1.074	0.206	0.247	0.462
	1.0	0.859	0.107	1.166	0.262	0.125	0.225
Douglas fir	1.0	0.646	0.041	1.095	0.147	0.147	1.026
			Hypothesis No. 3				
Tulip tree	0.5	1.000	0.092	1.480	0.197	0.248	0.359
	1.0	0.560	0.045	1.277	0.148	0.454	0.651
Oak	0.5	1.030	0.186	0.767	0.216	0.202	0.314
	1.0	0.750	0.177	0.823	0.242	0.239	0.298
Beech	0.5	0.790	0.073	1.112	0.174	0.342	0.577
	1.0	0.900	0.102	1.244	0.251	0.146	0.261
Douglas fir	1.0	0.360	0.008	0.867	0.046	0.568	1.364
		Values Deduced from Ultrasonic Technique Reported by Preziosa et al. (1981) and Fidler (1983)					
Oak	0.5	0.390	0.050	0.370	0.090	0.650	0.380
Douglas fir	1.0	1.232	0.145	0.052	0.009	0.314	0.456
		Values from Static Measurements Reported by Keylwerth (1951)					
Beech	—	0.518	0.043	0.449	0.070	0.359	0.707

From Bucur, V., *Holzforschung*, 40(5), 315, 1986. With permission.

the computed slowness curves and the experimentally derived values for a typical case of beech. The agreement between theoretical and measured values is quite good in *LR* and *LT* planes (less so in the *RT* plane), probably because of the distorted wavefront induced by the anatomic filtering structure in this plane (see also Chapter 6).

Table 5.7 gives experimental and theoretical slowness values for yellow poplar, using different hypotheses. The best agreement was observed between the second and the third hypotheses.

The technical constants deduced from inverting the [C] matrix in the [S] matrix in the third hypothesis are mentioned in Table 5.3. The calculation of the [S] matrix presented some special problems, at least for some species. It turns out that for most conifers the matrix [C] is very close to being singular; e.g., the determinant, |C|, approaches zero. In order to exercise some control over this problem we chose to test the sensitivity of the inversion by introducing an interval of values for the off-diagonal terms. In each case the lowest, the average, and the highest of the values for C_{12}, C_{13} and C_{23} were used in the calculation of the [S] matrix in the various combinations. Approximately 27 combinations of [C] and inverse [S] matrices were calculated for each species. This sensitivity check enabled us to focus on the range of technical constant values of practical interest (e.g., Young's moduli and Poisson's ratios). The choice of the engineering constants was made by selecting those values corresponding to the highest E_L modulus. The argument in favor of this depends upon previous measurements of axial wave propagation in cylindrical rods for the same species, using the same ultrasonic equipment. We found that the E_L on cylindrical rods was considerably larger than the range of E_L in the present case, where cubes were used in the direct transmission method.

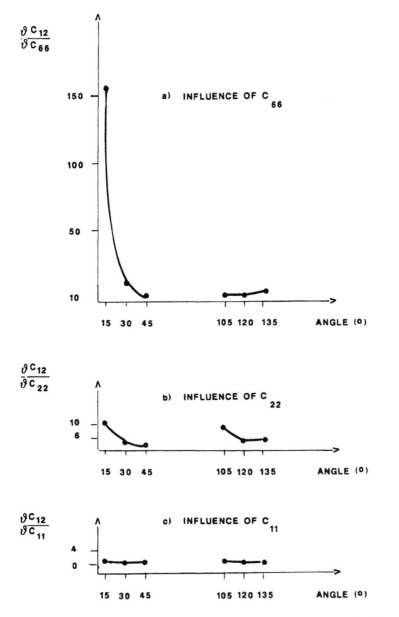

Figure 5.3. The influence of the propagation angle α on C_{12} and the related terms C_{66}, C_{22}, and C_{11} expressed by partial derivatives. Second hypothesis. (From Bucur, V., *Holzforschung*, 40(5), 315, 1986. With permission.)

Discrepancies in the set of observations from Table 5.3 may be attributed to different factors such as the inherent natural variability of wooden species, the measurement errors or scatter, and the type of testing method.

The fourth hypothesis for the averaging procedure is based on the assumption that both values of quasilongitudial and quasitransverse velocities of redundant measurements could be used for the calcula-

Table 5.5. Absolute (10^8 N/m^2) and relative (%) errors for the nondiagonal terms C_{LR} of the stiffness matrix related to the propagation angle α

	Propagation angle					
	15°	**30°**	**45°**	**105°**	**120°**	**135°**
Value (10^8 N/m^2)	60.19	52.46	54.77	47.32	46.69	45.34
Absolute error (10^8 N/m^2)	33.57	11.56	6.52	12.94	7.51	8.52
Relative error (%)	55.77	22.03	11.90	27.34	16.07	18.79

Note: Measurements in *LR* plane on beech at 1 MHz.

From Bucur, V. and Archer, R. R., *Wood Sci. Tech.*, 18(4), 255, 1984. With permission.

tion of nondiagonal stiffness terms. The relationships between those velocities and the diagonal terms of the stiffness matrix could be deduced from Christoffel's equation as:

$$(\Gamma_{11} - \rho V^2)(\Gamma_{22} - \rho V^2) - \Gamma_{12}^2 = 0 \tag{5.11}$$

$$(\rho V^2)^2 - \rho V^2(\Gamma_{11} + \Gamma_{22}) + \Gamma_{11}\Gamma_{22} - \Gamma_{12}^2 = 0 \tag{5.12}$$

$$(\rho V^2)^2 - \rho V^2[C_{11}n_1^2 + C_{22}n_2^2 + C_{66}(n_1^2 + n_2^2)] + (C_{11}n_1^2 + C_{66}n_2^2)(C_{66}n_1^2 + C_{22}n_2^2)$$
$$- (C_{12} + C_{66})^2\, n_1^2\, n_2^2 = 0 \tag{5.13}$$

Several equations of interest relate the roots V_{QL} and V_{QT} and the coefficients of Equation (5.13) as:

$$\rho V_{QL}^2 + \rho V_{QT}^2 = C_{11}n_1^2 + C_{22}n_2^2 + C_{66}(n_1^2 + n_2^2) \tag{5.14}$$

$$V_{QL}^2 + V_{QT}^2 = V_{11}^2 n_1^2 + V_{22}^2 n_2^2 + V_{66}^2(n_1^2 + n_2^2) \tag{5.14'}$$

and

$$\rho V_{QL}^2 \times \rho V_{QT}^2 = (C_{11}n_1^2 + C_{66}n_2^2)(C_{66}n_1^2 + C_{22}n_2^2) - (C_{12} + C_{66})^2 n_1^2 n_2^2 \tag{5.15}$$

which gives us

$$C_{12}/\rho = (n_1 n_2)^{-1}[(V_{11}n_1^2 + V_{66}^2 n_2^2)(V_{66}^2 n_1^2 + V_{22}n_2^2) - V_{QL}^2 V_{QT}^2]^{1/2} - V_{66}^2 \tag{5.16}$$

Experimental measurements of V_{QL}^* and V_{QT}^* do not satisfy Equation (5.14). From this complementary equation we deduce the correction k:

$$k = [V_{11}^2 n_1^2 + V_{22}^2 n_2^2 + V_{66}^2(n_1^2 + n_2^2)]/V_{QL}^{*2} V_{QT}^{*2} \tag{5.17}$$

This correction must then be introduced in Equation (5.16) for the term:

$$V_{QL}^2 \times V_{QT}^2 = k^2 V_{QL}^{*2} V_{QT}^{*2} \tag{5.18}$$

corresponding to the smallest value deduced for all the experimental redundant measurements determined on separate specimens cut at convenient angles to the principal direction. This optimization procedure leads to a new matrix, [C].

The slowness surface in the *RT* plane for maple, following the fourth hypothesis, is shown in Figure 5.5. Very good agreement between theoretical and experimental points was observed for this species when provided with the longitudinal and transverse mode velocity data. This method can be applied to all three anisotropic planes, with accurate results.

These considerations of the averaging procedure for wood material obviously do not cover all aspects concerning the calculation of technical constants from ultrasonic measurements, but it seems that for

Table 5.6. Off-diagonal terms (10^8 N/m^2) of the stiffness matrix for different species

Plane	Frequency (MHz)	Values deduced from measurements		Optimum term selected
		Calculated	Error	
		Tulip Tree		
LR	0.5	50.72	2.434	47.28
	1.0	42.38	6.278	36.10
LT	0.5	32.96	4.022	28.94
	1.0	27.22	4.316	22.91
TR	0.5	9.37	1.128	10.49
	1.0	12.07	5.395	12.07
		Oak		
LR	0.5	31.49	2.060	29.43
	1.0	29.53	1.962	27.57
LT	0.5	29.14	4.513	24.62
	1.0	25.99	4.022	21.97
RT	0.5	9.61	0.785	10.40
	1.0	10.10	0.491	10.59
		Beech		
LR	0.5	50.03	4.708	45.32
	1.0	57.09	5.199	51.89
LT	0.5	30.61	2.747	27.86
	1.0	31.20	5.297	25.89
RT	0.5	13.24	0.589	13.83
	1.0	9.52	0.785	10.30
		Pine		
LR	0.5	33.16	4.513	28.64
	1.0	22.61	5.591	22.66
LT	0.5	12.95	4.905	17.85
	1.0	15.50	2.747	12.75
RT	0.5	13.93	0.441	13.93
	1.0	13.54	0.294	13.24
		Spruce		
LR	0.5	29.63	2.453	27.17
	1.0	34.73	3.041	31.68
LT	0.5	14.13	0.000	14.13
	1.0	13.05	0.589	13.64
RT	0.5	9.23	0.422	9.23
	1.0	12.56	1.569	10.99
		Douglas Fir		
LR	0.5	39.73	1.472	38.26
	1.0	36.30	6.867	29.43
LT	0.5	25.99	6.377	25.99
	1.0	23.54	5.886	17.66
RT	0.5	17.02	0.638	17.02
	1.0	21.04	5.248	15.79

From Bucur, V. and Archer, R. R., *Wood. Sci. Tech.*, 18(4), 255, 1984. With permission.

each species a suitable combination of theoretical and experimental data allow the determination of nine engineering constants. The choice of one set of technical constants could be made by selecting those values corresponding to the highest E_L modulus. The argument in favor of this selection depends on previous measurements with longitudinal waves in cylindrical bars done on the same species using the same ultrasonic equipment.

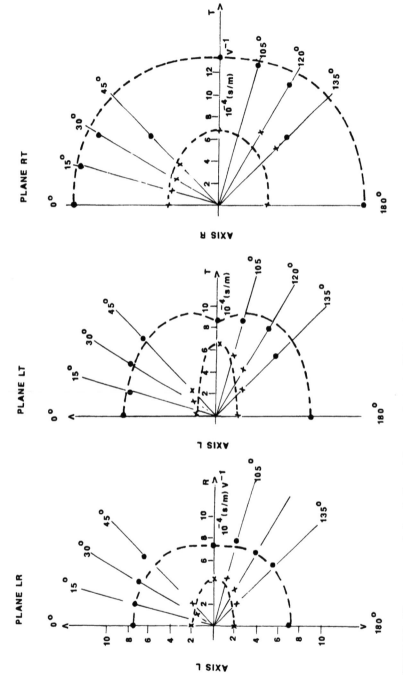

100

Figure 5.4. Slowness curves for beech. Computation according to the third hypothesis, the most conservative error prediction. (From Bucur, V. and Archer, R.R., Wood Sci. Tech., 18(4), 255, 1984. With permission.)

Table 5.7. Theoretical and experimental values of slowness on yellow poplar at 0.5 MHz, using different hypotheses

Plane	Wave	Angle	Experimental	No. 1	No. 2	No. 3
				Slowness (m/s)		
				Calculated in hypothesis		
LT	QL	15°	2.06	4.62	1.81	1.85
		30°	2.19	3.12	1.99	1.99
		45°	2.31	2.37	2.38	2.38
	QT	15°	8.87	7.48	8.36	8.88
		30°	9.62	7.36	8.49	8.89
		45°	10.50	7.08	8.67	9.23
TR	QL	15°	5.62	4.89	4.69	4.90
		30°	4.72	4.72	5.04	5.17
		45°	5.87	4.63	5.59	5.60
	QT	15°	15.82	6.79	13.11	13.72
		30°	12.82	7.21	11.35	12.99
		45°	10.06	7.39	10.18	12.45

From Bucur, V., *Holzforschung*, 40(5), 315, 1986. With permission.

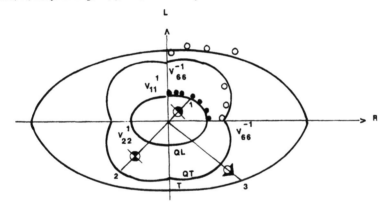

Figure 5.5. Slowness surface in maple, following the fourth hypothesis. Rotation *T* in *LR* plane. *QL* = Quasilongitudinal wave; *QT* = quasitransverse wave; *T* = transverse wave. (Bucur, 1987).

5.1.2. STIFFNESSES AND MODE CONVERSION PHENOMENA FROM BULK TO SURFACE WAVES

Chapter 4 showed that for the determination of the nondiagonal terms of the stiffness matrix numerous, off-axis bulk wave velocity measurements, as well as an appropriate fastidious optimization procedure, are called for. To simplify the procedure the potential of the surface waves must be analyzed to permit access to the off-diagonal terms of the stiffness matrix using very few experimental measurements. The properties of surface waves enable us to deduce the off-diagonal terms of the stiffness matrix, when the surface wave velocity is known, using the general equation:

$$C_{ij} = f(\rho, \quad V_{surface}, C_{ii}, \quad n_i) \tag{5.18'}$$

For the specific case of the orthotropic solid and surface wave propagation along axis 1, with the free surface corresponding to the (1,2) plane and when C_{13} must be calculated, the following bicubic equation (Deresiewicz and Mindlin, 1957) is used:

$$(1 - d)q^3 - (1 - e + 2f)q^2 + f(f + 2)q - f^2 = 0 \tag{5.19}$$

where q is a function of the mass density and surface wave velocity. The coefficients of this equation

Table 5.8. Coefficients of Equation (5.19) and relationships between elastic constants and surface wave velocities on an orthotropic solid with the free surface of the (1,2) plane

Propagation direction	C_{ij}	Coefficients of Equation (5.19)			
		q	d	e	f
1	C_{13}	$\rho V^2/C_{55}$	C_{55}/C_{33}	C_{11}/C_{33}	$(C_{11} - C_{13}^2/C_{33})/C_{55}$
2	C_{23}	$\rho V^2/C_{44}$	C_{44}/C_{33}	C_{44}/C_{33}	$(C_{22} - C_{23}^2/C_{33})/C_{44}$

From Bucur, V. and Rocaboy, F., *Ultrasonics*, 26, 344, 1987. With permission of the publishers, Butterworth Heinemann Ltd.

are given in Table 5.8. For the other terms C_{ij}, similar equations can be derived. Equation (5.19) is thus of greater interest as far as direct determination of nondiagonal terms is concerned. [Under specific, in-axis propagation conditions the equation governing surface wave propagation involves a single off-diagonal term.] In other words the following equation must be solved:

$$M\,C_{ij}^4 + N\,C_{ij}^2 + P = 0 \tag{5.20}$$

where M, N, P are real coefficients involving diagonal stiffness terms and in-axis surface wave velocities. Table 5.9 gives the expressions for the coefficients in the specific case of propagation along axis 1, with the free surface in the (1,2) plane.

Obviously, the agreement between C_{ij} terms, determined from bulk and surface velocities, could be used as an argument for the development of a methodological approach based on a mode conversion technique from bulk waves to surface waves (Bucur and Rocaboy, 1988).

The first step toward checking the validity of the surface wave method for wood characterization involves the following procedure: the stiffness terms obtained from bulk wave velocity measurements are substituted into Equation (5.19) and the corresponding surface wave velocities are computed according to the search procedure developed by Farnell (1970). The predicted values for surface wave velocities are then checked against the measured values. Good agreement between the two sets of velocities could be used to assert the validity of the experimental approach to surface wave velocity measurements.

The second step is a computation of off-diagonal stiffnesses from surface wave velocity measurements, following Equation (5.20). Experimentally, the coefficients M, N, and P are computed from the diagonal C_{ii} terms obtained from bulk wave velocity measurements, and the value of surface wave velocities is obtained from other sets of measurements. This second approach is introduced to check the consistency of the bulk wave procedure for the selection of C_{ij}. Obviously, agreement between C_{ij} terms computed from bulk and surface velocities must exist in order for our methodological approach to be valid.

Table 5.10 provides results of measured and computed surface wave velocities on beech and spruce. The highest discrepancy between values is related to the transversal anisotropic plane (2,3), or *RT*. The rather small differences between maximum and minimum values of surface velocities for beech can be explained easily from considerations of wood microstructure. Beech is one of the diffuse, porous hardwood species, presenting a smooth transition from earlywood to latewood in annual rings. This species, unlike spruce, is better modeled as a homogeneous anisotropic medium, assumed in the development of the continuum theory, which ignores the ring structure of wood. In spruce the transition between earlywood and latewood is very sharp. This layered structure induces dispersive propagation

Table 5.9. Coefficients of Equation (5.20)

Free surface orientation	1,2 plane
Propagation direction	1
Corresponding C_{ij} term	C_{13}
Coefficients	$M = (\rho V^2 - C_{55})/(C_{33}C_{55})$
	$N = 2[(\rho V^2)^2/C_{55} - \rho V^2\,(1 + C_{11}/C_{55}) + C_{11}]$
	$P = (\rho V^2)^3\,(C_{33} - C_{55})/C_{55}\,(\rho V^2)^2\,(C_{33} - C_{11} + 2C_{11}C_{33}/C_{55}) + \rho V^2(2 + C_{11}/C_{55})C_{11}C_{33} - C_{11}^2 C_{33}$

Note: The coefficients for the other planes are deduced from corresponding permutations.

From Bucur, V. and Rocaboy, F., *Ultrasonics*, 26, 344, 1987. With permission of the publishers, Butterworth Heinemann, Ltd.

Table 5.10. Comparison between measured and computed surface velocities on beech and spruce

Species	Surface velocities (m/s) in planes		
	12	13	23
	Computed Values		
Beech	1455	1238	900
Spruce	1294	1194	385
	Measured Values		
Beech	1404–1460	1260–1249	880–920
Spruce	1173–1300	1113–1210	380–440

From Bucur, V. and Rocaboy, F., *Ultrasonics*, 26, 344, 1987. With permission of the publishers, Butterworth Heinemann Ltd.

Table 5.11. Terms of the stiffness matrix used for the computation of the velocities of surface waves on beech and spruce

Species	Stiffness terms (10^8 N/m²)								
	C_{11}	C_{22}	C_{33}	C_{44}	C_{55}	C_{66}	C_{12}	C_{13}	C_{23}
Beech	173.25	32.63	16.40	6.21	10.87	15.18	30.36	16.89	7.42
Spruce	139.23	12.19	6.37	1.10	7.33	8.48	2.84	2.31	7.60

From Bucur, V. and Rocaboy, F., *Ultrasonics*, 26, 344, 1987. With permission of the publishers, Butterworth Heinemann Ltd.

of a probe pulse and reduces the accuracy of velocity measurements. Ultrasonic propagation is particularly dispersive in the (1,2) and (2,3) planes.

The next artful step in our analysis is to determine the three off-diagonal stiffness terms without resorting to a fastidious optimization procedure. For this purpose, the diagonal terms (obtained from on-axis bulk velocity measurements) and measured surface wave velocities were introduced in Equation (5.20) (see Table 5.11). The corresponding set of off-diagonal stiffness terms could then be computed easily using the same search procedure as was used for surface velocity computation. Table 5.12 shows that the two sets of stiffness terms lie very close to one another. This statement is better understood if the interaction between wood microstructure and ultrasonic waves is kept in mind, based on the consideration that surface wave particle motion is elliptical, and therefore may undergo modulation by wood microstructure. In spite of the fact that C_{ij} terms are much more difficult to obtain than C_{ii} terms, they are most representative of the anisotropy of the considered species. The combination of on-axis bulk wave velocities for C_{ii} terms and surface wave measurements for C_{ij} terms provides the technical constants listed in Table 5.13.

It is worth recalling, however, that the determination of C_{ij} terms is considerably eased with the surface wave method proposed. This is because it exclusively involves on-axis velocity measurements. A single wood specimen cut along the axis is therefore required. The motivation for a simple procedure

Table 5.12. Computed C_{ij} terms (10^8 N/m²) derived from Equation (5.20)

Species	Off-diagonal terms of stiffness matrix					
	Bulk waves			Surface waves		
	C_{12}	C_{13}	C_{23}	C_{12}	C_{13}	C_{23}
Beech	30.36	16.89	7.42	34.42	16.16	10.67
Spruce	2.84	2.31	7.60	3.78	3.58	7.88

From Bucur, V. and Rocaboy, F., *Ultrasonics*, 26, 344, 1987. With permission of the publishers, Butterworth Heinemann Ltd.

Table 5.13. Engineering constants from bulk- and surface-wave velocities and static tests for beech and spruce

Species	Young's moduli (10^8 N/m²)		
	E_L	E_R	E_T
Ultrasonic Tests			
***Bulk Waves*[a]**			
Beech	138.22	25.95	14.02
Spruce	138.39	3.03	1.59
***Surface waves*[b]**			
Beech	134.71	22.11	12.70
Spruce	137.05	2.43	1.26
Static Bending Tests[c]			
Beech	140.00	22.80	11.60
Spruce	137.69	9.19	4.90

[a] Measured parameters: bulk velocities in and out of axes.
[b] Measured parameters: surface in axis velocities.
[c] Measured parameters: displacements in static bending. (Data from Kollmann and Côté, 1968.)

From Bucur, V. and Rocaboy, F., *Ultrasonics*, 26, 344, 1987. With permission of the publishers, Butterworth Heinemann Ltd.

to search for C_{ij} terms came from the necessity to determine all engineering constants (three Young's moduli and six Poisson's ratios). However, it is well known that the measurements of Poisson's ratios for wood is a very difficult experimental task in the static regime (Sliker, 1989). A knowledge of surface wave velocities could simplify the measurement methodology. The reasons underlying this observation are presented below.

Mason (1958) noted the possible use of Rayleigh wave velocity measurements to determine Poisson's ratios. Because Rayleigh wave velocity, which is always less than the shear wave velocity, depends on Poisson's ratios, one can plot the surface velocity: shear velocity ratio as a function of Poisson's coefficients. These suggestions can be used in future research programs.

It seemed appropriate in this chapter to summarize the possibilities of ultrasonic measurements related to all off-diagonal constants using a mode conversion technique on an orthotropic solid (Table 5.14). Motivating the development of this technique was the need for a nondestructive approach to the determination of all technical constants on living trees or wood structures, for both mechanical behavior evaluation and defect detection purposes. Surface velocity measurements are thus expected to make the procedure easier. Within fairly well-defined practical limits, the mechanical characterization, and therefore defect detection, within standing trees may be answered already if an appropriate signal processing would remove the ambiguity that occurs when trying to distinguish between the longitudinal and shear bulk wave components and surface wave components.

Table 5.14. Mode conversion technique and corresponding surface and bulk velocity measurements (Bucur, 1989)

Free surface	Off-diagonal terms	Bulk velocities		Specimens
		Longitudinal	Shear	
12	C_{13}	V_{11}	V_{55}, V_{66}	Sample, block
	C_{23}	V_{22}	V_{44}, V_{66}	Sample, block
13	C_{12}	V_{11}	V_{66}, V_{55}	Standing trees or
	C_{32}	V_{33}	V_{44}, V_{55}	logs
23	C_{21}	V_{22}	V_{66}, V_{44}	Logs structural
	C_{31}	V_{33}	V_{55}, V_{44}	elements

Table 5.15. Comparison between calculated E_L from static and resonance tests and C_{LL} from ultrasonic tests, for Douglas fir

Test method	Static test E_L	Resonance test E_L	Ultrasonic C_{LL}
Values (10^8 N/m^2)	73.4–117.4	59.0–118.0	149.7

From Sinclair, A. N. and Farshad, M., *J. Test. Eval.*, 15(2), 77, 1987. With permission.

The reader should note that mode conversion phenomena arise as a result of structure-induced scattering mechanisms. Initially it appears that mode conversion and energy transfers from one to several wave types would lead to unwanted complications when interpreting velocity readings. On the other hand, mode conversion processes complicate theoretical treatment of the scattering problem, but offer the possibility of extracting additional information on the structure-governed mechanical properties of wood material.

5.1.3. YOUNG'S MODULI, SHEAR MODULI, AND POISSON'S RATIOS FROM DYNAMIC (ULTRASONIC AND RESONANCE) TESTS AND STATIC TESTS

As in Chapter 4, using the ultrasonic method accedes to the stiffness terms [C], whereas in using static test the compliances [S] are measured. Thus, if the coefficients of either the [C] or the [S] matrices are known, the coefficients of the inverse matrix, respectively [S] or [C], can be calculated by simple inversion.

For practical purposes it is sometimes convenient to express the elastic behavior of wood structure in terms of engineering elastic coefficients (Young's moduli, shear moduli, and Poisson's ratios) that can be deduced from the terms of the [S] matrix. Various wood elastic moduli, for a variety of species originating from the temperate or tropical zones, were deduced from static and dynamic tests by Carrington (1923), Campredon (1935), Pouligner (1942), Doyle et al. (1945), Hearmon (1948), Kollmann (1951), Kollmann and Côté (1968), Schumacher (1988), Bolza and Kloot (1963), Jayne (1972), Mullins and McKnight (1981), Bodig and Jayne (1982), Guitard (1985) and the U.S. Forestry Service (1987).

Pioneering work related to elastic constants determined by ultrasonic and static tests was done by Hearmon in 1965. Musgrave (1961) was the first to calculate all terms of the stiffness matrix of spruce. Using the static data on spruce reported by Hörig (1935), Musgrave deduced the corresponding nine terms of the stiffness matrix and computed slowness and wave surfaces. Hearmon emphasizes the need for fundamental studies of vibrational techniques in the audio-frequency or ultrasonic ranges, for the estimation of the elastic behavior of wood material.

Regarding the comparison between data in the ultrasonic and static regimes Sinclair and Farshad (1987) reported results on Douglas fir (Table 5.15) using only eight specimens. They point out the difficulty of producing valuable specimens for all tests, and note that the time of propagation of an ultrasonic signal can be measured more accurately than the resonance frequency or static strain. Table 5.15 shows that E_L has the highest value in ultrasonic tests.

Comparison of the results obtained from a large number of ultrasonic and static test specimens can be made on the basis of the examination of the theoretical relationships among stiffnesses, compliances, and engineering constants. The results of the experiments undertaken on spruce are given in Table 5.16 (Bucur et al., 1987). Furthermore, comparison among ultrasonic, resonance, and static tests was made

Table 5.16. Engineering constants for Sitka spruce at 12% moisture content, deduced from static and ultrasonic tests (Bucur et al., 1987)

Technical constants (10^8 N/m^2)						Poisson's ratios (10^{-3})		
E_L	E_R	E_T	G_{TR}	G_{LT}	G_{LR}	vLR	vLT	vRT
Static Test								
90.36	8.17	4.03	—	—	—	57	29	438
Ultrasonic Test								
95.64	10.37	4.87	0.91	10.95	11.96	45	49	43

Table 5.17. Young's modulus E_L for fir at 12% moisture content, deduced from static bending, resonance, and ultrasonic tests (From Haines et al., 1993)

Parameters	Units	Values Average	Coefficient of variation (%)
Density	kg/m³	465	11
E_L static	10^8 N/m²	126	19
E_L resonance	10^8 N/m²	139	17
C_{LL} ultrasonic	10^8 N/m²	154	15

Note: E_L resonance/E_L static = 1.10; C_{LL}/E_L resonance = 1.22.

on fir (Table 5.17). Table 5.16 demonstrates that the ultrasonic values of Young's moduli are slightly higher than the corresponding static measured moduli under compression on cubic specimens. The same conclusion was reached by Pluvinage (1985) and Launay et al. (1988) on beech and Sitka spruce wood, respectively, when static bending and static compression tests were compared to ultrasonic tests (goniometric technique). From Table 5.17 we note that the Young's modulus, E_L, from the resonance test is about 10% higher than the static test. The stiffness term C_{LL} is 22% higher than the static E_L because no correction with Poisson's ratio was made.

Coming back to the engineering constants available in the literature (especially those determined from static tests), it can be noted that only Young's modulus, E_L, is mentioned for the majority of commonly used species. The other moduli (Young's moduli E_R, E_T, or shear moduli) and Poisson's ratios have not been extensively examined, probably because of the difficulty of making precise small strain measurements (due to the equipment limitation) and of producing specimens on which the spatial inhomogeneity of wood, induced by the annual ring curvature, was limited or avoided.

Important progress in the field of static measurements was made by Sliker (1988, 1989), Ben Farhat (1985), and El Amri (1987). Improved accuracy in measurements was achieved by using low modulus strain gauges having little or no sensitivity perpendicular to the gauge axis, by making strain measurements with a resolution of 0.1×10^{-6} microstrains per meter, and by having a proper loading device.

To outline the difficulties mentioned, attempts were made to predict off-diagonal elastic constants from the well-known physical properties of specific gravity (Bodig and Goodman, 1971; Guitard, 1985) or diagonal compliances for a large group of species, using statistical regression analysis. This empirical approach has a very speculative character and could aid in estimating the behavior of wooden structures in studies connected with finite element techniques or in strain and stress distribution in wooden members.

Despite the difficulties, the concept of the ultrasonic velocity method is simple and attractive in application. It is relatively easy to measure the required velocities. The ultrasonic procedure is far less tedious than the static testing procedure for the measurement of Poisson's ratios. The static procedure has the disadvantage of being complicated, requiring a special, heavy apparatus and a considerable length of time to perform.

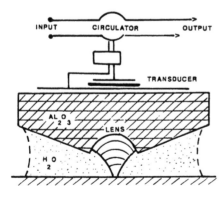

Figure 5.6. Typical configuration of the scanning acoustic microscope lens in the reflection mode (Quate et al., 1979).

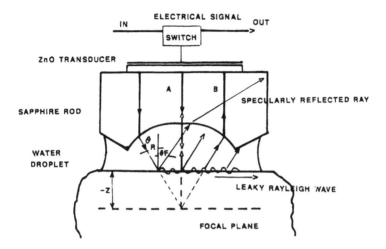

Figure 5.7. Signal reflected by the specimen as a result of interference between a surface Rayleigh wave (B) and a longitudinal wave reflected at normal incidence (A). (From Attal, J., in *Ultrasonic Signal*

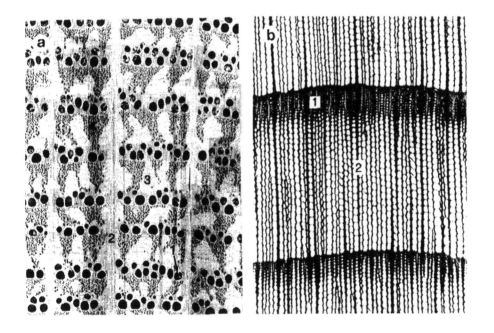

Figure 5.8. Microscopic description of oak, spruce, and pine. (Microscopic sections courtesy of Simone Garros; photographs courtesy of François Thiercelin.)

5.2. LOCAL CHARACTERIZATION

Wood is always inhomogeneous and anisotropic at all structural levels because of the juxtaposition of its anatomic elements. The cell walls give wood its mechanical resistance, and the principal framework constituent is cellulose.

To go a step further in understanding the relation between wood structure and its mechanical behavior, a layered model was proposed to be satisfactory for representing the wood structure at macroscopic

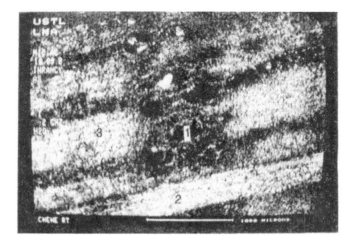

Figure 5.9. Acoustic micrograph of oak in the radial-tangential plane. The round ring is porous (1). Ray (2) and fiber zone (3) are well defined (Bucur, et al., 1992).

Figure 5.10. Acoustic micrograph of a multiseriate oak ray (Bucur et al., 1992).

and microscopic scales. On the macroscopic scale the layered system simulates the presence of earlywood and latewood in annual rings. On the microscopic scale the layered system is related to the organization of the cellular walls, which is composed of a finite number of homogeneous orthotropic layers (Schniewind, 1972). This basic cell wall organization is the same for all wood species. At the molecular level the physical organization of the cellulose molecules is also orthotropic. Depending on the nature of the investigation, all models acknowledge the precise determination of the elastic constants of the elements of the system.

The recent development of acoustic microscopy or of photoacoustics could provide very refined tools for the characterization of the structural elements of wood at the microscopic and submicroscopic scales.

5.2.1. ACOUSTIC MICROSCOPY

Spectacular development in microscopic techniques using different types of radiation (X-rays, microwaves, infrared, lasers, etc.) has been achieved recently in the study of the structural organization of

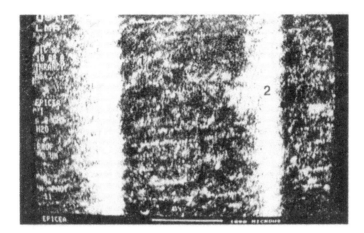

Figure 5.11. Acoustic micrograph of spruce in the radial-tangential plane. The transition is sharp between the earlywood (dark zone) and latewood (light zone) (Bucur et al., 1992).

Figure 5.12. Acoustic micrograph of pine in the longitudinal-radial plane. Detail of tracheids (1), pits (2), and rays (3) (Bucur et al., 1992).

new materials such as superconductors and composites. Acoustic microscopy is part of this development, and would appear to be one of the most promising nondestructive tools for the study of the physical properties of materials (Lemons and Quate, 1974, 1979; Attal, 1979, 1989; Jones, 1987).

Acoustic microscopy is most interesting in its direct interaction between the wave and the elastic properties of the material through which it propagates. Resolution occurs on the millimeter or micron scale for ultrasonic frequencies in the megahertz or gigahertz ranges. The principal difference between acoustical and optical images lies in the restriction of the optical microscope to reveal only surface features. In contrast, acoustic waves propagate into the specimen and permit the exploration of the region beneath the surface.

Determining the tenor of the relationships between an anatomical element of wood and its elastic constants via acoustic microscopy has not been reported in the literature. This section is an attempt to do so.

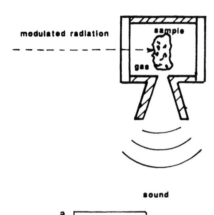

Figure 5.13. Photoacoustic effect induced by a laser beam that generates sound (Busse, 1987).

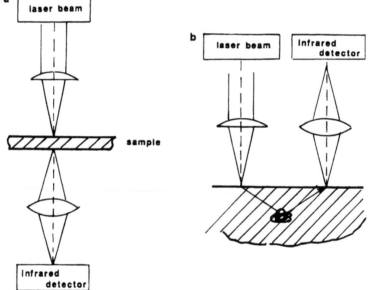

Figure 5.14. The device for thermal wave detection in (a) transmission and (b) reflection modes (Busse, 1987).

5.2.1.1. Operating Principle

Construction of the scanning acoustic microscope is described in detail by Attal (1979), Quate et al. (1979), and Briggs (1985), so only the elements of its operation are given here. Figure 5.6 represents typical configuration of a scanning acoustic microscope head in the reflection mode. The transducer is excited by an electrical pulse of appropriate frequency, which generates an ultrasonic beam in the sapphire rod (Figure 5.7). This beam (A) is focused at the distance ($-z$) beneath the surface of the sample through a drop of liquid, and this forms the acoustical coupling medium between the cavity of the lens and the specimen. The ultrasonic pulse is reflected, first at the interface and subsequently by structures located at the focal plane. Thus, the echoes produced traverse the system in the reverse order and are detected via video signal by a receiver connected to the transducer. The signal reflected by the specimen is the sum of two types of waves: the surface (Rayleigh) acoustic wave generated and then re-emitted in the liquid (B) at the critical angle Θ_R, and the longitudinal wave (A) reflected at normal incidence. The interference of these waves can be observed by varying the focal distance. If the focus is very near the surface, the amplitude of the detected pulse is related directly to the acoustic reflectance function at the specimen-liquid interface, which is in turn directly related to the local acoustic properties

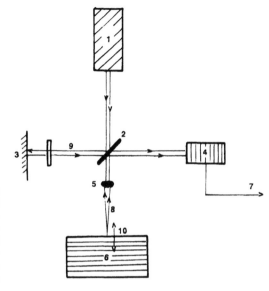

Figure 5.15. Laser interferometer used for the measurement of elastic constants of materials. (1) Laser; (2) beam splitter; (3) mirror; (4) photodetector; (5) neutral density filter; (6) specimen; (7) output; (8) probe beam; (9) reference beam; (10) surface displacement. (From Scruby, C. B., *Ultrasonics*, 27, 195, 1989. With permission of the publishers, Butterworth-Heinemann Ltd.©)

of the sample. By scanning the local point over the desired field of view an acoustic image can be built up in a few seconds and displayed on a television monitor.

The experiments reported here (Bucur et al., 1992) were performed at 200 MHz with a lens having a 1.7-mm focal length, focused approximately 10 μm below the surface.

Acoustic images were obtained on small blocks of wood of 20 × 20 × (5 to 10) mm following the *L*, *R*, and *T* natural directions of the wood specimen. The specimen surfaces were carefully prepared using a microtome, used currently to cut thin sections of wood for studies in optical microscopy. The moisture content of wood was <8% in all cases. The effective surface scanned on the specimens was 5 × 5 mm².

Using acoustic reflectometry the special staining techniques necessary to optical or other microscopies are completely avoided.

5.2.1.2. Acoustic Images

Images obtained using optical and acoustical techniques are shown for purposes of comparison. Figure 5.8 presents the anatomical features of oak, spruce, and pine revealed via optical microscopy. Further detail of this technique is given by Bosshart (1974) and Jacquiot et al. (1955, 1973).

Acoustic micrographs of oak, spruce, and pine, obtained in the reflection mode and explored in one of the three main anisotropic planes, are shown in Figures 5.9 to 5.12. Contrast in the acoustic micrographs is intrinsically apparent. The anatomical features of oak in the *RT* plane are illustrated in Figures 5.9 and 5.10, in which the scale is 1000, 500, and 250 μm, respectively. Details of the oak round ring-porous growth pattern are shown in Figure 5.9. The earlywood pores are large and concentrated in a narrow dark band at the beginning of the annual ring growth. Latewood pores, which are much smaller, are quite difficult to observe in the latewood zone. Note the zone of fibers (bright) and that of an oak-type multiseriate ray. The contrast observed is due to the difference in acoustic impedance (product of density and velocity) corresponding to each type of structural feature. It is well known (Polge et al., 1973) that oak rays have a higher density than do pores or the parenchymal zone. This fact is clearly seen in the acoustic micrographs, in which the rays appear as a brilliant zone as compared to the dark zone of the vessels. Details of the ray structure are shown in Figure 5.10, at a scanning scale of 250 μm. The constitutive cells of the multiseriate ray can be observed easily.

Figures 5.11 and 5.12 illustrate the anatomical details of two softwoods, spruce and pine. One of the most important anatomical features that readily identify wood species is the sharp difference between the latewood and the earlywood in the annual ring. Note in Figure 5.11 an abrupt transition between the two zones in an *RT* plane. The dark zone corresponds to the earlywood. This acoustic micrograph

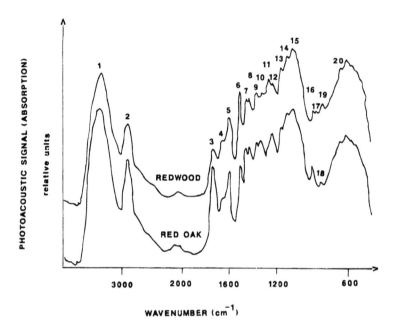

Figure 5.16. Infrared–photoacoustic spectra of redwood and red oak. Measurements of 400-μm thick radial sections. (From Kuo et al., *Wood Fiber Sci.*, 20(1), 132, 1988. With permission.)

techniques to X-ray densitometry for the definition of the elastic behavior of the anatomical elements in every annual ring.

Anatomic details in the longitudinal-radial plane in pines are given in Figure 5.12. Long tracheids and their common wall sculpturing due to pitting appear as additional evidence in this acoustic micrograph. Some details of the pitting architecture are also revealed.

The challenge in using acoustic microscopy, and particularly reflectometry, is to find a means of visualizing the anatomical features and acoustic properties of the wood at the highest possible resolution. Using a lens focused at 200 MHz, it is possible to obtain acoustic micrographs with a resolution of 6 μm in a scanning dimension ranging between 50 μm and 5 mm. A second aspect of the technique lies in its subsurface imaging capability. When the acoustic beam is focused inside the wood sample, penetrating ultrasonic waves are scattered by any existing microstructural detail. A compromise must be established between resolution and subsurface imaging depth, keeping in mind that the attenuation increases with frequency. At the moment the most promising frequency range for wood species appears to be between 1 and 200 MHz, where penetration up to a few millimeters can be attained easily.

The intrinsic contrast in acoustic micrographs of wood is quite good. The special staining technique, which is necessary in optical microscopy, is not needed here. The very fact that an acoustic microscope visualizes directly the acoustic and elastic properties of the wood may be a chief attribute in the development of a new nondestructive procedure for very fine quantitative anatomical studies. We refer here to the possibility of elastic constants measurements of all anatomic elements (fibers, vessels, tracheids, rays, cellular walls, microfibrils, etc.).

5.2.2. PHOTOACOUSTICS IN WOOD SCIENCE

The development of photoacoustics for the study of wood samples is due to the availability of laser sources and to the development of the corresponding electronics for the detection of thermal and acoustic waves. Laser beams together with ultrasonic waves afford the opportunity to perform noncontact measurements in hostile environments or in geometrically difficult-to-reach locations. Rosencwaig (1980) and Pao (1977) discuss this rapidly growing field which aims to obtain, through a non-contact techique, supplementary information to the existing non-destructive methodology.

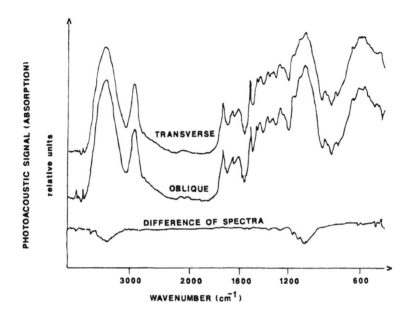

Figure 5.17. Infrared–photoacoustic spectra of ponderosa pine on transverse and oblique (45°) sections. (From Kuo et al., *Wood Fiber Sci.*, 20(1), 132, 1988. With permission.)

5.2.2.1. Principles

Photoacoustics (optoacoustics) was developed as the result of the effect observed when a modulated light beam was focused on the surface of an absorbing solid, which produced a local temperature modulation. In Figure 5.13 illustrates this effect. The light beam intensity modulated at the frequency (ω) produces a temperature modulation, ΔT, on the solid. The temperature of the air adjacent to the solid also changes, and a pressure modulation is achieved. This pressure modulation can be detected with a microphone as a sound at the modulation frequency ω. This is the photoacoustic effect associated with three fundamental processes: the absorption of the incident energy, the generation and propagation of thermal waves, and the generation and propagation of acoustic waves. The intensity of the detected sound depends on optical input power, the modulation frequency, and the thermal and acoustical properties of the specimen. The success of the technique is related also to the efficiency of the "photoacoustic cell" containing the microphone and the sample.

For inhomogeneous specimens (in addition to the previous factors), temperature modulation depends on the coordinates at which the light beam hits the solid. Investigation of the relationships between the temperature variation and acoustic wavelength is related to the photoacoustic spectroscopy.

This technique may be applied to a very wide range of materials: metals (Bourkoff and Palmer, 1985), reinforced composites (Busse, 1987), polymers (Busse and Eyerer, 1983), anisotropic solids (Castagnede and Berthelot, 1992), very thin films, textiles (Flynn et al., 1985), food (Belton and Tanner, 1983), wood (Kuo et al., 1988), and drugs.

The advantages of photoacoustic spectroscopy are manifest. No physical contact, no coupling medium, no sample preparation, and no potentially hazardous radiation is involved. Inspection can be carried out when the specimen is accessible from one side only.

Imaging techniques developed using thermal waves when laser scanning experiments are performed are the basis of a new microscopy method known as scanning photoacoustic microscopy. Some results were reported by Busse (1985) for crack detection in ceramic materials, delamination in composites, and defect detection in metals.

5.2.2.2. Instrumentation

The modulation of average surface temperature is produced by laser-generated thermal waves in photoacoustic detection monitors. The experimental arrangement for thermal wave transmission and reflection

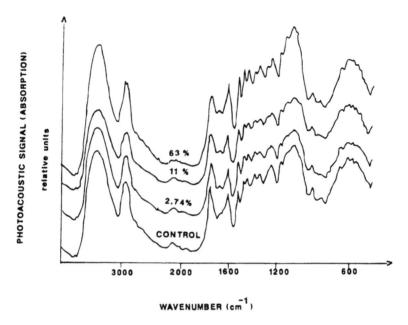

Figure 5.18. Infrared–photoacoustic spectra of Eastern cottonwood samples with various degrees of decay (2.74%; 11%; 63%) caused by brown rot fungus (*Gleoephyllum trabeum*). (From Kuo et al., *Wood Fiber Sci.*, 20(1), 132, 1988. With permission.)

is shown in Figure 5.14. Infrared detectors are used to avoid difficulties related to the complicated laser alignment. The lenses are coupled in such a way that infrared emission is received only from the region around the modulated optical spot. The parameters studied are the amplitude and the phase angle of temperature modulation with respect to the modulated laser beam. Photoacoustic spectra can be obtained by recording the signal from the microphone as a function of the wavelength of the incident light beam.

The infrared photoacoustic spectroscopy technique (Kuo et al., 1988) uses the newly developed fast Fourier transform and operates on wood specimens with a thermal diffusivity of 2×10^{-3} J/cm²/s. Modulation frequencies vary from 20 to 200 Hz. The thermal sensing depth ranges from 18 to 56 μm. The wood spectra are adjusted for carbon black. The photoacoustic cell atmosphere used was helium gas, which was selected for its optimal signal generation efficiency. The wood specimens were microscopic sections of 400 μm thickness.

Recently developed laser interferometers (Scruby, 1989) are shown in Figure 5.15. This material characterization equipment was used to measure longitudinal and shear wave speeds.

5.2.2.3. Applications

Infrared photoacoustic spectroscopy (Kuo et al., 1988) has been used recently to determine the characteristic absorption bands of softwood and hardwood. Chemical differences between species were observed in the regions between 900 and 800 cm^{-1} (Figure 5.16).

Figure 5.17 shows the spectra derived from ponderosa pine specimens in a transverse section and in a section cut at 45°. Spectral differences are due to the differences in cellolosic microfibrillar orientation.

Measurements performed on minute quantities of brown-rot decayed wood samples (Figure 5.18) showed that photoacoustic spectroscopy can overcome the experimental difficulties encountered with other analytical methods such as the X-ray diffraction technique, and give an accurate image of the intensity of the biological attack due to brown-rot decay.

REFERENCES

Attal, J. (1979). The acoustic microscope, a tool for nondestructive testing. in *Nondestructive Evaluation of Semiconductor Materials and Devices*. Zemel, J. N. (Ed.). Plenum Press, New York, 631–676.

Attal, J. (1989). Scanning acoustic microscopy: general description, field of applications and quantitative measurements. in *Ultrasonic Signal Processing*. Alippi, A. (Ed.). World Scientific, Singapore, 69–94.

Belton, P. S. and Tanner, S. F. (1983). Determination of the moisture content of starch using near infrared photoacoustic spectroscopy. *Analyst.* 108 (May), 591–596.

Ben Farhat, M. (1985). Contribution à la Mesure des Constantes Rhéologiques du Bois par Compression d'Échantillon Cubiques, Ph.D. thesis, Institut National Polytechnique de Lorraine, Nancy.

Bodig, J. and Goodman, J. R. (1973). Prediction of elastic parameters for wood. *Wood Sci.* 5(4), 249–264.

Bodig, J. and Jayne, B. (1982). *Mechanics of Wood and Wood Composites*. Van Nostrand Reinhold, New York.

Bolza, E. and Kloot, N. H. (1963). The Mechanical Properties of 174 Australian Timbers. Technol. Pap. No. 25, Division of Forest Products, CSIRO, Sydney, Australia.

Bosshard, H. H. (1974). *Holzkunde. Mikroskopie und Makroskopie des Holzes*. Vol. 1. Birkhäser Verlag, Basel.

Bourkoff, E. and Palmer, C. H. (1985). Low-energy optical generation and detection of acoustic pulses in metals and nonmetals. *Appl. Phys. Lett.* 46, 143–145.

Briggs, A. (1985). *An introduction to Scanning Acoustic Microscopy*. Oxford University Press, Oxford.

Bucur, V. (1985). Ultrasonic velocity, stiffness matrix and elastic constants of wood. *J. Catgut Acoust. Soc. Ser.* I(no. 44), 23–28.

Bucur, V. (1986). Les termes non diagonaux de la matrice des rigidités du bois. *Holzforschung.* 40(5), 315–324.

Bucur, V. (1987). Wood characterization through ultrasonic waves. in *Ultrasonic Methods in the Evaluation of Inhomogeneous Materials*. NATO ASI Ser. E, Applied Science No. 126. Alippi, A. and Mayer, W. G. (Eds.). Martinus Nijhoff, Dordrecht, 323–333.

Bucur, V. (1989). Mode conversion ultrasonic testing on wood. in *Ultrasonic Signal Processing*. Alippi, A. (Ed.). World Scientific, Singapore, 463–473.

Bucur, V. and Archer, R. R. (1984). Elastic constants for wood by an ultrasonic method. *Wood Sci. Tech.* 18(4), 255–265.

Bucur, V., Carminati, M., Perrin, J. R., and Perrin, A. (1987). La Variabilité du Bois d'Épicea de Sitka et sa Reponse Acousto-Élastique. Station Qualité des Bois, Centre de Recherches Forestières, Nancy, France.

Bucur, V. and Rocaboy, F. (1987). Surface wave propagation in wood: prospective method for the determination of wood off-diagonal terms of stiffness matrix. *Ultrasonics.* 26, 344–347.

Bucur, V., Saied, A., and Attal, J. (1992). Identification of wood anatomical elements by acoustic microscopy, unpublished data.

Busse, G. (1985). Imaging with optically generated thermal waves. *IEEE Trans. Sonics Ultrasonics.* SU-32(2), 355–364.

Busse, G. (1987). Characterization through thermal waves: applications of optoacoustic and photothermal methods. in *Ultrasonic Methods in the Evaluation of Inhomogeneous Materials*. Alippi, A. and Mayer, W. G. (Eds.). Martinus Nijhoff, Dordrecht, 105–122.

Busse, G. and Eyerer, P. (1983). Thermal wave remote and non-destructive inspection of polymers. *Appl. Phys. Lett.* 43(4), 355–357.

Campredon, J. (1935). Contribution à l'étude des propriétés élastiques des bois. *Ann. Ecole Eaux For. Nancy.* (5), 253–283.

Carrington, H. (1923). The elastic constants of spruce *Philos. Mag.* (45), 1055–1065.

Castagnede, B., Jenkins, J. T., and Sachse, W. (1990). Optimal determination of the elastic constants of composite materials from ultrasonic wave speed measurements. *J. Appl. Phys.* 67(6), 2753–2761.

Castagnede, B., Kim, K. Y., and Sachse, W. (1991). Determination of the elastic constants of anisotropic materials using laser-generated ultrasonic signals. *J. Appl. Phys.* 70(1), 150–157.

Castagnede, B. and Berthelot, Y. (1992). Photoacoustic interactions by modulation and laser impact: applications in mechanics and physics of anisotropic solids (review article). *J. Acoust. Soc. Am.* in press.

Deresiewicz, H. and Mindlin, R. D. (1957). Waves on surface of a crystal. *J. Appl. Phys.* 28, 669–671.

Doyle, D. V., Drown, J. T., and McBurney, R. S. (1945). The Elastic Properties of Wood. USDA Forest Service Rep. No. 1528, Forest Product Laboratory, Madison, WI.

El Amri, F. (1987). Contribution à la Modélisation Élastique Anisotrope du Matériau Bois. Ph.D. thesis. Institut National Polytechnique de Lorraine, Nancy, France.

Every, A. G. and Sachse, W. (1990). Determination of elastic constants of anisotropic solids from acoustic wave group velocity measurements. *Phys. Rev.* 42(13), 8196–8205.

Every, A. G., Castagnede, B., and Sachse, W. (1991). Sensitivity of Inverse Algorithms for Recovering Elastic Constants of Anisotropic Solids from Longitudinal Wave Speed Data. Paper presented at Ultrasonics International '91, Le Touquet, France, July 1 to 4, 459–462.

Farnell, G. W. (1970). Properties of elastic surface waves. in *Physical Acoustics*. Vol. 6. Mason, W. P. (Ed.). Academic Press, New York, 106–166.

Flynn, I. J., Miller, R. M., and Spillane, D. E. M. (1985). The examination of textile fibers and fabrics by photoacoustic spectroscopy. *Proc. Int. Conf. Acoustic Emission and Photo-Acoustic Spectroscopy and Applications*. Stephens, R. W. (Ed.). Chelsea College, London, 35–42.

Guitard, D. (1985). *Mécanique du Matériau Bois et Composites*. Cepadues Editions, Toulouse.

Haines, D., Bucur, V., Leban, J. M., and Herbe, C. (1993). Resonance Vibration Methods for Young's Modulus Measurements of Wood. Station Qualité des Bois, Centre de Recherches Forestières, Nancy, France.

Hearmon, R. F. S. (1948). The Elasticity of Wood and Plywood. Forest Products Res. Spec. Rep. No. 7, His Majesty's Stationery Office, London.

Hearmon, R. F. S. (1965). The Assessment of Wood Properties by Vibrations and High Frequency Acoustic Waves. Paper presented at 2nd Symp. Nondestructive Testing of Wood. Washington State University, Olympia, 49–66.

Hörig, H. (1935). Anwendung der Elastizitätstheorie anisotroper Körper auf Messungen an Holz. *Ingen. Arch.* (VI), 8–14.

Hosten, B. and Castagnede, B. (1983). Mesures des constantes élastiques du bois à l'aide d'un interféromètre ultrasonore numérique et leur optimisation. *C.R. Acad. Sci. Paris II.* 296, 1761–1764.

Jacquiot, C., Trenard, Y., and Dirol, D. (1955). *Atlas d'Anatomie des Coniféres*. Centre Technique de Bois, Paris.

Jacquiot, C., Trenard, Y., and Dirol, D. (1973). *Atlas d'Anatomie des Bois des Angiospermes*. Centre Technique du Bois, Paris.

Jayne, B. A. (1972). *Theory and Design of Wood and Fiber Composite Materials*. Syracuse University Press, Syracuse, NY.

Jones, H. W. (1987). *Acoustic Imaging*. Plenum Press, New York.

Kollmann, F. F. P. and Côté, W. A. (1968). *Principles of Wood Science and Technology*. Springer-Verlag, Berlin.

Kollmann, F. F. P. (1951). *Technologie des Holzes und der Holzwerkstoffe*. Springer-Verlag, Berlin.

Kriz, R. D. and Stinchcomb, W. W. (1979). Elastic moduli of transversely isotropic graphite fibers and their composites. *Exp. Mech.* February, 41–49.

Kuo, M., McClelland, J. F., Luo, S., Chien, P., Walker, R. D., and Hse, C. (1988). Applications of infrared photoacoustic spectroscopy for wood samples. *Wood Fiber Sci.* 20(1), 132–145.

Launay, J. et al. (1988). Comparison between Six Methods to Estimate Elastic Constants of Sitka Spruce Wood. Actes du Coll. Comportement Mécanique du Bois, Bordeaux, June 8 to 9, 13–24.

Lemons, R. A. and Quate, C. F. (1979). Acoustic microscopy. in *Physical Acoustics*. Vol. 14. Masson, W. P. and Thurston, R. N. (Eds.). Academic Press, New York, 1–92.

Lemons, R. A. and Quate, C. F. (1974). Acoustic microscope scanning version. *Appl. Phys. Lett.* 24(4), 163–165.

Mason, W. P. (1958). *Physical Acoustics and the Properties of Solids*. Van Nostrand Reinhold, Princeton, NJ.

Mullins, E. J. and McKnight, T. S. (1981). *Canadian Woods—Their Properties and Uses*. 3rd ed. University of Toronto Press, Toronto.

Musgrave, M. P. J. (1961). Calculation Relating Propagation of Elastic Waves in Anisotropic Media. Basic Phys. Div. Rep. No. 7. Natl. Physical Laboratory, London.

Pao, Y. H. (1977). *Optoacoustic Spectroscopy and Detection*. Academic Press, New York.

Pluvinage, G. (1985). Etude Critique de la Détermination Expérimentale des Constantes Élastiques dans le Bois. Rep. Contr. DGRST 81G1058, Université de Metz, Metz, France.

Polge, H., Keller, R., and Thiercelin, F. (1973). Influence de l'élagage de branches vivantes sur la structure des accroissements annuels et sur quelques caractéristiques du bois de Douglas. *Ann. Sci. For.* 30(2), 127–140.

Pouligner, J. (1942), *Contribution à l'Étude de l'Élasticité du Bois.* Gauthier Villars, Paris.

Quate, C. F., Atalar, A. and Wickramasinghe, H. K. (1979). Acoustic microscopy with mechanical scanning. A review. *Proc. IEEE.* 67, 1092–1114.

Roklin, S. I. and Wang, W. (1989a). Ultrasonic evaluation of in-plane and out-of-plane elastic properties of composite materials. in *Review of Progress in Quantitative Nondestructive Evaluation.* Thompson, D. O. and Chimenti, D. E. (Eds.). Plenum Press, New York, 1489–1496.

Roklin, S. I. and Wang, W. (1989b). Critical angle measurement of elastic constants in composite material. *J. Acoust. Soc. Am.* 86(5), 1876–1882.

Rosencwaig, A. (1980). *Photoacoustics and Photoacoustic spectroscopy.* John Wiley & Son, New York.

Schniewind, A. P. (1972). Elastic behaviour of the wood fiber. in *Theory and Design of Wood and Fiber Composite Materials.* Jayne, B. A. (Ed.). Syracuse University Press, Syracuse, NY.

Schumacher, R. T. (1989). Compliances for wood for violin top plates. *J. Acoust. Soc. Am.* 84(4), 1223–1235.

Schruby, C. B. (1989). Some applications of laser ultrasound. *Ultrasonics.* 27, 195–209.

Sinclair, A. N. and Farshad, M. (1987). A comparison of three methods for determining elastic constants of wood. *J. Test. Eval.* 15(2), 77–86.

Sliker, A. (1988). A method for predicting non-shear compliances in RT plane of wood. *Wood Fiber Sci.* 20(1), 44–55.

Sliker, A. (1989). Measurement of the smaller Poisson's ratios and related compliances for wood. *Wood Fiber Sci.* 21(3), 252–262.

U.S. Forestry Service (1987). in *Wood Handbook.* Agric. Handb. No. 72. Washington, D.C.

Wu. E. M., Jerina, K. L., and Lavengood, R. E. (1973). Data Averaging of Anisotropic Composite Material Constants. ASTM STP 521, American Society for Testing and Materials, Philadelphia, 229–252.

Chapter 6

Structural Anisotropy and Ultrasonic Parameters of Wood

Normally, the anisotropy of material properties is defined as the variations in the material physical response to the applied stress along different specimen axes. As far as elastic properties of materials are concerned, the response to the applied stress tensor is investigated in terms of the elastic strain tensor. For biological materials, the anisotropy results from the non-random distribution and orientation of the structural components. Most biological materials are heterogeneous; however, a nonrandom organization of structural elements enables us to consider these materials as homogeneous anisotropic media at the macroscopic level for overall mechanical behavior investigation. This means that the material response to the applied stress is characterized by a set of linear relationships between stress and strain components; in other words, the material elastic behavior is fully defined by its stiffness tensor. An accurate estimation of wood mechanical behavior requires simultaneous views of its structure and wave propagation phenomena. Clearly, wave parameters are affected by wood structure which acts as a filter. This interaction sharply reveals the anisotropy of this material. In what follows, the interaction between ultrasonic waves and wood anisotropy induced by its structure, resulting from the specific disposition of anatomical elements during the life of the tree, is examined.

6.1. FILTERING ACTION INDUCED BY THE ANATOMIC STRUCTURE OF WOOD

Wood anisotropy can be considered at different macroscopic levels owing to:

- the specific disposition of anatomical elements, leading to the necessity to consider three mutually perpendicular planes of elastic symmetry
- the successive deposition of earlywood and latewood layers within the annual ring
- the presence of fibers running along the growth axis of the tree and the array of rays running from the heart to the bark

In order to relate the anatomical structure to the acoustic behavior of wood it is necessary to understand some mechanisms which, during the propagation of vibration, would separate the cells and allow them to act independently (Bucur, 1980). Study on structural features of the cell walls (speaking very generally) of wood shows that tracheids are "tubes" of a cellulosic crystalline substance embedded in an amorphous matrix (lignin). This particular cellular organization was initially modeled by Price (1928). From an acoustical point of view, the wood structure can be considered as a rectangular system of cross-homogeneous closed "tubes" embedded in a matrix. The longitudinal orientation of tracheids or fibers is partially disturbed by "horizontal tubes", or medullary rays.

It is interesting to note that in the longitudinal direction the dissipation of acoustical energy takes place at the limit of the "tubes". Accordingly, the continuous and uniform structure of softwoods, built up by long anatomical elements, provides high acoustical constants values. Experimental measurements in the ultrasonic range sustain this model. The strong correlation between fiber length and ultrasonic velocity V_{11} (Polge, 1984), $r = 0.90$, might open the possibility that a specific, nondestructive methodology for fiber length measurements may be developed.

In the radial direction the acoustical waves once more find a tubular structure in the presence of rays, but in the tangential direction an acoustical conducting structure is completely absent. Indeed, the presence of rays in hardwoods could be a major factor in wood anisotropy in planes containing the R axis (RT and LR), however, other anatomical attributes play a role in wood anisotropy. Two other microstructural aspects may be considered: the annual ring structure (with latewood having almost solid, thick-walled cells, and with earlywood having thin-walled cells) and the difference between the R and T directions in cell arrangement (McIntyre and Woodhouse, 1986). Longitudinal cells, tracheids, and fibers tend to be aligned in the R direction and randomly distributed in the T direction. This disposition of anatomical elements may also have a significant influence on the propagation of shear waves. The modulation of shear propagating ultrasonic waves by the structure of wood must be

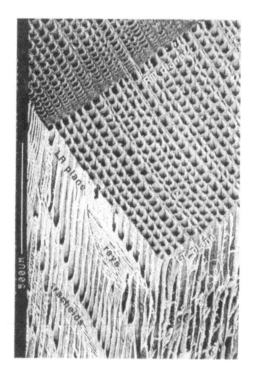

Figure 6.1. Resonance spruce. The regular structure of tracheids is evident in the *RT* plane. The rays are visible in *LR* plane. (Photo courtesy of Dr. Claire Barlow.)

understood in terms of both propagation and polarization direction. Figure 6.1 shows details of the anatomic organization of spruce in different anisotropic planes.

The ultrasonic energy injected into a fibrous material couples several modes (longitudinal, flexural, and torsional) on each fiber. The physical properties of the cellular wall, density, rigidity modulus, etc., and the shape and size of the fibers or other elements affect the transmitted ultrasonic field. Each structural element acts independently as an elementary resonator. The spatial distribution of velocities and frequencies that matched the frequency of the natural fibers could explain the acoustical behavior of wood as illustrated by its overall parameters.

6.2. ESTIMATION OF ANISOTROPY BY VELOCITIES OF LONGITUDINAL AND TRANSVERSAL BULK WAVES

The first step in estimating wood anisotropy is to relate bulk velocities to symmetry axes. The continuum theory, which ignores the ring structure of wood, can be used safely only if the wavelength is long as compared to the ring spacing, the proportion of the latewood in the annual ring, and the fiber dimensions.

This section describes some acoustical properties of 15 species:

• Broadleaved species

Acer campestre L., common maple
Acer pseudoplatanus L. (of curly grain), curly maple
Aesculus hippocastanum L., horse chestnut
Caesalpinia brasiliensis S., pernambuco
Fagus sylvatica L., beech
Liriodendron tulipifera L., tulip tree
Quercus petraea Liebl., oak

Table 6.1. Average values of ultrasonic velocities (m/s) in different wood species

Species	Density (kg/m³)	Velocities (m/s)					
		V_{11}	V_{22}	V_{33}	V_{44}	V_{55}	V_{66}
Hardwoods							
Poplar	326	5074	2200	1210	642	1250	1536
Horse chestnut	510	4782	2311	1382	536	1166	1549
Tulip tree	574	5625	2047	1511	566	1272	1413
Oak	600	5071	2148	1538	683	1252	1546
European plane	620	5060	2178	1646	840	1234	1460
Common maple	623	4695	2148	1878	630	1148	1305
Beech	674	5074	2200	1560	960	1270	1510
Curly maple*	700	4350	2590	1914	812	1468	1744
Pernambuco	932	4935	2435	2034	1006	1280	1294
Softwoods							
Silver spruce*	352	5500	2225	1850	325	1386	1361
Spruce*	400	5600	2000	1600	298	1425	1374
Douglas fir	440	5500	2330	1990	560	1660	1662
Sitka spruce*	430	5550	2300	1500	350	1480	1500
Sitka spruce common	450	5200	2200	1500	450	1560	1630
Red spruce*	485	6000	2150	1600	330	1240	1320
Spruce common	485	5353	1580	1146	477	1230	1322
Pine	580	5000	2100	1200	600	1030	1050

* Denotes resonance woods.
From Bucur, V., *Int. Assoc. Wood Anat. Bull.*, 9(1), 67, 1988. With permission.

Platanus acerifolia Willd., European plane
Populus spp., poplar

• Coniferous species

Picea abies (L.) Karst., spruce
P. engelmannii Parry, silver spruce
P. rubens Sarg., red spruce
P. sitchensis (Bong) Carr, Sitka spruce
Pseudotsuga menziesii (Mirb.) Franco, Douglas fir
Pinus spp., pine

The velocity measurements were carried out using a classical direct transmitting pulse technique at 1 MHz frequency on small, clear specimens (disks and cubes). All measurements were performed at 12% moisture content. Ultrasonic velocity values, which were measured along the symmetry axes, are presented in Table 6.1.

As a general comment on longitudinal waves, V_{11} is always greater than V_{22} and V_{33}. The longitudinal orientation of cells along the *L* axis is the best explanation of the specific ordering of velocity values as cell walls provide a continuous wave path. It is interesting to note that in the longitudinal direction the dissipation of acoustic energy probably takes place at the limit of tracheids or fibers. Accordingly, the continuous and uniform structure of softwoods built up of long anatomical elements provides high values of acoustical constants.

Table 6.2 presents the measured values of shear velocities. Within the same symmetry plane these values are different if the direction of propagation is changed (e.g., *LR* and *RL;* the first index is for propagation direction, the second for polarization). For softwoods with a pronounced annual ring structure, the differences between V_{TR} and V_{RT} used for V_{44} computation are 10 to 15%; for ring-porous hardwoods such as oak, the difference is 17%. For the diffuse-porous woods which show only small differences between early- and latewoods, such as the beech or tulip tree, the discrepancy between the two values is only about 5%. The modulating action of wood structure must be understood in terms of both propagation and polarization directions. Furthermore, the discrepancies observed in the shear velocities are probably due to the waveguide effect induced by annual ring layering and by alternation

Table 6.2. Some measured shear velocities (m/s) on disk-type specimens

Velocity	Measured values on disks (m/s)						
	Tulip tree	Beech	Maple	Oak	Douglas fir	Spruce	Sitka
For V_{66}							
V_{LR}	1455	1517	1835	1559	1638	1322	1669
V_{RL}	1472	1498	1869	1482	1630	1320	1628
For V_{55}							
V_{LT}	1275	1273	1550	1252	1691	1229	1586
V_{TL}	1273	1267	1587	1250	1618	1239	1563
For V_{44}							
V_{RT}	682	981	892	647	560	477	450
V_{TR}	710	941	948	782	500	405	400

Note: The measured velocities Vxy are noted with two indices: x corresponds to propagation direction and y corresponds to polarization direction.

From Bucur, V., *Int. Assoc. Wood Anat. Bull.*, 9(1), 67, 1988. With permission.

of earlywood and latewood (Bucur and Perrin, 1988). The material behaves like a filter with alternating passband and stopband (Woodhouse, 1988). This phenomenon is shown consisely in Figure 6.2. The influence of the annual ring is evident on V_{TR} where the propagation is parallel to the layering.

In hardwoods such as beech or tulip tree, which are characterized by a very small amount of latewood in the annual ring, the differences between V_{TR} and V_{RT} are relatively small. The diffuse, porous hardwoods, which have a very small proportion of latewood in the annual ring, seem to be much closer to the hypothesis of the continuum theory than are the softwoods and oak. Similarities between the behaviors of oak and softwoods in the acoustic field also may be deduced if the corresponding acoustic emission responses under the four-points bending test are compared (Vautrin and Harris, 1987).

Of particular interest in the anisotropy estimation are the relationships between several bulk velocities observed on the velocity surface (Figure 6.3). The spatial filtering action of wood structure is easily connected with longitudinal and shear velocities or with quasilongitudinal and quasitransversal velocities in different anisotropic planes. One of the possible approaches to the estimation of wood anisotropy is to compute the ratios between velocities of longitudinal and transversal waves in the three main symmetry directions (Table 6.3). Such an approach affords the means of estimating how far wood behavior departs from a purely orthotropic behavior. Extrapolating from data presented in the table we note that the ratios between longitudinal velocities in the *L, R,* and *T* axes are roughly $V_{11}:V_{11} = 1$, $V_{11}:V_{22} = 0.5$, and $V_{11}:V_{33} = 0.33$ for hardwoods, and $V_{11}:V_{11} = 1$, $V_{11}:V_{22} = 0.4$, and $V_{11}:V_{33} = 0.28$ for softwoods. These ratios are in agreement with those of the mechanical properties cited by Kollmann and Côté (1968).

Another point to consider is the ratio of shear velocities which shows a lack of symmetry between V_{55} and V_{66}. These two values are related to the longitudinal planes. Although it may be convenient to look at the ratios between shear velocities in the same axis (e.g., in axis 3, $V_{55}:V_{44}$, or, V_{TL} and V_{TR}), it

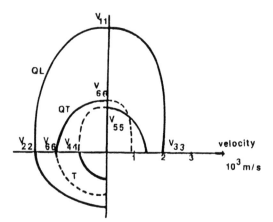

Figure 6.2. Velocity surface on curly maple calculated from optimized values of the off-diagonal terms of the stiffness matrix (Bucur, 1987).

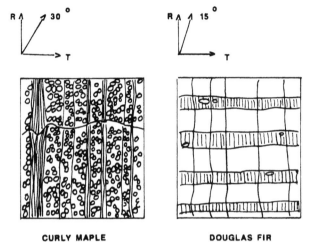

Figure 6.3. Spatial filtering action of wood structure in transverse plane *RT*. The optimum angle for C_{ij} calculation is 15° for Douglas fir with few rays, and 30° for curly maple with many rays (Magnification × 25) (Bucur and Rocaboy, 1988a).

Table 6.3. Acoustic anisotropy expressed by the velocity ratios

	Acoustic anisotropy expressed						
	Between axes				In axis		
	$\dfrac{V_{11}}{V_{22}}$	$\dfrac{V_{11}}{V_{23}}$	$\dfrac{V_{44}}{V_{55}}$	$\dfrac{V_{44}}{V_{66}}$	$\dfrac{V_{66}}{V_{55}}$	$\dfrac{V_{66}}{V_{44}}$	$\dfrac{V_{55}}{V_{44}}$
Hardwoods							
Poplar	2.31	4.19	0.51	0.42	1.23	2.39	1.95
Horse chestnut	2.07	3.46	0.46	0.35	1.33	2.89	2.17
Tulip tree	2.75	3.72	0.44	0.40	1.11	2.49	2.25
Oak	2.36	3.30	0.55	0.44	1.23	2.26	1.83
European plane	2.32	3.07	0.68	0.58	1.18	1.74	1.47
Common maple	2.18	2.50	0.55	0.48	1.14	2.07	1.82
Beech	2.30	3.25	0.76	0.64	1.18	1.56	1.32
Curly maple*	1.68	2.27	0.55	0.47	1.19	2.15	1.81
Pernambuco	2.02	2.43	0.79	0.78	1.01	1.29	1.27
Softwoods							
Silver spruce*	2.47	2.97	0.23	0.24	0.98	4.19	4.26
Spruce*	2.80	3.50	0.21	0.24	0.96	4.19	4.78
Douglas fir	2.39	3.67	0.24	0.23	1.01	4.29	4.23
Sitka spruce*	2.36	2.76	0.34	0.34	1.00	2.97	2.96
Red spruce*	2.79	3.75	0.27	0.25	1.06	4.00	3.76
Spruce common	3.39	4.68	0.39	0.36	1.07	2.77	2.58
Pine	2.38	4.17	0.58	0.57	1.02	1.75	1.72

* *Denotes resonance woods.*
From Bucur, V., *Int. Assoc. Wood Anat. Bull.*, 9(1), 67, 1988. With permission.

is obvious that softwoods exhibit a very high ratio of anisotropy (4.7 for *Picea abies*), whereas hardwoods show a relatively small anisotropy (1.27 for pernambuco).

As shown previously, a simultaneous view into the three symmetry planes of the anisotropic behavior of a wood species is presented on the velocity surface. It is also conceivable that for velocity surface drawing the optimum off-diagonal terms of the stiffness matrix need to be known and, consequently, the optimum value of velocity needs to be measured out of the principal axes. Of interest is the examination of Table 6.4. A remarkable result is the optimum value of velocities at 45°, which was observed in all three symmetry planes of curly maple. Examination of micrographs of this species (see Wagenfurh and Scheiber, 1974; Schweingruber, 1978) led to the conclusion that the obvious structural

Table 6.4. Corresponding angle of propagation for optimum off-diagonal terms (10^8 N/m^2) of the stiffness matrix (Bucur and Rocaboy 1988a)

Species	Plane	Off-diagonal stiffnesses			
		Label	Value	Axis	Angle
Beech	1,2	C_{12}	30.36	1	75°
	1,3	C_{13}	16.89	1	15°
	2,3	C_{23}	7.42	2	30°
Tulip tree	1,2	C_{12}	24.61	1	60°
	1,3	C_{13}	28.35	1	45°
	2,3	C_{23}	8.20	2	30°
Curly maple	1,2	C_{12}	33.84	1	45°
	1,3	C_{13}	18.58	1	45°
	2,3	C_{23}	22.31	2	45°

disorder induced homogeneity, which is reflected in the acoustical properties of this species. It is reasoned that this is the basis for the empirical uses of this peculiar wood structure, such as violin backs.

Detailed analyses of hardwood micrographs related to the optimum C_{ij} values emphasize the influence of rays on the acoustic anisotropy of wood. In the *RT* plane the corresponding propagation angle for C_{ij} optimum is 30° vs. *R* axis for the tulip tree, beech, and horse chestnut, and the angle for Douglas fir and Sitka spruce is 15° (Figure 6.3). The other species studied show explicitly that methods involving a single off-axis measurement of 45° are not appropriate for an accurate definition of anisotropy for most wood species.

Using the example of the 1,2 (*LR*) plane, the value of the angle (75°) for beech and for tulip tree (60°) (both taken with respect to the *L* axis) clearly reinforces the idea that cell wall density is greater for rays than it is for fibers. This agrees with the results obtained from the X-ray and image analysis of Thiercelin and Keller (1975). A more detailed analysis of the corresponding micrographs (Figure 6.4) shows that tulip tree rays are more uniformly distributed than those of beech. The presence of isolated, very wide rays in beech is a possible explanation for the 75° angle. This ray distribution also reinforces the strength of the structure in the vertical direction in 1,3 (*LT*) plane (e.g., 15° for beech).

The spatial filtering action of wood structure is easily connected to the surface velocities through the calculation of the three corresponding C_{ij} terms. The results are given by Bucur and Rocaboy (1988b). They obtained good agreement between bulk and surface C_{ij} terms. Thus, the consistency of the optimization procedure for bulk C_{ij} terms was assessed.

Another point worth considering is the intersection between *QT* and *T* curves that occurs in the *1,3* or in the *L,T* plane of the velocity surface for all wood species. At that particular intersection both shear types merge into a single wave type. In other words some structural singularity modulates ultrasonic waves and forces their polarization. Measurements of the corresponding polarization direction are not currently accessible, but microfibril angle may be advanced as a possible microstructural parameter of influence.

6.3. ESTIMATION OF ANISOTROPY BY INVARIANTS

The invariants of tensors are parameters used for the characterization of the elastic or viscoelastic behavior of anisotropic solids. Investigations were undertaken by Betten (1982), Musgrave (1985), Roy and Tsai (1992), and Hosten (1992).

The stability of calculated invariants vs. different angles of propagation confirms the validity of the structural model (orthotropic or other) chosen for the tested material. The invariants can be associated with the anisotropy of the material. The objective of this section is to illustrate basic concepts related to the estimation of wood anisotropy through acoustic and elastic invariants. The division into two parts seems artificial because stiffnesses are deduced from acoustic measurements, but at the same time, the stiffnesses are considered to be elastic constants. The discussion is also intended to aid in comprehension of the approach related to wood anisotropy and invariants.

6.3.1. ACOUSTIC INVARIANTS

The acoustic invariants, deduced from Christoffel equations, can be related to bulk velocities measured in principal directions of elastic symmetry. This seems rather simple. A more sophisticated anisotropic estimation requires the computation of Voigt and Reuss moduli. Henceforth, this approach was considered for the assignment of elastic invariants, deduced from all nine stiffnesses and compliances.

The first approach to wood anisotropy interpretation was to compute acoustic invariants and to average them. Because those quantities are insensitive to the direction of propagation, they can act as references for anisotropy investigation. Combining invariant values in the three main symmetry planes we obtain a single global value that characterizes each species (Bucur, 1988). To better understand this approach, it is useful to analyze the theoretical considerations that enable us to compute the acoustic invariants. The starting point is the theoretical velocity surface. We saw previously that three sheets can be obtained by plugging the optimum stiffness terms into Christoffel's equations. In the particular case of propagation taking place in the 1,2 plane, we have:

$$2\rho V_{QL}^2 = (\Gamma_{11} + \Gamma_{22}) + \{(\Gamma_{11} - \Gamma_{22})^2 + 4\,\Gamma_{12}\}^{1/2} \tag{6.1}$$

$$2\rho V_{QT}^2 = (\Gamma_{11} + \Gamma_{22}) - \{(\Gamma_{11} - \Gamma_{22})^2 + 4\,\Gamma_{12}\}^{1/2} \tag{6.2}$$

$$\rho V_T^2 = C_{55}n_1^2 + C_{44}n_2^2 \tag{6.3}$$

Equation (6.3) is explicitly that of an ellipse.

The velocity curves corresponding to the quasilongitudinal and quasitransversal waves lie exterior and interior, respectively, to the following ellipse:

$$\Gamma_{11} + \Gamma_{22} = (C_{11} + C_{66})n_1^2 + (C_{22} + C_{66})n_2^2 = 2\rho V^2 \tag{6.4}$$

where

$$V^2 \text{ ellipse} = 1/2(V_{QL}^2 + V_{QT}^2) \tag{6.5}$$

A complete analysis involves simultaneously plotting the velocity curves that correspond to the three planes of elastic symmetry. The respective position of each curve is governed by the ratios between the diagonal stiffness terms. An analysis of possible intersections between the three curves in each plane and of discontinuities that may appear at the junction between two planes provides information on the departure of the test material from the truly orthotropic model. In the specific case of propagation taking place in the 12 plane, this departure arises from the following theoretical considerations.

- At angle α:

$$(\Gamma_{11} + \Gamma_{22}) = C_{11}n_1^2 + C_{22}n_2^2 + C_{66}(n_1^2 + n_2^2) \tag{6.6}$$

- At angle $\beta = \pi - \alpha$

$$(\Gamma_{11} + \Gamma_{22}) = C_{11}n_2^2 + C_{22}n_1^2 + C_{66}(n_1^2 + n_2^2) \tag{6.7}$$

Consequently, adding Equations (6.6) and (6.7) leads to an expression invariant with the angle α under consideration:

$$\Sigma\Gamma = C_{11} + C_{22} + 2C_{66} \tag{6.8}$$

Expressed in terms of velocities, a velocity invariant I_{12} can be introduced:

$$I_{12(V)} = 1/2(V_{11}^2 + V_{22}^2 + 2V_{66}^2)^{1/2} \tag{6.9}$$

This expression corresponds to the sum of the four possible values when ultrasonic waves are made

Table 6.5. Ratio of acoustic invariants for different species of European and Australian origin

Hardwoods	Ratio of invariants	Softwoods	Ratio of invariants
Poplar	0.47	Silver spruce*	0.474
Horse chestnut	0.51	Spruce*	0.418
Tulip tree	0.43	Douglas fir	0.45
Oak	0.49	Sitka spruce*	0.497
European plane	0.51	Sitka spruce	0.446
Common maple	0.55	Red spruce*	0.415
Beech	0.52	Spruce*	0.355
Curly maple*	0.63	Pine	0.468
Pernambuco	0.61	Huon pine[+]	0.416
Cedar[+]	0.47		
Queensland maple[+]	0.476		
Silky oak[+]	0.49		
Queensland walnut[+]	0.511		
Sassafras[+]	0.446		
Black wood[+]	0.504		

Denotes resonance woods; [+] *Australian woods.*
From Bucur (1988) and Bucur, V. and Chivers, R.C., *Acoustica,* 75(1), 69, 1991. With permission.

to propagate in the axes of the considered plane of elastic symmetry. For a propagation direction out of axis, the corresponding value of invariant $I_{12(V^*)}$ is obtained from complementary angles velocity readings, following the relation:

$$I_{(V^*)} = 1/2(V^2_{QL(\alpha)} + V^2_{QT(\alpha)} + V^2_{QL(\beta)} + V^2_{QT(\beta)})^{1/2} \tag{6.10}$$

Considering all the symmetry planes of wood and, consequently, all the values of velocities to be measured, the simplest expression of invariants of velocity in the notation matrix, are

- In plane 12 or *LR:*

$$I_{12} = ((V^2_{11} + V^2_{22} + 2V^2_{66})/4)^{1/2} \tag{6.11a}$$

- In plane 13 or *LT:*

$$I_{13} = ((V^2_{11} + V^2_{33} + 2V^2_{55})/4)^{1/2} \tag{6.11b}$$

- In plane 23 or *RT:*

$$I_{23} = ((V^2_{22} + V^2_{33} + 2V^2_{55})/4)^{1/2} \tag{6.11c}$$

Combining the values of invariants as a ratio between the invariant in the transversal plane (*RT*) and the average invariant in the longitudinal plane (*LR* and *LT*) we obtain a unique value R_I:

$$R_I = 2 \cdot I_{23}/(I_{12} + I_{13}) \tag{6.12}$$

This synthetic treatment of invariants may be considered in terms of defining a global parameter representative of the overall acoustical parameters of a wood species. This approach is similar to that used in composite materials characterization, in which the need for data averaging (Wu et al., 1973) arises from handling experimental data of a tensorial nature.

The methods described below compute invariants and average them. Consequently, the voluminous experimintal data are compactly reduced into invariants that are easy to handle. This procedure provides useful physical insight into acoustic parameters and could serve in nondestructive research of wood properties.

Numerical results are presented in Table 6.5. It is well known that for isotropic solids the invariants ratio must be considered 1. In the case of the timber species studied the values of invariants range from 0.355 to 0.610.

Table 6.6. Wood anisotropy expressed by the Voigt and Reuss moduli (10^8 N/m^2)

Species Density (kg/m^3)	Voigt moduli		Reuss moduli (10^8 N/m^2)		Anisotropy		
	KV	GV	KR	GR	AC	AS	Ratio
1	200	7.67	3.83	1.49	0.87	0.68	0.63	1.07
2	352	28.80	7.64	8.27	0.54	0.55	0.87	0.64
3	370	23.00	10.00	7.71	1.27	0.50	0.77	0.64
4	400	23.80	8.90	8.39	0.82	0.48	0.83	0.57
5	400	31.00	8.43	9.84	1.15	0.52	0.76	0.68
6	430	30.50	11.00	8.85	1.84	0.55	0.71	0.77
7	437	32.10	10.10	6.59	1.17	0.65	0.79	0.82
8	440	27.72	14.26	15.77	4.22	0.27	0.54	0.51
9	485	35.50	13.60	9.86	1.27	0.57	0.83	0.68
10	485	20.36	13.05	5.81	1.68	0.56	0.77	0.72
11	510	27.37	11.28	9.66	3.64	0.48	0.51	0.93
12	560	23.70	12.90	17.60	7.28	0.15	0.28	0.53
13	574	37.90	15.03	10.20	4.74	0.58	0.52	1.11
14	580	30.26	11.74	7.53	4.47	0.60	0.45	1.34
15	600	26.00	15.80	13.20	8.85	0.33	0.28	1.16
16	600	23.67	17.61	11.14	6.84	0.36	0.44	0.82
17	626	26.70	15.30	11.60	6.90	0.39	0.38	1.04
18	670	41.10	17.10	21.90	9.96	0.30	0.26	1.16
19	674	36.88	17.61	13.75	9.91	0.46	0.28	1.63
20	700	46.00	15.70	21.00	6.62	0.37	0.41	0.91
21	700	39.39	16.89	25.03	9.36	0.22	0.29	0.78
22	720	36.20	17.60	13.30	8.19	0.46	0.37	1.27
23	740	51.50	18.50	21.50	11.70	0.41	0.23	1.82
24	932	80.98	17.52	40.12	11.93	0.34	0.19	1.78

Note B: The species studied were

(1) balsa, (2) *Picea engelmanii**, (3) *P. sitchensis**, (4) *P. abies**, (5) *P. rubra**, (6) *P. sitchensis**, (7) *P. sitchensis**, (8) Douglas fir, (9) red spruce*, (10) Norway spruce, (11) horse chestnut, (12) *Acer rubrum**, (13) tulip tree, (14) pine, (15) *A. macrophyl**, (16) oak, (17) *A. macrophyllum*, (18) *A. pseudoplatanus*, (19) beech, (20) *A. saccharum**, (21) curly maple*, (22) *A. saccharum**, (23) *A. platanoides**, and (24) pernambuco. Species denoted with * are resonance woods.

Examination of the microscopic sections of the species under consideration leads to the conclusion that the structural disorder in curly maple probably induces an acoustic homogeneity reflected in the highest value of the invariant ratio, 0.63. A related ratio is that of pernambuco wood:invariants, 0.61.

Figure 6.4 illustrates the experimental relations between the acoustic invariants and the mass density ($r = 0.678***$). The weight of evidence suggests that wood species having high density and any important organized structure in the millimeter length scale in the *RT* plane exhibit high values for the ratio of invariants. The variation in density has a corresponding variation in ultrasonic velocity.

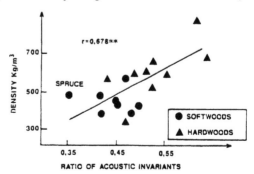

Figure 6.4. Correlation between ratio of acoustic invariants and wood density (Bucur, 1990).

a)

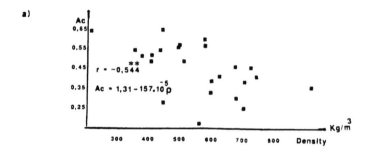

b)

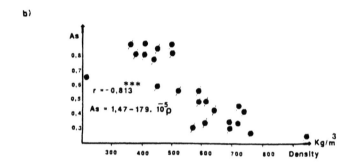

c)

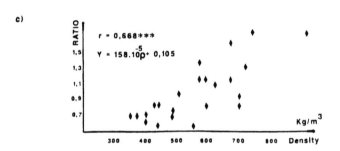

Figure 6.5 Relationships between elastic anisotropy and density. (a) Compression anisotropy and density; (b) shear anisotropy and density; (c) anisotropy ratio and density. (From Bucur, V., *Mater. Sci. and Eng.*, A 122, 83, 1989. With permission.)

Table 6.7. Principal components analysis; explanation of species variability following three principal statistical axes

| Parameters | Explanation of variability (%) following the axis: | | | |
	1	2	3	Total
Voigt and Reuss moduli, stiffnesses anisotropy ratio	72.1	14.0	8.6	94.7

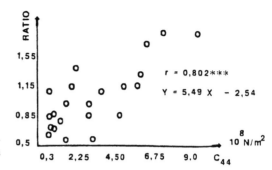

Figure 6.6. Relationship between anisotropy ratio and shear terms C_{44}. (From data in Bucur, 1989.)

Consequently, the acoustic behavior of those species is less anisotropic than that of species having low density and typical softwood structure or high density and a ring-porous structure.

6.3.2. ELASTIC INVARIANTS

Elastic anisotropy estimation, using the procedure suggested by Katz and Meunier (1989) and based on the Voigt and Reuss moduli, is delineated below. The nine terms of the stiffness or the compliance matrices are related by linear relations to the Voigt and Reuss moduli as follows:

- For Voigt moduli:

$$C_{11} + C_{22} + C_{33} = 3A \tag{6.12a}$$

$$C_{12} + C_{13} + C_{23} = 3B \tag{6.12b}$$

$$C_{44} + C_{55} + C_{66} = 3C \tag{6.12c}$$

- For Reuss moduli:

$$S_{11} + S_{22} + S_{33} = 3A' \tag{6.13a}$$

$$S_{12} + S_{13} + S_{23} = 3B' \tag{6.13b}$$

$$S_{44} + S_{55} + S_{66} = 3C' \tag{6.13c}$$

Using these expressions, the Voigt bulk modulus KV and the shear modulus GV and the equivalent Reuss bulk KR and shear GR moduli, are

- For Voigt moduli:

$$3KV = A + 2B \tag{6.14a}$$

$$5GV = A - B + 3C \tag{6.14b}$$

- For Reuss moduli:

$$1/KR = 3A' + 6B' \tag{6.15a}$$

$$5/GR = 4A' - 4B' + 3C' \tag{6.15b}$$

Following the assumption that the Voigt modulus represents the upper bound on the elastic properties of a multiphase system (strain is uniform across the interface), whereas the Reuss modulus represents the lower band of the elastic properties (a uniform stress distribution exists across the interface), the

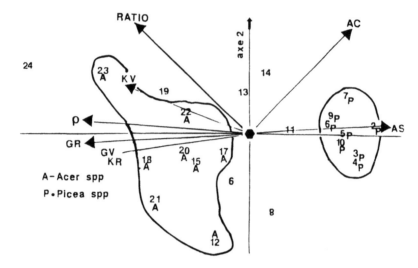

Figure 6.7. Principal components analysis in 1,2 statistical plane (Bucur, 1990). The cluster of experimental points in the principal components plane and the correlation circle are superimposed. The vector of the anisotropy ratio divides the species into two groups. In the positive sense of axis 1 are located the *Picea* spp. specimens, whereas the *Acer* spp. specimens are located in the negative sense. The vector of the anisotropy ratio divides the specimens into two groups corresponding to each group of species, (*Picea* and *Acer* spp.). The species studied were (1) balsa, (2) *Picea engelmanii**, (3) *P. sitchensis**, (4) *P. abies**, (5) *P. rubra**, (6) *P. sitchensis**, (7) *P. sitchensis**, (9) red spruce*, (10) Norway spruce, (11) horse chestnut, (12) *Acer rubrum**, (13) tulip tree, (14) pine, (15) *A. macrophyl**, (17) *A. macrophyllum*, (18) *A. pseudoplatanus*, (19) beech, (20) *A. saccharum**, (21) curly maple*, (22) *A. saccharum**, and (23) *A. platanoides**. Species denoted with * are resonance woods.

difference between them provides a measure of the compressive and shear elastic anisotropies (Hill, 1963; Watt, 1979; Hearmon, 1956, 1961, 1965; Katz and Meunier, 1989).

For convenience, the following expressions of compressive (AC) and shear (AS) elastic anisotropy were suggested:

$$AC = (KV - KR)/(KV + KR) \tag{6.16a}$$

$$AS = (GV - GR)/(GV + GR) \tag{6.16b}$$

Obviously, for an isotropic solid both parameters (AC and AS) are zero because $KV = KR$ and $GV = GR$.

Some results of the anisotropy of wood species derived from Voigt and Reuss moduli are given in Table 6.6. The data cover fairly wide ranges of density (from 200 to 932 kg/m^3) and anisotropy ratios (from 0.50 to 1.78). Two species, balsa and tulip tree, have an anisotropy ratio near 1.0. This means that compressive and shear anisotropy are quite identical. Furthermore, Douglas fir (ratio, 0.51) and pernambuco (ratio, 1.78) are in opposite positions. The relevance of numbers obtained for Douglas fir in shear anisotropy (0.54), which is much larger than that corresponding to pernambuco (0.19), may be connected to the structural organization of those species. However, the pernambuco is the unique species, exhibiting such a low shear anisotropy.

Figure 6.5 shows the relationships between elastic anisotropy and density. Generally, an increase in density is followed by a decrease in AC and AS. This may be of interest as the explanation for the propagation phenomena in inhomogeneous solids.

Judging from the observations on the mean values of compression and shear anisotropy, the relationships between density and the shear term C_{44} or the ratio of nondiagonal terms $C_{13}:C_{23}$, connected with the propagation in the RT plane, were studied in Figure 6.6. A strong correlation was observed between the anisotropy ratio and shear terms in the RT plane. Beyond this statement lies a whole hierarchy of questions concerning the interaction of ultrasonic waves with the complex structural organization of wood species, which is related to the influence of latewood proportion in the annual ring, the width of growth ring, the thickness of the cellular wall, etc.

In order to appreciate the relationships between the variabilities of the 24 species considered in Table 6.7 and the anisotropy expressed by Voigt and Reuss moduli, a principal component analysis (Morrison, 1967) was performed. A linear combination of eight variables which progressively explain the variance of the population under consideration was adopted. The multivariate system summarizes 95% of the variability of the population in a three-axis system (Table 6.7). In the 1,2 plane of statistical analysis 86% of the variability is explained.

The relationships between the initial variables and the principal component axes are given on the correlation circle superimposed on the scatter swarm of observation points in the plane of components 1,2; 1,3 and 2,3 (Figure 6.7). In the plane 1,2 the anisotropy ratio vector divides the species into two groups: in the positive sense of axis 1, *Picea* spp., and in the negative sense, *Acer* spp. Of the total variability, 86% in this plane is explained. The two groups are divided by the *KV* vector in the 1,3 plane, and 81% of the variability is explained. The third plane explains only a small proportion of the total variance (23%), and is not analyzed here.

Note that the use of the anisotropy ratio derived from the simultaneous analysis of the Voigt and Reuss moduli may provide an idea about the structural organization of solid wood.

REFERENCES

Barlow, C. (1993). Scanning electron micrograph of resonance spruce. Personal communication.

Betten, J. (1982). Integrity basis for a second-order and a fourth-order tensor. *Int. J. Math. Sci.* 5(1), 87–96.

Bucur, V. (1980). Anatomical structure and some acoustical properties of resonance wood. *Catgut Acoust. Soc. Newslett.* No. 33, May, 24–29.

Bucur, V. (1990). About the anisotrophy of resonance wood for violins. *Fortschritte der Akustik, DAGA '90*, Vienna, 553-557.

Burcur, V. (1987). Varieties of resonance wood and their elastic constants. *J. Catgut Acoust. Soc.* No. 47.

Bucur, V. (1988). Wood structural anisotropy estimated by acoustic invariants. *Int. Assoc. Wood Anat. Bull.* 9(1), 67–74.

Bucur, V. (1989). Bulk and surface waves for wood anistropy characteristics. *Mater. Sci. and Eng.* A122, 83-85.

Bucur, V. and Rocaboy, F. (1988a). Anisotropy of Biological Orthotropic Structures, Estimated from Ultrasonic Velocity Measurements. Applications to Wood. No. 1. Centre Recherches Forestières de Nancy, France.

Bucur, V. and Rocaboy, F. (1988b). Surface wave propagation in wood: prospective method for the determination of wood off-diagonal terms of stiffness matrix. *Ultrasonics.* 26(5), 344–347.

Bucur, V. and Perrin, J. R. (1988). Ultrasonic waves-wood structure interaction. *Proc. Inst. Acoust. Edinburg.* 10 (Part 2), 199–206.

Bucur, V. and Chivers, R. C. (1991). Acoustic properties and anisotropy of some Australian wood species. *Acustica.* 75(1), 69–74.

Hearmon, R. F. S. (1956). Elastic constants of anisotropic materials. II. *Adv. Phys.* 5, 323–382.

Hearmon, R. F. S. (1961). *An Introduction to Applied Elasticity.* Oxford University Press.

Hearmon, R. F. S. (1965). The assessment of wood properties by vibrations and high frequency acoustic waves. *Proc. 2nd Symp. Nondestructive Testing of Wood.* Washington State University, Olympia, 49–65.

Hill, R. (1963). Elastic properties of reinforced solids: some theoretical principles. *J. Mech. Phys. Solids.* 11, 357–372.

Hosten, B. (1992). Stiffness matrix invariants to validate the characterization of composite materials with ultrasonic methods. *Ultrasonics.* 30(6), 365–371.

Katz, L. and Meunier, A. (1989). A generalized method for characterizing elastic anisotropy in solid living tissue. *J. Mater. Sci. Mater. Med.* 1, 1–8.

Kollman, F. and Côté, W. A. (1968). *Principles of Wood Science and Technology.* Springer-Verlag, Berlin.

McIntyre M. E. and Woodhause, J. (1986). On measuring wood properties. III. *J. Catgut Acoust. Soc.* No. 45, 14–23.

Morrison, D. F. (1967). *Multivariate Statistical Methods.* McGraw-Hill, New York.

Musgrave, M. P. J. (1985). Personal communication.

Polge, H. (1984). Essais de caracterisation de la veine verte du merisier. *Ann. Sci. For.* 41, 45–58.

Price, A. T. (1928). A mathematical discussion on the structure of wood related to its structural properties. *Philos. Trans. R. Soc. London.* 228(A), 1–62.

Roy, A. K. and Tsai, S. W. (1992). Three-dimensional effective moduli of orthotropic and symmetric laminates *ASME J. Appl. Mech.* 59(1), 39–47.

Schweingruber, F. H. (1978): Microscopic Wood Anatomy. Zurcher A. G., CH-6301 ZUG.

Thiercelin, F. and Keller, R. (1975). Influence des gros rayons ligneux sur quelques propriétés du bois de hêtre. *Ann. Sci. For.* 32(2), 113–129.

Vautrin, A. and Harris, B. (1987). Acoustic emission characterization of flexural loading damage in wood. *J. Mater. Sci.* 22, 3707–3711.

Wagenfurh, R. and Scheiber, C. (1974). *Holzatlas.* VEB Fachbuchverlag, Leipzig.

Watt, J. P. (1979). Hanshin-Shtrikman bounds on the effective elastic moduli of polycrystals with orthotropic symmetry. *J. Appl. Phys.* 50, 6290–6295.

Woodhouse, J. (1985). Personal communication (letter).

Wu, E. M., Jerina, K. L.; and Lavengood, R. E. (1973). Data Averaging of Anisotropic Composite Material Constants. ASTM STP 521, American Society for Testing and Materials, Philadelphia, 229–252.

Part III
Quality Assessment

Wood Species for Musical Instruments

That wood is a unique material used in the art of musical instrument craftsmanship is an understatement. After a long period of evolution, the skill and devotion of luthiers established the most appropriate wood species for typical instruments. This chapter examines principal wood species used for the most popular wood instruments employed in today's classic symphonic orchestras. The organization of instruments in an orchestra is based on four standard groups: strings, woodwinds, brass, and percussion. While considering them, it must be borne in mind that wood is used for strings, woodwind, and percussion instruments.

As regards its construction and other elements, the violin is the most fascinating among all instruments. It is therefore natural to consider the selection of resonance wood, analyzing knowledge related to the structural organization or to the acoustical and mechanical properties of this material.

7.1. ACOUSTIC PROPERTIES OF WOOD SPECIES

Of the "resonance woods", spruce (*Picea abies*), was the first considered for the top plate (Figure 7.1), second was curly maple, used for the back plate, ribs, and neck. It was included in this group because of its role in modern violin acoustics. Finally, all species with remarkably regular anatomic structure and high acoustic properties are included under the heading "resonance wood". Table 7.1 and Color plates 1 to 6* give several anatomical characteristics of resonance woods (spruce, fir, and curly maple). More details on the anatomical structure of different wood species for musical instruments are given by Richter (1988). Attempts to model the structure of resonance wood, using a simple theoretical model consisting of an array of tubes (wood cells), were made by Bucur (1980a,b), Woodhouse (1986), and Kahle and Woodhouse (1989). Marok (1990) compared the wood to a stochastic medium. Niedzielska (1972) investigated the relationships between anatomic elements of spruce and its acoustic properties. Statistical relationships between some mechanical properties of resonance wood were deduced by Dettloff (1985).

Violin makers (luthiers) have traditionally selected their boards according to the simplest anatomical criteria, such as straight grain, fine texture, and low density, and they supplement these visual cues with rather crude bending tests. Other criteria concern the constitution of the annual ring: violins and violas, 1 mm average ring width (0.8 to 2.5 mm is the limit); cellos, 3 mm; and double bass, 5 mm; the proportion of latewood in the annual ring is typically on the order of 1/4 and the discrepancy between the respective densities of latewood and earlywood is as wide as possible (typically 900 and 280 kg/m^3, respectively), so that the overall density is maintained at around 400 kg/m^3. The transition between earlywood and latewood must be as smooth as possible. Compression wood is completely rejected. The luthier's general criterion is that a regular structure is the primary requirement for soundboards. Final, sophisticated selection is achieved through bending tests, which are performed by scientists. The result is that matched soundboards exhibit similar elastic behaviors (Ono et al., 1983, 1984; Norimoto et al., 1984, 1986). The most important criterion of selection of curly maple is connected with the beauty of the wavy grain structure. At first glance this criterion is obviously concerned with aesthetics, however, Chapter 6 showed that the very complex structure of curly maple plays an important role in its acoustic behavior.

7.1.1. ACOUSTIC PROPERTIES OF RESONANCE WOOD FOR VIOLINS

Schelleng (1982) described the most important acoustical parameters for the characterization of wood for violins, using small, thin specimens in a resonance frequency method. Physical explanations of violin function (Cremer, 1983; Fletcher and Rossing, 1991; Meyer, 1983, 1984; Caussé, 1992; Weinreich and Caussé, 1991; Gough, 1980, 1981, 1987; Hancock, 1989; Hutchins, 1962, 1975, 1981, 1983, 1991; Beldie, 1975) revealed specific behaviors and provided a well-defined foundation for the discussions of instrument quality characteristics, which are intrinsically connected with wood and string parameters.

---

* Color plates follow page 140.

FRONTAL VIEW   LONGITUDINAL SECTION

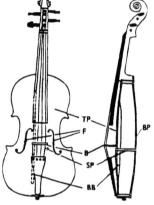

TRANSVERSE SECTION

Figure 7.1. Components of the violin. TP, Top plate of spruce; BP, back plate of maple; R, ribs of maple; SP, soundpost of spruce; B, bridge of maple; BB, bass bar of spruce; F, f holes. (From Janson et al., *J. Acoust. Soc. Am.*, 95(2), 1100, 1994. With permission.)

The acoustic behavior of wood material during the vibration of violin plates is related to the elasticity of the material along or across the grain, under extensional or bending vibrations, and to the internal friction phenomena caused by the dissipation of vibrational energy.

7.1.1.1. Spruce

Theoretical understanding of linear vibration propagation in solids and technological advances in the last 40 years have permitted the development of a frequency resonance method, using thin strips of wood for the measurement of elastic constants along and across the grain (Young's moduli E_L and E_R and shear moduli G_{LR} and G_{LT}), and for the corresponding damping constants, expressed as the logarithmic decrement, commonly noted in the literature as $2\pi \tan \delta_L$ or $2\pi \tan \delta_R$ or quality factor Q_L or Q_R (Barducci and Pasqualini, 1948; Beldie, 1968; Holz, 1967, 1968, 1973, 1974; Ghelmeziu and Beldie, 1972; Harajda and Poliszko, 1971; Ek and Jansson, 1985; Dunlop, 1988; Haines, 1979, 1980; Sobue et al., 1984; M. Hutchins, 1981, 1983, 1984, 1991). The frequency resonance method also allows access to the corresponding sound velocity of extensional or bending waves. Knowing the density of the material (ρ), acoustic impedance ($V \cdot \rho$) and acoustic radiation ($V:\rho$) can be deduced. The latter parameter can assist in matching two violin plates with different stiffnesses and densities, but on which the ratio $V:\rho$ is identical (Schelleng, 1982).

Data on resonance wood elastic constants that were measured on strips are given in Table 7.2. From this table it is evident that the damping constants are highly influenced by the range of frequency. Holz (1967b) noted a dramatic influence of frequency in the range 2 to 10 kHz. Figure 7.2 shows the internal friction and the lack of influence of frequency on E_L.

Another approach to answering the question about the influence of frequency on the measured constants of spruce resonance wood was proposed by Rocaboy and Bucur (1990). Specimens were selected from parts used for top and back plates of ten conventional violins made by C. M. Hutchins and numbered 261 to 271. Ultrasonic and frequency resonance methods were used to measure Young's moduli and stiffnesses on the same samples (Table 7.3). (The reader is reminded that the number of the specimen corresponds to the number of Hutchins' instruments). The velocities, measured by the ultrasonic technique, are greater than those measured by the frequency resonance method by about 10%. As expected, the differences between stiffnesses and Young's moduli are higher (25% in the

Table 7.1. Anatomical description of resonance wood (Bucur, 1980b)

Characteristics	Species
Tracheids	**Spruce** 20–40 μm in diameter in radial plane in earlywood and smaller in latewood. Small, bordered pits in one row on the radial walls. Pits leading to ray parenchyma piceoid, small, quite uniform in size, with distinct border. Generally 2/ cross-field, in a single row (Color Plates 1, 2, and 3). **Fir** 25–65 μm in diameter in radial plane in earlywood 10–25 μm in latewood. Bordered pits in one row or very rarely paired on the radial walls. Pits leading to ray parenchima taxodioid, small, quite uniform in size, with distinct border, 1–4/cross-field. **Curly maple** Fiber tracheids, thick-walled medium to very coarse (Color Plate 4). Very variate in length (0.01–3 mm).
Raw	**Spruce** Two types, uniseriate and fusiform; uniseriate rays numerous (80–100/cm) and 1–26 cells in height. Fusiform rays scattered, with a transverse resin canal, 3–5 seriate through the central position, tapering above and below to uniseriate margins like the uniseriate rays, up to 16 cells in height; end walls nodular with indentures; ray tracheids in both types of rays usually restricted to one row on the upper and lower margins, nondentate. **Fir** Uniseriate, very variable in height (1–50 cells), consisting wholly of ray parenchyma, rays numerous: 80/cm. **Curly maple** Up to 5 mm in height along the grain, homocellular, 1–20 (mostly 2–10) seriate, ray-vessel pitting similar to intervessel type.
Resin canal	**Spruce** Thick-walled epithelium, longitudinal canals, small, with 125 μm diameter, usually rare (22/cm²); transverse canal much smaller. **Fir** Normal resin canals wanting; longitudinal traumatic canals sometimes present, sporadic in widely separated rings, arranged in a tangential row which frequently extends for some distance along the ring. **Curly maple** Wanting.
Vessels	**Spruce and Fir** Wanting **Curly maple** Solitary, 25–150 μm in diameter, 30–50/mm²; perforation plats simple for the most part, occasionally scalariform, a few bars, intervessel pits oval, orbicular, widely spaced—50 μm in diameter

longitudinal direction and much more in the radial direction) than the differences between the velocities deduced previously. Theoretically a correct comparison between these methods could be made only when the same constants are compared. If Young's moduli are considered, then the stiffnesses must be corrected with Poisson's ratio values, and consequently the differences observed previously may be reduced.

From this description of the experimental methods used here it should be clear that the main influence of frequency is expected to be on parameters that are connected to the internal friction phenomena. The velocity or the real part of elastic constants are less influenced by the frequency range of measurement.

Among the European spruce resonance varieties, the spruce known by luthiers as "haselfichte" has a special place because of its peculiar indented ring pattern (Bonamini et al., 1991; Corona, 1990a; Schmidt-Vogt, 1986). Chiesa (1987) determined the ultrasonic velocity in the L, R, and T directions and noted the following values: $V_{LL} = 4450$ m/s, $V_{RR} = 2410$ m/s, and $V_{TT} = 1290$ m/s, at 8% moisture content. The longitudinal V_{LL} is very low when compared to values measured in perfectly straight

Table 7.2. Elastic constants of resonance wood

Real Part of Elastic Constants and Corresponding Velocities, Measured with Frequency Resonance Method

Density (kg/m³)	Velocities (m/s)				Young's moduli 10⁸ N/m²		Shear moduli (10⁸ N/m²)	
	V_{LL}	V_{RR}	V_{TL}	V_{TR}	E_L	E_R	G_{TL}	G_{TR}
Spruce								
480	5,600	1,200	1,307	359	150	7.4	8.2	0.62
440	6,000	1,100	1,215	316	160	5.0	6.5	0.44
Sitka Spruce								
480	5,200	1,700	1,581	309	130	13	12.0	0.46
460	5,200	1,500	1,062	242	130	11	5.1	0.27
Red Spruce								
480	6,300	950	1,060	277	90	4.8	5.4	0.37
450	5,700	1,300	1,192	305	150	7.9	6.4	0.42
White Spruce								
480	5,200	1,600	1,241	306	130	12	7.4	0.45
460	5,700	1,600	1,224	339	150	12	6.9	0.53

Damping Constants (Logarithmic Decrement), Measured with Frequency Resonance Method

Density (kg/m³)	Low frequency				High frequency			
	$2\pi \tan \delta_L$	fr_L (Hz)	$2\pi \tan \delta_R$	fr_R (Hz)	$2\pi \tan \delta_L$	fr_L (Hz)	$2\pi \tan \delta_R$	fr_R (Hz)
Spruce								
480	0.022	642	0.069	1,046	0.084	16,587	0.098	13,133
440	0.021	779	0.058	753	0.075	13,025	0.077	12,008
Sitka Spruce								
480	0.030	245	0.063	190	0.049	14,551	0.071	11,311
460	0.032	552	0.059	1,156	0.081	11,332	0.070	12,042
Red Spruce								
480	0.022	873	0.074	553	0.052	9,931	0.120	8,074
450	0.022	797	0.063	696	0.052	9,613	0.072	8,718
White Spruce								
480	0.023	547	0.063	445	—			
460	0.022	591	0.066	437	0.064	12,817	0.082	14,527

From Haines, D., *Catgut Acoust. Soc. Newslett.*, 31, 23, 1979. With permision.

resonance wood of the same density (440 kg/m³). The extreme scarcity of this raw material makes it very precious for violin makers. The indented spruce is difficult to carve, but has a rich and attractive color and seems to give a brilliant tone to the violins. Acoustically, indented spruce is interesting in that an understanding can be gained of the relationships of wave paths with anatomical features (see below).

Referring briefly to different varieties of spruce that can replace *P. abies* for musical instrument parts, Ono (1983a) listed the dynamical properties of *P. glehni* Mast. (akaezomatsu) and *P. jezoensis* Carr. (kuroezomatsu) and compared them to Canadian *P. sitchensis* Carr. and European *P. abies*. The main properties of the corresponding specimens are given in Table 7.4. It is evident that the density and the dynamic properties of Japanese species are in the same range as those for European resonance spruce.

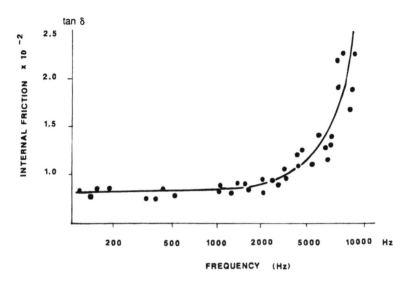

Figure 7.2. Resonance spruce, internal friction vs. frequency. (From Holz, D., *Holztechnologie*, 8(4), 221, 1967. With permission.)

Table 7.3. Elastic constants from ultrasonic and frequency resonance method on spruce resonance wood

	Density	Ultrasonic method				Resonance method			
Sample	ρ (kg/m³)	V_{LL} (m/s)	V_{RR} (m/s)	C_{LL} (10^8 N/m²)	C_{RR} (10^8 N/m²)	V_{LL} (m/s)	V_{RR} (m/s)	E_L (10^8 N/m²)	E_R (10^8 N/m²)
261	420	5810	1489	141.8	9.3	5597	1318	131.6	7.3
262	420	5527	1554	128.3	10.1	5550	1354	129.4	7.7
264	400	5852	1489	137.0	8.9	5878	1414	138.2	8.0
265	400	5830	1384	150.3	8.4	—	1373	—	8.3
266	400	5085	1560	103.4	9.7	4888	1423	95.6	8.1
267	490	5626	1572	155.1	12.1	5229	1324	134.0	8.6
268	420	5697	1625	136.3	11.1	5354	1423	120.4	8.5
269	440	5776	1379	146.8	8.4	6560	1261	189.4	7.0
270	380	5600	1589	119.2	9.6	5767	1550	126.4	9.1
271	450	5359	1575	128.3	11.2	5706	1592	1465	11.4

Note: The number of specimens corresponds to the number of C. M. Hutchins' instruments.

From Rocaboy, F. and Bucur, V., *Catgut Acoust. Soc. J.* Ser. 2, 1(6), 21, 1990.

Table 7.4. Acoustic and elastic properties of resonance wood of different origins

Species	Origin	Age (Years)	Diameter (cm)	Density (kg/m³)	E_L (10^8 N/m²)	Q_L
Picea glehnii	Japan	325	62	427	119	164
P. jezoensis	Japan	150	52	428	129	171
P. sitchensis	Canada	315	137	424	109	137
P. abies	Germany	200	60	445	142	177

From Ono, T., *Holzforschung.* 37, 245, 1983. With permission.

Table 7.5. Acoustic and elastic constants of maple (*Acer* spp.)

Density (kg/m³)	Velocities (m/s)		Young's moduli (10⁸ N/m²)		Shear moduli (10⁸ N/m³)		Logarithmic decrement $2\pi\delta$ (10⁻²)	
	V_{LL}	V_{RR}	E_L	E_R	G_{TL}	G_{TR}	L	R
Maple (*Acer platanoides*)								
750	3800	1700	110	20	17	0.89	3.7	6.1
Silver Maple (*Acer saccharinum*)								
760	3800	1900	110	26	13	0.49	4.1	6.7
Sycamore (*Acer pseudoplatanus*)								
630	3700	1600	87	16	—	—	4.7	7.0
Curly Maple (*Acer pseudoplatanus*)[a]								
580	4491	2379	117	13.8	—	—	—	—

[a]Ghelmeziu and Beldie, cited by Bucur, 1977.
From Haines, D., *Catgut Acoust. Soc. Newslett.*, 31, 23, 1979. With permission.

The elastic parameters in the ultrasonic range for resonance wood of different origins are given in the appendix at the end of this chapter.

7.1.1.2. Curly Maple

In his excellent book devoted to spiral grain and wave phenomena in wood formation Harris (1988) analyzes the effect of the environment and of genetics on different types of grain deviation in wood.

The wavy grain in maple (*Acer pseudoplatanus* L.) is considered to be a natural defect that enormously enhances the commercial value of timber and is important to the aesthetics and acoustics of the violin. The distinctive pattern, also called "fiddleback figure", has sinuosity along the annual ring. Changes in grain direction on the annual ring and close and abrupt corrugation of ray cells on surrounded fibers may be observed locally or over the tree stem. The highly decorative pattern of curly maple is very easily observed on a longitudinal radial surface in which the reflection of light and the color distribution is different on a smooth surface and on a surface with irregularly oriented cells. Arbogast (1992) and Bucur (1983, 1992) explain this aspect further. Leonhardt (1969) tried to classify curly maple used in the manufacture violins by basing the classification upon criteria related to the texture. The appearance of this species falls into four groups: wide, wavy grain arranged in narrow zones or in large zones, or short, wavy grain (flames) arranged also in narrow and large zones.

Vintoniv (1973) spoke to the effect of growth conditions on the acoustic, physical, and mechanical properties of sycamore (*Acer pseudoplatanus*). Harvesting trees for musical instruments should be done at altitudes of 800 to 1200 m as these trees have the best acoustic properties. Variations in density induced by the effects of altitude had only a small influence on the values of Young's moduli.

Table 7.5 provides some acoustic and elastic constants of maple (*Acer* spp.). However, sound velocity in the longitudinal direction is less than that measured in spruce.

Holz (1974) reported comparative data on curly maple and common maple. The relationship between density and modulus of elasticity showed that wavy structure has a higher density and a slightly higher E_L than plain structure (respectively, 550 to 700 kg/m³ and 100 to 120 × 10⁸ N/m²). The damping factor is different only in the frequency range 200 to 500 Hz, on which "wavy textured maple has a damping factor of by 0.1 × 10⁻² higher", being 0.9 × 10⁻² for plain structure and 1.0 × 10⁻² for wavy structure.

The scarcity of curly maple motivated scientists to undertake studies on wood species to replace it. Bond (1976) proposed the tropical wood, *Intsia bakeri*, commercially called merbau or mirabou, for the back of the violin. Another promising tropical species is mansonia, and other tropical species from South America (Delgado et al., 1983; Souza, 1983) and Australia (Bamber, 1964, 1988; Abbott, 1987; Bariska, 1978; O'Toole and Gilet, 1983; Dunlop, 1988, 1989; Bucur and Chivers, 1991; Dunlop and Shaw, 1991). Further investigation of the acoustic properties and related anatomical characteristics of these timbers at the microscopic and submicroscopic scales is vital if we are to obtain a deeper understanding of their physical and acoustic properties.

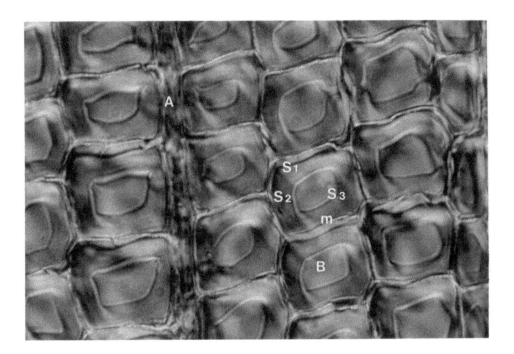

Plate 1 Spruce. Transverse section (original magnification × 400). (A) Uniseriate ray and (B) tracheids in latewood. The structural organization of the cellular wall can be observed; (m) intercellular layer; (S$_1$, S$_2$, S$_3$) are the outer, middle, and inner layers of the secondary wall.

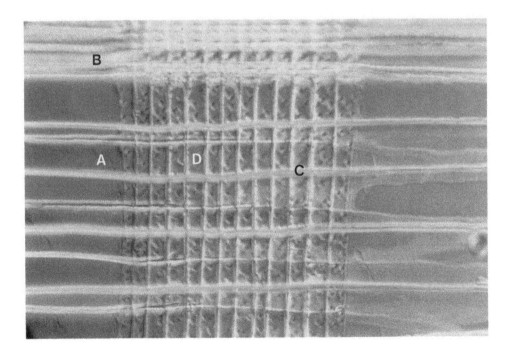

Plate 2 Spruce. Section in LR plane (original magnification × 200). Tracheids in (A) earlywood zone, (B) the latewood zone, and (C) rays. Note the tracheids cross field and (D) piceoid pits pairs.

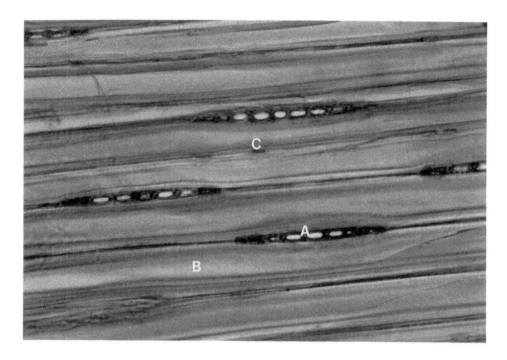

Plate 3 Spruce. Section in LT plane (original magnification × 200). (A) Heterogeneous uniseriate rays, (B) tracheids, and (C) pits in cellular tracheid walls.

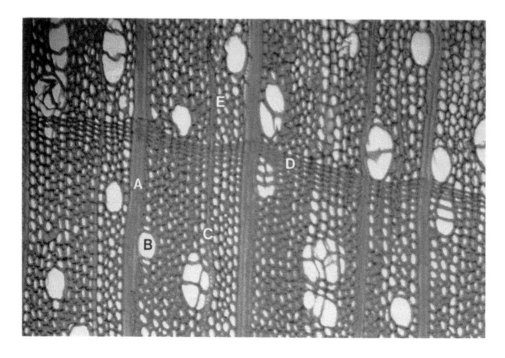

Plate 4 Curly maple. Section in RT plane (original magnification × 200). Two sizes of rays are found in this species: the larger pluriseriate ray (A) being as wide or wider than the vessels (B) and the narrow uniseriate ray (C). Note the abrupt transition in the diameter of latewood (D) to earlywood (E) fibers.

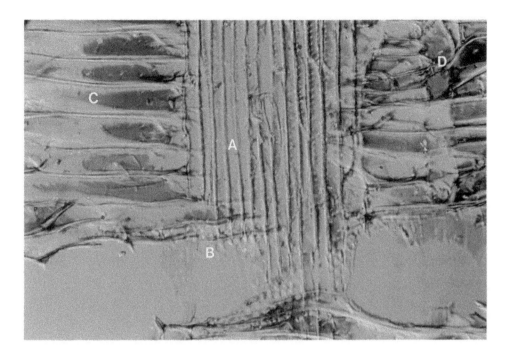

Plate 5 Curly maple. Section in LR plane (original magnification × 200). (A) Rays and (B) heavily pitted wall of vessel. Two kinds of fibers can be observed: (C) vertically aligned fibers and (D) deviated or corrugated fibers.

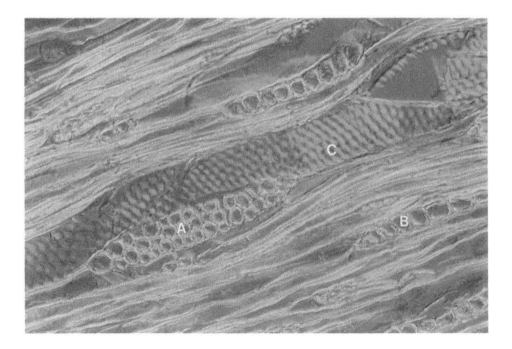

Plate 6 Curly maple. Section in LT plane (original magnification × 200). Zone of less ordered anatomic structure, with (A) multiseriate and (B) uniseriate rays and fibers. (C) Vessel with pits.

The photomicrographs were taken in transmitted polarized light by V. Bucur from wood specimens of spruce and curly maple originated in France. Specimens were provided by Professor R. Keller, Ecole du Génie Rural, des Eaux et des Forêts de Nancy. The microscopic sections were produced by Madame Simone Garros at Centre de Recherches Forestières de Nancy.

Table 7.6. Mechanical properties of European spruce selected for guitar soundboards

Density (kg/m³)	Young's moduli (10⁸ N/m²)		Logarithmic decrement	
	E_L	E_R	$2\pi\delta_L$	$2\pi\delta_R$
406	130	(3.8)	0.020	0.067
420	111	11.0	0.022	0.058
403	121	9.1	0.021	0.057
518	136	(2.4)	0.026	0.008
460	150	7.6	0.021	0.064

From Richardson, B. E., in *Proceedings of the Institute of Acoustics*, Edinburgh, U.K., 8(1), 107, 1986. With permission.

7.1.1.3. Wood for the Bow

Pernambuco wood, *Guilandesia echinata*, also called Brazilian wood, is the preferred species for the manufacture of violin bows, its excellent flexibility and strength. Delune (1977) made reference to the successful utilization of *Manilkara bidentata*. Lancewood was proposed and tested, but the results were far from those of pernambuco. *Piratinera guyanensis* (*bois d'amourette*) seems very appropriate for manufacturing baroque-type instruments, whereas beechwood is used for mass-produced instruments.

No data exist that are related to stiffness or to damping of wood species for bow stick, a gap that must be filled in the future. Modal analysis could be a powerful tool to explain the influence of material characteristics on flexural modes of the bow stick.

7.1.1.4. Wood for Other Components

Other, small pieces that comprise the violin (the fingerboard, the tailpiece, the nut, the end button, and the pegs) have less influence on the acoustic quality of the instrument than do the top and back plates. The fingerboard is generally made of ebony (Mauritius ebony is often preferred), but other African and Indian hardwood species may be used. For mass-produced instruments colored hornbeam is used. The tailpiece and the pegs are most often made of ebony, but boxwood and rosewood are also used. Mountain mahogany (*Cerocarpus ledifolius*) is considered the American alternative for ebony (Abbott, 1987). *Acer campestre* is considered by Europeans to be one of the best woods for bridges; U.S. manufacturers use Canadian rock maple.

7.1.2. ACOUSTIC PROPERTIES OF WOOD FOR GUITARS

The growing interest in the classical guitar is an argument to include this instrument in this discussion, despite that it is not commonly used in the modern symphonic orchestra. A detailed description of the steps in guitar craftsmanship and acoustics is given by Jahnel (1981), Janson (1983), and Rossing (1981), but a brief overview is provided here.

Scientific methodology has tried to explain some peculiar requirements for wood quality, bearing in mind that an objective definition of quality criteria for the selection of the material may aid in reproducing instruments of good tonal quality. The choice of spruce resonance wood for the top plate is made carefully, according to the empirical rules established by the very conservative guild of instrument makers. A narrow grain spruce is preferred (2 mm annual ring width), although Cedar of Lebanon or Canadian red cedar (*Tsuga heterophylla*) is sometimes used (Delune, 1977). The requirements for structural regularity of the guitar are not as stringent as for violins. This is easily understood because the differences between the top plates of violins and guitars are mainly in size and in the amount of string tension. The violin top is carved whereas the guitar top is flat. Consequently, the stiffness requirements are different.

Table 7.6 lists some characteristics of spruce soundboards. The parameters were measured on narrow strip samples using the frequency resonance method. The analyzed wood was characterized by high stiffness, low internal friction and density, and by a nice wood texture appearance. Wood with a high Q is necessary to produce a sustained "sing" or decay time. By contrast, violins need more damping.

For the back and the sides of guitars, Brazilian rosewood is considered to be the best wood, although it is well known that guitar makers (Jahnel, 1981; Douau, 1986; Eban, 1981, 1991; Fisher, 1986; Richardson, 1986) use a wide variety of high-density hardwoods for backs, sides, and bridges. Alternative

Table 7.7. Parameters of some species used for backs and sides of guitars

Species	Density (kg/m³)	Young's moduli (10⁸ N/m²) E_L	Young's moduli (10⁸ N/m²) E_R	Logarithmic decrement $2\pi\delta_L$	Logarithmic decrement $2\pi\delta_R$
Cordia trichotoma	793	82.40	46.70	0.035	0.04
Swartzia	838	178.30		0.025	
Zollernia illicifilia	1095	267.80	33.20	0.015	0.03
Piptadenia macrocarpa	824		20		0.04
Machaeriam villosar	909	150.90	35.60	0.022	0.05
Dalbergia cearenis	1012	76.70		0.026	
Ferreira spectabilis	892	230.00		0.019	
Dalbergia nigra	1025	167.70		0.018	

From Richardson, B. E., in *Proceedings of the Institute of Acoustics*, Edinburgh, U.K., 8(1), 107, 1986. With permission.

tropical wood species are proposed in Table 7.7. All these species are characterized by high density and low damping.

7.1.3. ACOUSTIC PROPERTIES OF WOOD FOR WOODWIND INSTRUMENTS

The woodwinds of the classical orchestra are the flute, oboe, clarinet, and bassoon. Flutes and clarinets are now frequently made from metal or plastic materials, however, flutes and piccolos used in military bands could be made of African blackwood, rosewood, or macassar ebony. The main requirements in the selection of a wood species are to present a high dimensional stability, a high density, and a fine structure. For those instruments still made of wood, the effect of the wood species is observed mainly in the aesthetics and tonal quality. The material used can influence the playing behavior of woodwind instruments (Benade, 1976) due to:

the alteration of the frequency of the air column
the vibrational damping phenomena at the walls induced by air friction or by the oscillatory temperature
the turbulence in the vibration of the air column

The working properties necessary to wood species selected for woodwind instruments are ease of accurate turning, boring, and drilling. These properties are connected with the acoustical behavior of the instrument, which is influenced strongly by the design of the tone hole system. The open holes determine the existence of a cutoff frequency which is peculiar to each instrument.

The finest quality clarinets or oboes are entirely handmade from African blackwood. Cocuswood and various rosewoods have also been used when African blackwood was difficult to obtain. Boxwood is also reported to be a satisfactory substitute in oboe manufacture. African blackwood is also considered the best wood species for the manufacture of bassoons, but other dense hardwoods such as *Acer pseudoplatanus* or *A. platanoides* may be utilized.

Reeds (single in the clarinet family and double in the oboe family) are composed of thin cane. One of the best types of cane is *la canne de Provence*, from Provence in the south of France (Heinrich, 1991).

Table 7.8 lists the acoustic properties of the major species used in making woodwind instruments, which are determined by the most popular measurement technique: narrow strip samples on which resonance frequency was measured.

Table 7.8. Acoustical properties of main species used for woodwind instruments

Density (kg/m³)	Velocity (m/s) V_{LL}	Velocity (m/s) V_{RR}	Young's moduli (10⁸ N/m²) E_L	Young's moduli (10⁸ N/m²) E_R	Shear moduli (10⁸ N/m²) G_{TL}	Shear moduli (10⁸ N/m²) G_{TR}	Internal friction $2\pi\delta_L$	Internal friction $2\pi\delta_R$
790	4600	1600	170	20	27	8.8	0.019	0.047
830	4400	1800	160	28	30	9.2	0.017	0.038

Indian Rosewood (*Dalbergia latifolia*)

Brazilian Rosewood (*Dalbergia nigra*)

From Haines, D., *Catgut Acoust. Soc. Newslett.* No. 31, 23, 1979. With permission.

Table 7.9. Honduras rosewood: acoustic properties measured in the range 130 to 350Hz via frequency resonance method

Size of specimen (mm)	Sound velocity (m/s)	tan δ_L (10^{-3})	Density (kg/m^3)	Radiation (m^4/kg/s^{-1})
500 × 71 × 30	4590	4.11	1040	4.41
420 × 62 × 25	4460	3.74	1060	4.20

Note: Radiation is calculated as the ratio of velocity and density.

From Hase, N., *Mokuzai Gakkaishi.* 33(10), 762, 1987. With permission.

7.1.4. ACOUSTIC PROPERTIES OF WOODS FOR PERCUSSION INSTRUMENTS

Percussion instruments have been part of culture since the earliest human societies, used in religious rites, dances, or in long-distance communication. Scholes (1972) and Leipp (1976) provide detailed descriptions of ancient and modern percussion instruments. Those in which wood is used are drums, tambourines, castanets, marimba, and xylophones.

Drum rims are usually composed of ash or beech, and even sometimes plywood. Drumsticks may be made from hickory, lancewood, rosewood, or hornbeam. Castanets, which give characteristic color to flamenco and other Latin dances, are composed of two round pieces of hardwood, either rosewood or ebony.

The xylophone is composed of a series of wooden bars of different size and wood species (maple, walnut, spruce, exotic species) (Bariska, 1984). The scale covered by this instrument is three or four octaves. The orchestral marimba (Ohta et al., 1984) is a deep-toned xylophone, with tubular metal resonators located under the keys, which sound an octave lower than the four-octave xylophone. This instrument is played usually with soft-headed beaters. Honduras rosewood (*Dalbergia Stevensonii* Standl.), a very durable species which has a brilliant tonal quality, is considered to be the best wood for the orchestral marimba. Table 7.9 lists some acoustical properties for this species. These properties have small internal friction and large sound radiation. A substitute for Honduras rosewood is onoore-kanba (*Betula schmidtii* Regel), a rather homogeneous and diffuse-porous species from Japan. Unfortunately, this wood presents with large deviations in grain orientation (15° or more). Such an anatomical structure affects the acoustical properties of wood, reducing the sound velocity and increasing the internal friction. For this reason, onoore-kanba is used for inexpensive, mass-produced marimba. Deschamps (1973) has identified by several wood species employed in the manufacture of xylophones used in traditional African music. These species could be used in the production of symphonic orchestra marimba.

7.1.5. ACOUSTIC PROPERTIES OF WOOD FOR KEYBOARD INSTRUMENTS: THE PIANO

The piano is the most popular of all instruments for domestic and concert hall use in occidental classical musical culture. Sound is produced by the vibration of strings struck by hammers. [An extensive description of the piano is given by Parmantier (1981). The acoustics of this instrument were studied by Suzuki and Nakamura (1990), Suzuki (1986), Boutillon (1986), Hall (1992), Chaigne and Askenfelt (1994), and Nakamura (1993)].

The soundboard, lying behind the strings in an upright piano and below them in a grand piano, "provides the means through which the energy of the vibrating string is transferred to the air, and it therefore amplifies the sound in comparison to that produced by the string alone, in the absence of the sound board. In doing this it draws energy away from the string and must therefore cause the string amplitude to decrease more rapidly than would be the case without the sound board" (Hancock, 1994). Pearson and Webster (1967), the best soundboards are produced from spruce (*Picea abies* harvested from the Austrian and Bavarian Alps and the Carpathians). Sitka spruce (*P. sitchensis* of Canadian origin) is currently recommended by the important piano manufacturers of Europe, North America, and Japan. Commercially known as "Romanian pine", the spruce from the Carpathian-Bucovina region of northern Romania, is considered one of the best woods for concert piano production because of its acoustic properties, its general appearance, and its pleasant, light yellow color. Resonance fir (*Abies alba*) also can be used.

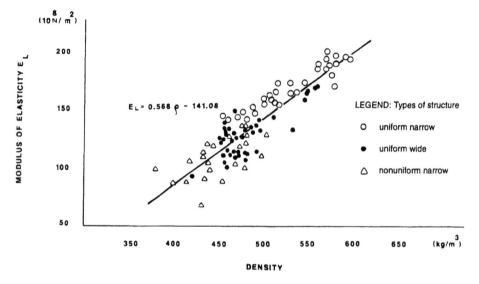

Figure 7.3. Relationship between density and Young's modulus E_L in Norway spruce with various annual ring structures. (From Holz, D., *Arch. Akustyki.*, 9(1), 37, 1974. With permission.)

The width of annual ring growth, 0.7 to 3 mm (Holz, 1967b), is an important factor in the selection of the raw material for soundboards. Closer-ringed soundboards are used under treble strings and wider-ringed boards are used under the bass. For acoustic considerations, the change in annual ring width should be gradual in adjacent soundboards.

Holz (1968 to 1974) and Ono and Kataoka (1979) extensively studied the acoustical properties of wood species used for piano soundboards. The frequency resonance method was used to determine the Young's modulus, the sound velocity, and the internal friction. The explored frequency range was 50 Hz to 10 kHz. The relationships between the density, the modulus of elasticity (E_L), and the corresponding sound velocity (V_{LL}) are given in Figures 7.3 and 7.4. Resonance wood with very fine structure is characterized by high values of E_L (maximum $E_L = 200 \times 10^8$ N/m^2) and of velocity ($V_{LL} = 5700$ m/s) whereas irregular and nonuniform structure produceded lower characteritics (minimum $E_L = 65 \times 10^8$ N/m^2 and $V_{LL} = 3900$ m/s).

Holz (1974) and Kataoka and Ono (1976) attached great importance to possibly using softwood species other than spruce in the manufacture of piano soundboards. Holz carefully selected fine structure specimens of Douglas fir (Oregon pine), fir, pine, and redwood. The relationship between E_L and the density is given in Figure 7.5, on which the regression line for spruce is shown. Holz stated that "the data for fir, Sitka spruce and redwood accomodate to the regression line for spruce". The resin content in pine (4 to 17%) and Douglas fir (5 to 7%) probably affects the behavior of these two species when compared to resonance spruce in which the resin content is very low (1.8 to 2.7%). Internal friction (tan δ_L) ranging from 0.7 to 0.9 $\times$ 10^{-2} is independent of density (Figure 7.6) and greatly influenced by the measurement frequency range, as can be seen in Figure 7.2. This means that the continuity and the uniformity in the longitudinal direction of the anatomic structure of resonance wood play more important roles in the acoustic behavior of the material than its density.

Other parts of the piano are less important than the soundboards for the acoustical quality of the instrument. Bars are produced from the same timber as the soundboards. Bridges, which must sustain the pressure of strings, are made of beech, Canadian rock maple, or, more recently, of laminated hardwood (birch or maple). Beech and Canadian rock maple are also used for wrest planks, which bear the load of tensioned strings. Tropical wood-laminated beams are sometimes recommended because of their strength and hardness. The casework could be comprised of mahogany or walnut with an ebonized finish, especially for concert pianos. For school pianos, natural oak finish is used. Several other tropical species such as wawa, Nigerian obeche, limba, Parana pine, makoré, and sapeli may be chosen as finishing veneers because of their aesthetic qualities. Action parts, those that comprise the mechanisms

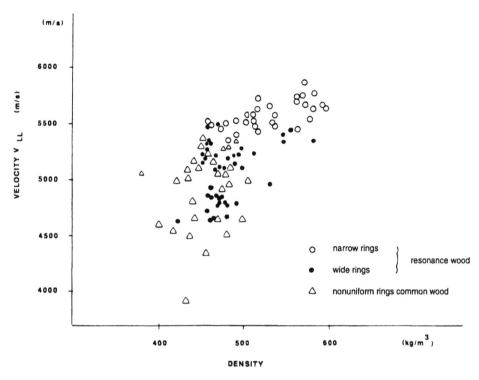

Figure 7.4. Relationship between sound velocity and density for piano soundboard resonance wood (Norway spruce) of various structural characteristics. (From Holz, D., *Arch. Akustyki.*, 9(1), 37, 1974. With permission.)

between the keys and the strings, are made from Canadian rock maple or beech. Japanese species also may be employed: Hase and Okuyama (1986, 1987) noted the use of *Betula maximowicziana* (E_L = 103×10^8 N/m²). Hornbeam or maple are used in small parts. Damper heads are sometimes made from Tasmanian myrtle or Indian rosewood. Hammers are made from Canadian rock maple, hornbeam, black walnut, or laminated birch (Hase et al., 1989) because of their highly elastic parameters (E_L = (174 to 211) $\times 10^8$ N/m²).

7.1.6. RELATIONSHIPS BETWEEN ELASTIC PROPERTIES OF RESONANCE WOOD AND ITS TYPICAL STRUCTURAL CHARACTERISTICS

The very useful data of Barducci and Pasqualini (1948) and Haines (1979) summarize the relationship between sound velocity and density for a very large variety of species. The general law for wood is to decrease velocity while increasing density. Previously, we noted that spruce resonance wood is characterized by high values of velocity of sound propagation in the longitudinal direction and by low density. This species is a special case of wood material having high mechanical performance. Establishing the relationships between elastic properties and typical structural characteristics must be oriented to the question, how can singular elastic properties be related to a specific set of structural characteristics? To encompass the fairly wide scope of the question, two criteria should be fulfilled. The first concerns the macroscopic structural parameters (width and regularity of annual rings, proportion of the latewood zone, density pattern in annual rings), and the second covers the microscopic and submicroscopic structural parameters (tracheid length, ray distribution patterns, microfibril angle, index of crystallinity).

7.1.6.1. Macroscopic Structural Parameters

Violin makers always insist on the importance of the regularity of the growth ring pattern for acoustical properties (Hutchins, 1978). The regularity index of wood is calculated as a ratio between the difference

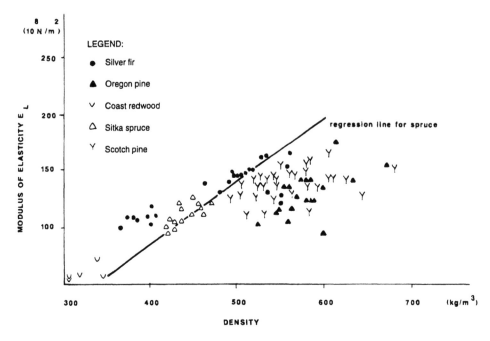

Figure 7.5. Comparison between Norway spruce resonance wood and other coniferous woods of fine and regular structure, apt to be used for piano soundboards. The regression relationship between Young's modulus and density for spruce is shown. (From Holz, D., *Arch. Akustyki.*, 9(1), 37, 1974. With permission.)

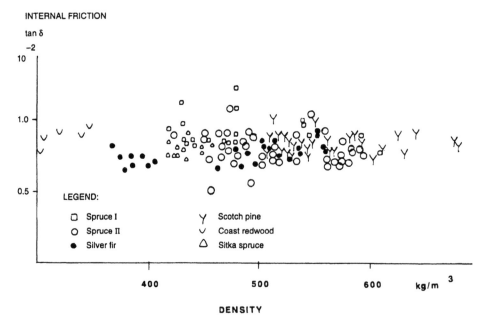

Figure 7.6. Internal friction and density in several coniferous species. (From Holz, D., *Arch. Akustyki.* 14(4), 195, 1973. With permission.)

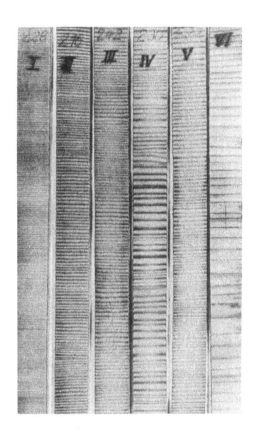

Figure 7.7. Transverse sections of resonance wood of various qualities. (I) Wood for guitars; (II) violins; (III) piano soundboard; (IV) other string instrument non-acoustic parts (viola, violoncello, etc.); (V) piano keyboard; (VI) uses other than musical instruments, i.e., boat construction. (From Holz, D., *Holztechnologie*, 25(1), 31, 1984. With permission.)

of maximum and minimum annual ring width to the maximum width. For very regular structure, this index could be 0.7. The following sections analyze several aspects connected with the growth ring pattern in spruce resonance wood and with the density pattern of the annual rings.

7.1.6.1.1. Growth Ring Pattern

Annual ring patterns can be observed easily in transverse sections (Figure 7.7), from which the width and the proportion of the latewood can be measured. Holz (1984) selected resonance wood specimens of different qualities on which ring structural parameters were measured together with elastic modulus E_R, the corresponding velocity and damping constant (Table 7.10). From Table 7.10 we observe that

Table 7.10. Radial characteristics of different quality Norway spruce resonance wood

Quality	Density (kg/m³)	E_R (10⁸ N/m²)	V_{RR} (m/s)	tan δ_R (10⁻²)	Annual ring parameters Width (mm)	Annual ring parameters Latewood (%)
Guitar	413	5.9	1190	1.8	1.08	21
Violin	528	10	1380	1.8	1.46	26
Piano	463	5.7	1110	1.6	1.64	25
Other*	456	12	510	2.8	2.65	41
Boat building	378	1.9	730	2.2	4.25	13

*Other = stringed instrument parts without any acoustical function.
From Holz, D., *Holztechnologie*. 25(1), 31, 1984. With permission.

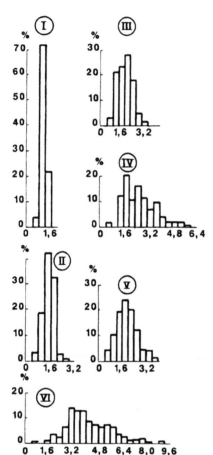

Figure 7.8. Frequency distribution of annual ring width in resonance wood. Wood for (I) guitars; (II) violins; (III) piano soundboards; (IV) other string instrument non-acoustic parts (viola, violoncello, etc.); (V) piano keyboard; (VI) uses other than musical instruments, i.e., boat construction. (From Holz, D., *Holztechnologie*, 25(1) 31, 1984. With permission.)

the relative proportion of the latewood, in soundboards for violins, guitars, and pianos, hardly exceeds 25%. Histograms of the annual ring width (Holz, 1984) show a very sharp distribution for violins and guitars. At the opposite end of the scale, structural spruce is characterized by a flat and nonsymmetric curve (Figure 7.8). Furthermore, for the same specimens, the distribution of sound velocity in the longitudinal direction and the distribution of acoustic radiation were presented (Figure 7.9). The difference between resonance and structural spruce is evident in the velocity histogram. The profile of the curve is similar, but the range of values is different. On the other hand, the curve corresponding to the acoustic radiation of resonance wood has a flat, parabolic profile, while that of structural wood is irregular.

Ultrasonic measurements of V_{LL} on resonance spruce and curly maple vs. the width of annual rings shows that an increase in annual ring width is associated with a decrease in velocity (Figure 7.10). Note from Table 7.10 that increasing the proportion of the latewood generated the increase in the elasticity modulus and the internal friction.

The different behaviors of resonance wood in the longitudinal and the radial directions provided information about the highly anisotropic nature of this material. Because we are interested in the overall anisotropy of the material, all acoustical parameters in all symmetry planes are needed. Ultrasonic methods may be used (Table 7.11), as demonstrated by Bucur (1987, 1988). The main differences between resonance wood and common wood structures are observed via shear waves. The ratio of acoustic invariants creates a difference of 11 to 13% between resonance and common structure.

The validity of ultrasonic measurements, when compared to measurements in the audio-frequency range, was demonstrated by Schumacher et al. (1988) by using values of elastic constants to model the

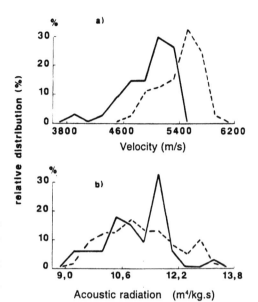

Figure 7.9. Frequency distribution of sound velocity on resonance spruce (---) and common spruce (——) of two parameters: velocity, V_{LL} (a) and resonant ratio, (V_{LL}/ρ) (b). (From Holz, D., *Holztechnologie*, 25(1), 31, 1984. With permission.)

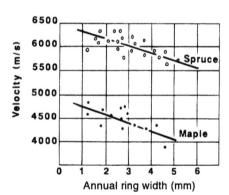

Figure 7.10. Ultrasonic velocity (V_{LL}) and annual ring width in resonance wood.

Table 7.11. Ultrasonic velocities on resonance wood

Species	Density (kg/m³)	Longitudinal velocities (m/s)			Shear velocities (m/s)		
		V_{LL}	V_{RR}	V_{TT}	V_{RT}	V_{LT}	V_{LR}
		Spruce					
Resonance	400	5600	2000	1600	298	1425	1374
Common	485	5353	1580	1146	477	1230	1322
		Maple					
Curly	700	4350	2590	1914	812	1468	1744
Common	623	4695	2148	1878	630	1148	1350

From Bucur, V., *J. Catgut Acoust. Soc.* Ser. 1, 47, 42, 1987. With permission.

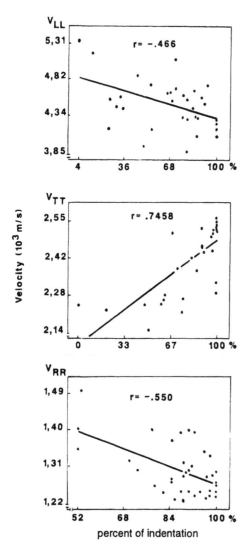

Figure 7.11. Influence of the proportion of indented rings on ultrasonic velocity. (From Chiesa, V., Thesis, University of Florence, 1987. With permission.)

behavior of violin plates. He stated that the real part of the elastic constants measured ultrasonically are well within the range of data reported in the literature using other acoustic methods. The influence of the frequency range on viscoelastic parameters is probably more important.

The particular pattern of indented rings in resonance wood has a wide influence on its acoustical characteristics. Figure 7.11 shows that the volumic proportion of indentations is strongly related to radial velocity V_{RR}. Longitudinal velocity V_{LL} and tangential velocity V_{TT} are inversely related to the presence of this peculiar tissue.

The relatively small V_{LL} in this resonance wood may be associated with the presence of indentations and to shorter tracheids than it is in normal wood. The distribution of tracheid length is wider in normal than in resonance wood (Fukazama and Ohtani, 1984). This may also affect the value of V_{LL}.

7.1.6.1.2. Densitometric Pattern of the Annual Ring in Resonance Wood
X-ray microanalysis was used to study the densitometric pattern of the annual rings in spruce and fir resonance wood. Radial specimens having both narrow and wide annual increments were selected from soundboards whose wood originated from the Val di Fiemme, Italy. Densitometric measurements were

Table 7.12. Density components in earlywood and latewood of spruce and fir resonance wood

Width (mm)	Density components (kg/m³)					Velocity (m/s), earlywood		Proportion (%)	
	D_{min}	D_{max}	D	D_{EW}	D_{LW}	V_{TT}	V_{LL}	EW	LW
Spruce, Wide Annual Rings									
4.1	323	838	433	380	700	1045	3399	69	31
4.0	323	838	433	351	652	1015	3875	75	25
4.9	275	957	476	339	734	976	3390	65	45
Fir, Wide Annual Rings									
4.8	269	995	521	410	825	1474	3639	71	29
4.2	3.39	072	441	415	864	1598	4197	79	21
Spruce, Narrow Annual Rings									
1.1	341	838	568	376	551	—	—	45	55
1.5	311	802	489	345	570	—	—	53	47
2.0	305	898	435	352	663	—	—	65	35
Fir, Narrow Annual Rings									
0.8	312	756	473	311	691	—	—	77	23
1.3	252	756	479	406	641	—	—	82	18
1.7	251	796	424	371	610	—	—	80	20

Note: EW = Earlywood; LW = latewood; D = average density or mass density.

completed with ultrasonic measurements (Bucur, 1984) of earlywood and latewood on specimens (Table 7.12) with wide annual rings. Due to equipment limitations ultrasonic velocity measurements were confined to specimens with large annual ring increments in which the internal zones of the annual rings were sufficiently large for the contact area with the ultrasonic transducer (about 1 mm²).

Figure 7.12 presents the radiographic image and the densitometric profile of resonance wood. Microdensitometric parameters are explained in Figure 7.13. The density components as well as the ultrasonic velocities of longitudinal waves in L and T directions are given in Table 7.12. The density of earlywood is half that of latewood, and the velocity V_{LL} in latewood is surprisingly small. As Table 7.12 demonstrates, microdensitometric analysis permits further inquiry into the annual ring organization expressed on a quantitative and a fine scale.

Microdensitometric measurements on new and old wood and ultrasonic measurements with shear waves revealed important differences between specimens (Table 7.13). The anisotropy expressed by microdensitometric parameters shows differences between the average and minimum densities in earlywood, D_{min} (22%), and the maximum density D_{max} in latewood (16%). More evident are the differences expressed by ultrasonic velocities. The anisotropy expressed by the ratio of shear velocities show 200 to 300% differences between early- and latewood. Moreover, insight into the annual ring pattern produced by the development of the X-ray technique through microdensitometric analysis permitted the development of dendrochronological investigations of musical instruments. Dendrochronological dating of the resonance wood used in violins was reported by Schweingruber (1983), Corona (1980 to 1990), and Klein et al. (1986). Other than dating the spruce tops, information can be obtained about storage time before wood utilization in instruments. Klein et al. (1986) reported 5 to 25 or even 50 years of wood storage for violins produced from 1563 to 1892 by Italian and German masters. Moreover, the presence of sapwood on some German instruments was proven. On the Italian-made instruments it seems that sapwood was always removed, as remarked by Leonhardt (1969), Ille (1975), Bariska (1978), Shigo and Roy (1983).

7.1.6.2. Microscopic and Submicroscopic Structural Parameters

These parameters complement the macroscopic parameters in tracing the structural inhomogeneities that produce the characteristic anisotropic features of resonance wood. Table 7.14 shows data on the

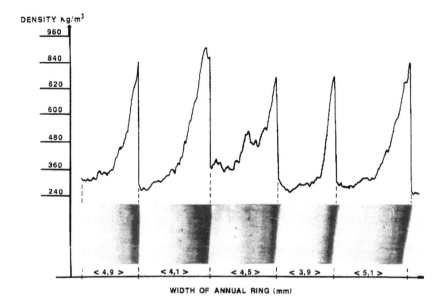

DENSITY kg/m³

WIDTH OF ANNUAL RING (mm)

Figure 7.12. Densitometric pattern of resonance wood.

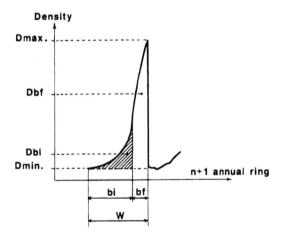

Figure 7.13. Microdensitometric components measured on the radiographic image of one annual ring. (Dbf) Average density of latewood zone, (Dbi) average density of earlywood zone, (W) annual ring width.

structural parameters of spruce resonance specimens for violins. Note the very small overall microfibril angle measured in the X-ray diffraction technique, and the very important difference in this parameter between earlywood and latewood. Also, it should be noted that the tracheid length for the specimens studied (Rocaboy and Bucur, 1990) is around 4 to 5 mm and that the length distribution around the average is very confined, as the histograms are very sharp.

Rays are important anatomic elements that induce anisotropy in spruce resonance wood (Schleske, 1990). The influence of the presence of rays on shear moduli on spruce resonance wood was demonstrated by Beldie (1968), as can be seen from Table 7.15. The values most affected are related to the transverse plane in which the difference between G_{RT} and G_{TR} is 50%.

Table 7.13. Microdensitometric and acoustic parameters of recent and old wood

Specimen	Year	Ring (mm)	D	Density components (kg/m³)			HT
				D_{min}	D_{max}	$D_{max} - D_{min}$	
Microdensitometric Analysis							
Picea rubens	1975	2.6	484	351	661	310	1.29
Picea abies	1756	1.0	398	292	551	258	1.46

Specimen	Year	Acoustic Parameters Shear velocity (m/s)				
		V_{TR}	V_{LT}	V_{LR}	V_{LR}/V_{TR}	G_{LR}/G_{TR}
Picea rubens	1975	552	1222	1228	2.35	5.53
Picea abies	1756	354	1284	1505	4.25	18.10

Note: HT = Heterogeneity = $(D_{max} - D)/(D - D_{min})$. Specimens were provided by C. M. Hutchins from boards and material from her collection.

Table 7.14. Structural parameters, at fine anatomic scale, of spruce resonance wood for violins

Sample	Ring width (mm)	Latewood (%)	Regularity	Crystallinity (%)	Microfibril angle		
					Overall	EW	LW
266	1.5	23	0.37	38.0°	2.94	11°–30°	2°–9°
269	1.9	25	0.52	35.5°	4.08	8°–15°	1°–8°
271	1.3	26	0.58	30.0°	7.95	12°–30°	4°–12°

Note: EW = earlywood; LW = latewood.

From Rocaboy, F. and Bucur, V., *Catgut Acoust. Soc. J.* Ser. 2, 1(6), 21, 1990. With permission.

Table 7.15. The influence of rays on shear moduli of spruce (10^8 N/m²) resonance wood

Young's moduli		Shear moduli					
E_L	E_R	G_{LR}	G_{RL}	G_{LT}	G_{TL}	G_{RT}	G_{TR}
151	11.3	5.73	2.50	6.21	5.64	0.31	0.60

Note: With shear waves the propagation direction is perpendicular to the polarization direction. The presence of rays strongly influences the shear moduli of spruce G_{RL} and G_{RT}. These moduli are smaller than G_{LR} and G_{TR}, respectively.

From Beldie, I. P., *Holz. Roh Werkst.* 26(7), 261, 1968. With permission.

Schleske (1990) analyzed the modification of the wood parameters during carving in the surface arching of the top of the violin. On arch contour the fibers and rays are shortened and strongly deviated from the principal directions L and R. The deviation of rays is higher than that of fibers. The new geometry induces a decrease in stiffnesses, and implicitly of sound velocity, in such a way that, as Schleske put it, "we can achieve an effective decoupling of the whole violin top plate at the edge". This very subtle way of working with the wood structure and anisotropy of violin plates is very well known to luthiers, who utilize it empirically to decrease the thickness of carved plates and to adjust the tone quality of instruments.

7.1.7. TONAL QUALITY OF MUSICAL INSTRUMENTS AND WOOD PROPERTIES

The sound of musical instruments depends on many factors, the most important of which are the acoustics of the room, the skill of the player and the ability of the instrument's sounding box to receive and transfer the energy of acoustic vibration. Quantitative relationships between the tone of the instrument and the construction parameters need to be established, as this is a discussion of the relationships between the tone and the properties of woods used for their manufacture. This approach requires simultaneously a deep

CALCULATED MODE

OBSERVED MODE

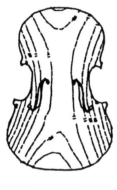

177.5 Hz   168 Hz

Figure 7.14. Mode shape and frequencies of a violin plate. Calculated data and observed patterns in holographic interferometry. (From Richardson, B. E. et al., 1987. *J. Catgut Acoust. Soc.*, No. 47, 12–16. With permission.)

understanding of the physical principles of sound production in the instrument and an understanding of the psychology of human perception (Meyer, 1983, 1984; Hutchins, 1975; Dünnwald, 1990).

The importance of the transient and steady-state behaviors of an instrument must be acknowledged. "This is a subject of great importance to the player because the way in which notes start to sound influences what the player calls the playability or the characteristic response of an instrument" (Hancock, 1994). The behavior of the instrument is influenced by its overall design as well as by the properties of the wood used in its construction (Woodhouse, 1993a, b, 1994).

Richardson (1988) noted that the most elusive aspect of the maker's art is his ability to produce instruments with a predetermined tone quality. The variability of wood used in the construction of the body plays an important role in the tone quality of the instrument. Objective selective criteria for the raw materials can assist the maker in his tasks.

To visualize the very complex modes of vibration of the body, different methods have been developed, ranging from vibration patterns of violin plates to modal analysis and holographic interferometry (Cremer, 1983; Reinicke and Cremer, 1970; Beldie, 1975; Bissinger and Hutchins, 1980; Kondo et al., 1980; Marshall (1985); Muller and Geissler, 1984; Fletcher and Rossing, 1991; Jansson and Niewczyk, 1989; Molin and Jansson, 1989; Rogers, 1986, 1991; Richardson et al., 1987). Figure 7.14 gives the calculated data and observed patterns of the mode shapes and frequencies of a violin top via the method of holographic interferometry. The use of computers permits modal analysis, which gives spectacular pictures of the soundbox vibrations (Figure 7.15).

The techniques used by luthiers to choose a quality wood by weighing the soundboards and by listening to the pitch and the timbre of the eigentone of the free plate held at its nodal point were discussed earlier in the chapter. Through this we are able to recognize a simple method of estimating the density of the material, the velocity of sound, and the damping constants.

7.1.8. CONCLUSION

The data presented in this chapter were an attempt to offer the prospect of understanding the logic and the scientific foundation of the art of the luthier, who invented skillful and ingeneous procedures to verify sufficiently "the physical properties of wood and plates without expensive equipment" (Muller, 1986). The scientific findings and testing equipment available in the 20th century can assist in obtaining more or less reproducible results, however, the skill of the master remains the unique contribution in the production of the art object that is the musical instrument.

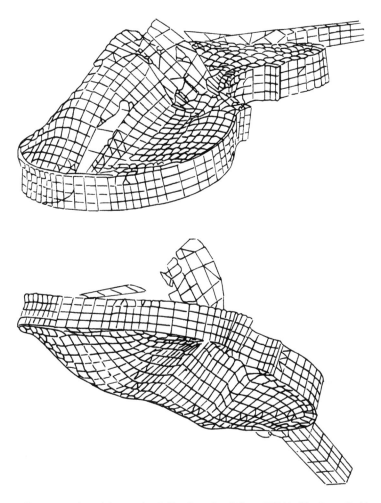

Figure 7.15. Reconstruction of the mode of vibration of a violin at 536 Hz (Knott as cited by Hutchins, C. M., *Catgut Acoust. Soc. J.* Ser. 2, 1(7), 36, 1991. With permission.)

7.2. FACTORS AFFECTING ACOUSTIC PROPERTIES OF WOOD FOR MUSICAL INSTRUMENTS

Thus far the discussion has provided a broad view of the elastical behavior of wood. The next pages deal with acoustic and other physical and chemical properties of wood.

7.2.1. INFLUENCE OF NATURAL AGING ON RESONANCE WOOD

The first study of the influence of natural aging of resonance wood on its acoustical properties was done by Barducci and Pasqualini (1948). Using a resonance frequency method they measured the sound velocity and the quality factor in longitudinal and radial directions of spruce and maple (Table 7.16). Bearing in mind the great variability of wood properties, not evaluated in this study, only the general trend of data can be considered. The best acoustical properties of wood could be expected to develop at 10 years of natural aging. Holz (1981) studied the influence of aging on acoustic and mechanical properties of resonance wood. He selected specimens aged 2 to 4, 60, and 180 years. The resonance method was used to determine longitudinal Young's modulus and internal friction.

Table 7.16. Influence of aging on acoustic properties of resonance wood (Barducci and Pasqualini, 1948)

Age (years)	Density (kg/m³)	Sound velocity (m/s)			Quality factor (Q_L)	Origin
		V_{LL}	V_{RR}	Ratio		
		Spruce				
1	460	5350	—	—	125	Italy
10	410	5700	1150	4.95	125	Italy
12	415	5600	—	—	135	Italy
52	440	5400	1500	4.70	130	Tyrol
67	450	5250	—	—	115	Tyrol
390	450	4200	950	4.40	95	Italy
		Maple				
1	720	4050	—	—	80	Italy
13	665	4300	—	—	105	Italy
17	785	4150	—	—	80	France

Figure 7.16 presents the relationships between E_L and density. The values of E_L for old wood are slightly smaller than those for fresh wood. The variations of the internal friction parameters in the longitudinal direction vs. frequencies in the range 100 Hz to 10 kHz show no important differences between old and fresh wood (Holz, 1981). By contrast, Yankovski (1967) noted a lower internal friction in old wood than in fresh wood; the values of tan δ are, respectively, 0.01 (old wood) and 0.025 (fresh wood) at 1.5 kHz. However, a simple plot of E_L and tan δ_L vs. frequency can be misleading. Some possible sources of confusion in interpreting data related only to longitudinal axes could be introduced by the very important natural variability of the wood material itself. To avoid the indistinguishable differences revealed, another methodological approach must be used. The requirements of the methods of analysis must be addressed at another structural level of the material, avoiding the overall properties of wood at the macroscopic scale (Bucur, 1975). Combining the nondestructive X-ray microdensitometric analysis, X-ray diffraction, and ultrasonic methods, and considering all anisotropic directions, as well as other physical and chemical properties of the cellular wall, we would, of course, obtain thorough results.

From the analysis of the function of the violin, we observe two main physical effects that may be included in the aging process of wood for musical instruments:

- Chemical and physical changes in wood against time, with its inevitable fluctuations of temperature, atmospheric relative humidity, etc.
- Effects due to the playing process, which consist mainly of the results of vibrations and static loading

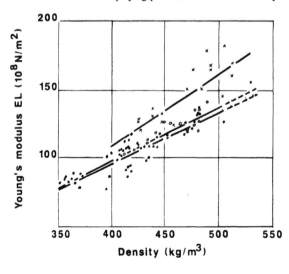

Figure 7.16. Relationships between E_L and density for old structural timber, old resonance wood, and fresh wood. (•) 180-Year old wood from a building; (×) 60-Year old wood from violin tops; (○) fresh wood from nonacoustical parts. (From Holz, D., *Holztechnologie*, 22(2), 80, 1981. With permission.)

induced by the strings. Detailed considerations of the static forces acting on the body of the violin, were developed by McLennan (1980). The effects of continuous resonance vibration on wood quality were analyzed by Sobue and Okayasu (1992).

Luthiers attach the greatest importance to thoroughly dry and well-seasoned resonance wood. To this end, the material is exposed to air for some years before it is used. Different periods of time are recommended, ranging from 3 to 100 years. The short periods (3 to 10 years) are necessary for dimensional stabilization of wood, when the material achieves perfect hygroscopic equilibrium. Zimmermann (1978) noted that the mechanism of water storage in wood is governed by the elasticity of tissues, the capillarity, and the cavitation phenomena. The primary loss of water was observed immediately after cutting the tree. The rupture of the sap column and most of the larger anatomic elements determines the pressure increase in the tissues. It appears that capillarity and cavitation play important roles in the stress distribution in wood, defined as when the moisture content drops from higher values to under the fiber saturation point. The small anatomic elements and/or some components of the cell wall hold the water a very long time and eliminate it only over a period of many years. This phenomenon is superimposed on the periodic fluctuations of wood moisture content due to the changing relative humidity in the atmosphere.

To avoid cavitation and consequently to produce raw material of very good quality, without cracks and internal tensions, very long air-drying periods are necessary. For the mass production of musical instruments, kiln drying procedures were developed based on a peculiar equilibrium between temperature (20° to 40°C) and relative humidity over a long period of time (3 to 4 months) (Koberle and Majek, 1978).

Violin makers often report that wood strengthens with age. This structural evolution has encouraged speculation, however, concerning the hypothesis of crystallographic structural modification as well as that of chemical and ultrastructural changes in lignin and hemicelluloses. Also note that the very long drying periods recommended in the literature (Leipp, 1965; Fukada et al., 1956) are limited to a century because of the losses of physical, mechanical, and, implicitly, acoustic properties of very old wood.

Several years ago, a hypothesis was advanced concerning the technique of the old Italian masters of long-time immersion of resonance logs in water. This treatment supposedly activates enzyme activity, which selectively destroys the borded pits and consequently produces a raw material with high acoustic properties. Barlow and Woodhouse (1990) used scanning electron microscopy to explore samples from many old Italian-made instruments, and found no evidence of pits degraded by bacterial attack. They noted that "most samples seemed indistinguishable from modern, air-dried spruce of musical instrument quality". This study showed once more that no significant differences in anatomical structure were observed between old and fresh wood.

Realistic information about wood behavior affected by aging are undoubtedly connected with chemical properties (Fengel and Wegener, 1989). Data on the chemical composition of old spruce and maple are given in Table 7.17. The main points deduced are

Table 7.17. Some chemical components of old and fresh spruce and maple (data from Pishik et al., 1971)

Age (years)	Chemical composition (%)				
	Cellulose	Lignin	Warm water extractives	Hemicelluloses	Ash
Spruce					
0	54.47	26.29	0.30	17.79	0.26
3	54.40	25.40	0.97	18.50	0.28
50–70	53.94	25.01	0.92	17.80	0.25
150–200	50.78	24.64	2.82	14.16	0.48
200–300	52.34	24.90	2.58	13.16	0.48
300–400	50.77	24.26	4.60	12.11	0.47
500–700	49.45	23.97	2.21	10.35	0.42
Maple					
0	46.53	24.18	0.32	22.51	0.56
70	46.53	23.97	1.64	19.80	0.51
100	47.71	21.91	1.60	18.98	0.52

Table 7.18. Index of cellulose crystallinity in fresh and old resonance wood, derived via X-ray diffraction technique (Bucur, 1975)

Natural aging of wood (years)									
1	4	10	30	40	50	60	90	100	120
Crystallinity Index on Spruce									
39	46.5	40.5	41.9	43.6	41.7	39.6	39.9	46.9	46.9
Crystallinity Index on Maple									
32	33.4	—	—	36.0	46.0	—	—	—	—

- Cellulose is the most stable chemical component of wood, quantitatively unchangeable with age
- Lignin decreases slightly with age due to oxidation, an effect evident in the color and perfumed odor of old wood
- Hemicelluloses are the most unstable components, easily hydrolyzable in oligosaccharides

Pishik et al. (1971) provided convincing evidence that the extractives from old wood absorb much more in the ultraviolet region than those from fresh wood when spectral analysis techniques were employed. The submolecular modifications in old wood (Pishik et al., 1971) op.cit) were illustrated by magnetic resonance imaging. The number of OH groups in old wood is higher than in fresh wood.

Other interesting physical parameters to be studied in relation to the aging of wood are the piezoelectric constants. As reported by Fukada et al. (1956) for wood specimens of 8, 200, 350, 530, 700, 900, 1200, and 1300 years of age, the piezoelectric constant related to the longitudinal axes, the longitudinal Young's modulus under static or dynamic tests increased with age. A maximum was observed near 200 years. This phenomenon is explained by the reorganization of the fine structure, namely recrystallization of cellulosic chains and slow dissociation of cellulose molecules. Using the X-ray diffraction technique, Fukada (1965) stated that the index of crystallinity in Japanese cypress reaches a maximum at 350 years and decreases gradually with age up to 1400 years. He also suggested that the excellent acoustical quality of Italian violins made 3 centuries ago could be related to the optimal crystallinity observed in the cellulosic components of the wood. Consequently, the acoustical properties of the material reach a maximum. Fukada's statement encouraged us to study the behavior of cellulosic crystals in resonance wood (aged from 4 to 120 years). From Table 7.18 we may deduce that the index of crystallinity is variable vs. time and that it reveals a maximum at 4 and 100 years. (Unfortunately, our specimens were not old enough.)

Further knowledge of the crystallographic organization of cellulose and about the molecular organization of other chemical components (lignin, hemicelluloses, etc.) of old wood would be of great value for the understanding of the differences between old and fresh specimens.

7.2.2. INFLUENCE OF ENVIRONMENTAL CONDITIONS
The acoustical response of musical instruments depends critically on the conditions in the environment: room temperature, 60 to 65% relative humidity, and 8 to 10% moisture content of wood. It is well known that under these conditions the acoustic and mechanical properties of wood are optimal; i.e., wave velocities and elastic moduli present maximum values because damping parameters due to internal friction are minimal. Continuous changes in temperature and relative humidity with the seasons is of great importance to all wooden parts, varnished or not, old or new. The thinner elements are affected more than the thicker elements (Fryxell, 1964, 1981).

Thompson (1979) demonstrated that the frequencies and mode shapes excited in instrument plates, and hence the tonal quality, are greatly influenced by the temperature and moisture content of the wood. Even small variations in relative humidity affect the amplitude and frequencies of the different modes of vibration of plates. As an example, for mode #2 of a free violin back plate, the frequency was 336 Hz at 15% relative humidity, and at 79% relative humidity the frequency was 313 Hz. For mode #5 of vibration of the same plate at the same relative humidities, the corresponding frequencies were 160 and 142 Hz. Problems related to the tuning of violin plates when wood moisture and air relative humidity changed were discussed by C. M. Hutchins (1982).

An interesting study of temperature changes in a musical instrument was undertaken by Firth (1988), who used infrared thermography. Graphs (Figure 7.17) showed isotherms of the top plate of a guitar when held both statically and after playing. The highest temperature was recorded under the player's arm (34.5°C). The lowest temperature was recorded near the border of the instrument (22.5°C).

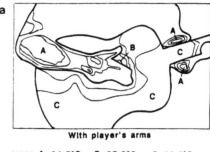

With player's arms

areas A: 34,5°C   B: 27,5°C   C: 22,5°C

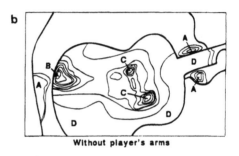

Figure 7.17. Isotherms of the top plate of a guitar, determined using infrared thermography. (a) Isotherms of the top plate of a guitar when held statically; (b) isotherms of the same plate without player's arm (Firth, 1983).

Without player's arms

areas A: 34,5°C   B: 30,5°C   C: 28,5°C   D: 22,5°C

7.2.3. INFLUENCE OF LONG-TERM LOADING

It is generally believed that a good violin improves with age through playing. To bring a violin to its optimal acoustic performance requires months or years of playing. When an instrument is not played for a long time, its acoustic qualities diminish due to the relaxation of stress induced by the static and dynamic loading of strings. This loading occurs through vibration of the instrument under a downbearing stress of about 20 lb (44 kg) from the strings and through the influence of the heat and humidity emanating from the violinist. Modification of the tonal quality of a new violin is described by makers and players in terms of "strengthening" of the wood. Examining the rheology of the system represented by the violin, a fatigue phenomena is observed under a superimposition of two regimes: one dynamic, when the instrument vibrates during play, and the other, static, induced by the stress of the strings. To simulate in a simplified manner the changes induced in a violin by playing it, specimens or violins loaded for a long period of time can be used.

The scientific literature lacks data concerning the influence of long-term mechanical loading on acoustical properties of resonance wood or on solid wood in general. Manasevici (1962) reported the effects on the behavior of common pine wood specimens ($20 \times 50 \times 1000$ mm) due to 51 days of continuous flexural resonance vibration. The overall amplitude of vibration was recorded, and the maximum value was recorded at the 15th day of the experiment. During the days following the amplitude decreased from the maximum value to a value higher than the value measured at the beginning of the experiments. Unfortunately, the author did not report data on resonance frequency or on quality factors for the assessment of the modifications induced in specimens.

Sobue and Okayasu (1992) studied the effect of continuous small amplitude vibration on Young's modulus and that of internal friction (tan δ) on different species of softwoods and hardwoods. The free-free flexural vibration with small amplitude (from 0.015 to 0.40 mm) and frequency (100 to 170 Hz) was applied during a period of 5 h. The Young's modulus E_L was not affected by the vibrational regime; rather the internal friction parameter tan δ_L decreased from 5 to 15%. This behavior is related to the modification of the normal alignment of cellulose chain molecules. The hydrogen bonds were broken under the continuous vibration, and this phenomenon is reflected in the diminishing of (tan δ). By

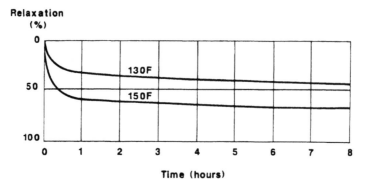

Figure 7.18. Relaxation expressed as the inverse of the ratio of strains at initial time and at time *t* (%) vs. the time of loading (hours) on small specimen of Sitka spruce under static bending loading, at 130° and 150°F. (From Gadd, C. W., *J. Catgut Acoust. Soc. J.*, 41, 17, 1983. With permission.)

applying the reaction rate process theory to the formation and rupture of the hydrogen bond, a relationship of (tan δ) vs. time was derived, and it was noted to be in good agreement with the experimental data.

The effect of 1500 h of vibration on a violin with strings up to pitch was investigated by Boutillon and Weinreich (cited by Hutchins and Rogers, 1992), and was found to produce a decrease in the B1 mode frequency from 580 to 550 Hz shortly after removal of vibrational field.

Gadd (1984) studied the influence of stress relaxation under static bending loading on Sitka spruce (Figure 7.18). When the constant load in the elastic domain is applied, the influence of temperature (130° to 150°F) is dependent on the behavior of wood during the first hour of loading. Gadd suggested that a treatment of soundboards for several hours at 150°F could stabilize the internal stresses in violin plates.

Modifications in the acoustic properties of resonance wood (aged 10 years under natural conditions) by long-term loading at a very low level of stress ($0.20 \ \sigma_{rupture}$) were made via the ultrasonic method (Bucur and Ghelmeziu, 1977; Bucur, 1980a). The reason for selecting this level of stress is given in Figure 7.19, in which the relationship between ultrasonic velocity and stress is given for spruce resonance wood, aged 10 years and having 8% moisture content (Bucur, 1978). This diagram describes the failure phenomena in wood. Four distinct zones were delineated: zone I, below $0.2\sigma/\sigma_{rupture}$; zone II, 0.2 to $0.7\sigma/\sigma_{rupture}$; zone III, $0.7\sigma/\sigma_{rupture}$; zone IV, over $0.9\sigma/\sigma_{rupture}$. An increasing tendency of velocity is observed in the first zone. In the second zone the velocity is almost constant. In the third and fourth zones the velocity decreases with increasing stress level. Nevertheless, the variation of velocity is generated by modifications of wood structure. Dinwoodie (1968), Keith (1971), and Cousins (1974) studied the anatomical and structural deterioration of wood generated by stress level. No microscopic modifications of anatomic structure were observed in the first zone. Chemical deteriorations of cellulosic chains probably occurred, as predicted by Cousins. In the second zone anatomical dislocations were observed, similar to those described by Keith as slip-lines. The coalescence of slip-lines finally determines the crack propagation in anatomic structure and macroscopic failure of wood.

Note that in the region corresponding to $0.2\sigma/\sigma_{rupture}$, the anatomical structure has not deteriorated. Modifications in the structure were probably produced at the finer structural scale [i.e., microfibrils, cellulosic chains, piezoelectric behavior of cellulosic crystal (Bazhenov, 1961)]. The model proposed by Mark (1967) for the cellulosic constituents of microfibrils is able to satisfy the observed phenomena. Accumulations of dislocations in cellulosic chains and coalescence of hemicelluloses due to hydrogen bond activation in more crystalline regions during loading could explain the increase in velocity under very low stress.

To determine the influence of long-term loading on acoustic properties of wood, the analysis of two types of loading was proposed: static bending and repeated bending shocks under Charpy pendulum at a very low stress level, namely $0.2\sigma/\sigma_{rupture}$ (Bucur and Ghelmeziu, 1977). The velocity and attenuation of longitudinal waves were measured with narrow band transducers of 50 kHz.

Typical samples of spruce, fir, and curly maple, $300 \times 20 \times 20$ mm, were cut from violin soundboards with corresponding long axes to the *L* and *R* directions. Specimens with long axes corresponding to the L direction developed stress in *LR* and *LT* planes, whereas specimens corresponding to the *R*

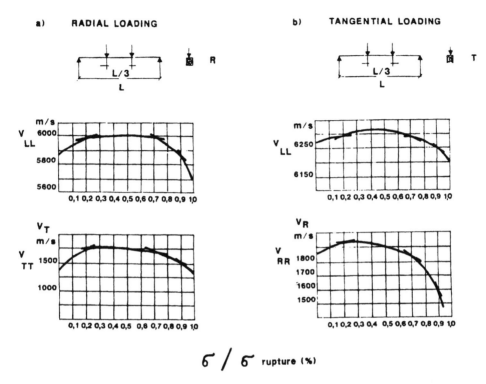

Figure 7.19. Relationship between ultrasonic velocity and stress in spruce resonance wood. (a) Specimen under static bending loading, in radial or tangential direction of the specimen; (b) longitudinal velocity (V_{LL}, V_{TT}) vs. the ratio of stress, at instant t and at the rupture ($\sigma_t/\sigma_{rupture}$) (Bucur, 1979).

direction developed stress in the *RT* and *RL* planes. All anisotropic planes may be analyzed in this manner. In each loading direction, three specimens were tested. Ultrasonic velocity and attenuation were measured as follows: for static bending loading after 1 h, 4 h, 1, 2, 4, 8, 10, 12, 20, 48, and 52 days; for dynamic bending shocks at 50, 100, and 200 to 2000 shocks, with 24 h of relaxation after the first cycle of 1000 shocks (Tables 7.19 and 7.20). Structural modifications were expressed by the variation of velocity and corresponding attenuation of longitudinal waves vs. time or number of shocks, as shown in Figure 7.20 and 7.21.

For static loading, the most affected anisotropic direction was that in which the stress was induced; e.g., for longitudinal loading, all velocities (V_{LL}, V_{RR}, V_{TT}) decreased as compared to initial values (Table 7.19). For radial and tangential loading only V_{LL} diminished, whereas V_{RR} and V_{TT} increased slightly. Attenuation presented an inverse variation with time, not velocity. The fluctuation in acoustic properties is important up to the 12th day of the experiment, after which all acoustical parameters become more or less stable. However, the ratio $(V/V_0)^2$ (where V is the velocity at instant t and V_0 is the velocity at $t = 0$) was chosen as a parameter for the synthetic expression of the structural modifications produced by different types of loading for all anisotropic axes (Figures 7.22 and 7.23). The profile of different curves illustrates the progressive structural transformation of wood. Such behavior observed in specimens could be connected to the fluctuation of the tonal timbre of a new instrument, induced by very fine structural modifications.

7.2.4. INFLUENCE OF VARNISHING

Protection against variable environmental conditions (humidity and temperature) is achieved by coating wood instruments with varnish. It is generally accepted that varnishing improves the appearance of instruments and benefits the acoustical characteristics by modifying (increasing) the mass, stiffness in the radial direction, and internal friction of plates. Measurements were performed on strips, rectangular

Table 7.19. Ultrasonic velocity and attenuation after 52 days of static bending loading at 0.2 $\sigma_{rupture}$ on spruce, fir, and curly maple specimens at 8% moisture content (Bucur and Ghelmezin, 1977).

		Spruce			Fir			Maple		
		Bending loading following the axis								
		L	R	T	L	R	T	L	R	T
		Velocities (m/s)								
V_{LL}	Initial	6150	6300	6200	5980	6000	6050	4800	4350	4400
	Final	6000	6100	6100	5800	5920	5840	4700	4230	4250
V_{RR}	Initial	2340	1940	2100	1680	1720	1700	2380	2360	2170
	Final	2200	1970	2170	1600	1800	1780	2330	2380	2270
V_{TT}	Initial	1500	1470	1470	1160	1300	1200	1680	1550	1520
	Final	1400	1530	1570	1100	1410	1300	1670	1630	1580
		Attenuations (neper/m)								
α_{LL}	Initial	0.46	0.50	0.45	0.57	0.60	0.59	1.26	1.20	1.30
	Final	0.50	0.52	0.52	0.62	0.63	0.69	1.32	1.23	1.36
α_{RR}	Initial	1.60	1.75	1.80	1.40	2.00	1.90	2.70	2.73	2.70
	Final	1.10	1.60	1.40	1.60	1.80	1.40	2.80	2.70	2.65
α_{TT}	Initial	2.40	2.80	3.50	3.90	3.00	3.60	3.20	3.83	3.50
	Final	2.50	2.00	2.60	4.20	2.00	2.90	3.70	2.92	3.30

Table 7.20. Ultrasonic velocity and attenuation after 2000 bending shocks at 0.2 $\sigma_{rupture}$ on spruce, fir, and curly maple specimens at 8% moisture content (Bucur and Ghelmezin, 1977).

		Spruce			Fir			Maple		
		Shocks following the axis:								
		L	R	T	L	R	T	L	R	T
		Velocities (m/s)								
V_{LL}	Initial	6000	6180	6080	5870	5900	5900	4800	4830	4530
	Final	5930	6000	6000	5700	5800	5800	4200	4750	5600
V_{RR}	Initial	1560	1460	1390	1330	1580	1600	2200	1890	1990
	Final	1460	1520	1450	1430	1670	1670	2110	1850	1850
V_{TT}	Initial	1260	1260	1260	1100	1100	1210	1350	1510	1440
	Final	1230	1340	1360	1080	1190	1290	1200	1470	1600
		Attenuation (neper/m)								
α_{LL}	Initial	0.58	0.58	0.63	0.50	0.50	0.59	0.80	0.78	0.70
	Final	0.76	1.02	0.70	0.57	0.57	0.88	1.50	1.12	1.26
α_{RR}	Initial	2.65	2.10	2.30	2.65	3.18	2.84	2.70	2.74	2.76
	Final	3.06	4.30	2.80	4.15	4.60	3.00	3.20	6.74	3.10
α_{TT}	Initial	5.40	5.50	4.70	5.00	4.96	4.10	2.00	3.00	2.00
	Final	5.50	4.90	2.40	5.60	4.30	3.30	6.00	3.80	2.67

plates, and carved violin plates by Schelleng (1968), Barlow et al. (1988), Haines (1980), C.M. Hutchins (1987), M. A. Hutchins (1991), and Ono (1991). As noted in Table 7.21 the internal friction in the R direction is increased by 100%. The anisotropy ratio E_R/E_L is also much increased by varnishing.

The frequency characteristics of plates are related to the nature of varnishing. The stiffness of the coating layers affects spruce more than maple. The effect of the sealer alone and of the sealer plus six layers of varnish can be easily proved by measuring the resonance frequency and the internal friction parameter in the R direction.

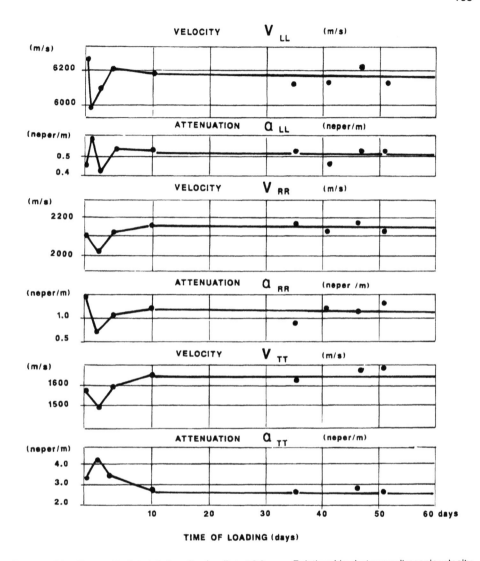

Figure 7.20. Spruce. Radial static bending loading at $0.2\sigma_{rupture}$. Relationships between ultrasonic velocity, attenuation, and time of loading. (From Bucur, *Ann. Sci. For.*, 37(3), 249, 1980a. With permission.)

7.3. CHEMICAL TREATMENTS FOR IMPROVING THE ACOUSTIC PROPERTIES OF COMMON SOLID WOOD USED FOR MASS-PRODUCED INSTRUMENTS

Such treatments (Norimoto et al., 1984 to 1992; Minato et al., 1990; Yano et al., 1988 to 1993) must act on the structural organization of the wood, producing a new material able to support a high acoustic energy conversion rate during playing. Yano and Minato (1992) demonstrated that the efficiency of acoustic energy conversion is proportional to $[(E/\rho^3)Q^{-1}]^{1/2}$ where E is Young's modulus, ρ is density, and Q^{-1} is the quality factor related to internal friction. This means that the requirements (in the longitudinal direction for the top plate of violins and guitars, and piano boards) for high-quality material are high: sound velocity, low internal friction, low density, and high dimensional stability vs. moisture content variation.

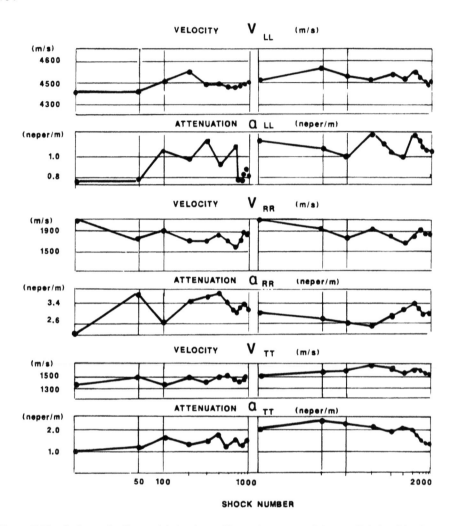

Figure 7.21. Curly maple. Tangential shocks on Charpy hammer at $0.2\sigma_{rupture}$. Relationships between ultrasonic velocity or attenuation and shock number. (From Bucur, V., *Ann. Sci. For.*, 37(3), 249, 1980a. With permission.)

Chemical treatments can act at the cellular and molecular levels. At the cellular level a chemical deposit can be observed on the internal faces of the lumen. At the molecular level we observe the formation of oxymethylene bridges between hydroxyl groups of wood constituents and chemical agents. The second approach seems to be more suitable for the treatment of parts of musical instruments. Acetylation and formaldehyde cross-linking induced the important dimensional stabilization of wood and modified acoustic properties, reducing the internal friction and creep deformation when relative humidity increased from 60% to 90%.

Table 7.22 lists numerical results of the modifications in the physical properties of chemically treated wood. Measurements were performed using the resonance frequency method on strips 120 mm (*L*) × 12 mm (*R*) × 3 mm (*T*) for longitudinal measurements and on 12 mm (*L*) × 120 mm (*R*) × 8 mm (*T*) strips for radial measurements. In the longitudinal direction the Young's modulus did not change significantly, whereas in the radial direction the increase could be 22 to 24%. Decreasing parameters are the internal friction expressed by tan δ (−60%) and the rupture modulus (−20%). The dependence

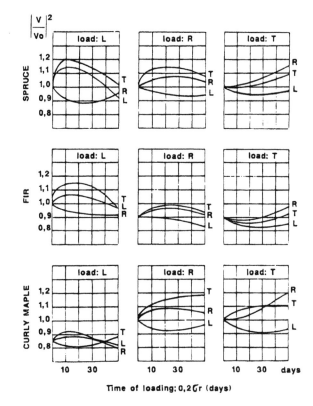

Figure 7.22. Evolution of structural modifications in wood induced by static bending loading at $0.2\sigma_{rupture}$. Relationships between $(V/V\sigma)^2$, where V is the velocity at instant t and $V\sigma$ is the initial velocity, and time of loading (days). (From Bucur, V., *Ann. Sci. For.*, 37(3), 249, 1980a. With permission.)

of acoustic properties on microstructure is evident from the analysis of these numbers. The chemical treatment acts more on amorphous constituents of the cellular wall than it does on crystalline constituents. Also, chemical treatment reduced the shrinkage coefficients by about 40%.

Formaldehyde was applied to the top plate, the soundpost, and the bridge of the violin (Yano and Minato, 1992). Several instruments were built with these parts and spectral analysis was performed to evaluate the violin tonal quality. It was concluded that formaldehyde treatment improves the tonal quality, especially for mediocre instruments. Remarkable results have recently been obtained with resorcin/formaldehyde treatment and with saligenin/formaldehyde treatment (Yano et al., 1994).

Another interesting approach to improve the quality of mass-produced instruments was proposed by Fulton (1991), who treated green wood with ammonia followed by the arching and bending of violin plates. For the proposed procedure the author emphasized several advantages related particularly to the shape of the arch. This shape is strictly related to the elastic properties of the material because of the continuity of the structure (the fibers or other anatomical elements are not cut as they are when the plates are carved). The differences in the vibrational behavior of the bent and carved plates are mainly in the Q factor of different modes, which could be 30% higher in bent than in carved plates. Treatment of violin parts at 140° to 180°C in vacuum for several hours can also markedly improve the acoustic quality (Hix, 1991).

7.4. COMPOSITES AS SUBSTITUTES FOR RESONANCE WOOD

From the previous section and Chapter 6 we have seen that the main requirement of resonance wood for the soundboards of musical instruments are high anisotropy, low density, high value of sound velocity, acoustic radiation and Young's modulus in the longitudinal direction and small damping coefficients in the same longitudinal direction, and relative, small values for other elastic constants.

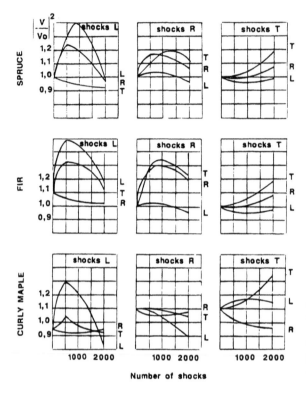

Figure 7.23. Evolution of structural modifications in wood induced by dynamic loading with bending shocks on Charpy pendulum. Relationships between $(V/V\sigma)^2$, where V is the velocity at instant t and $V\sigma$ is the initial velocity, and the shock number. (From Bucur, V., *Ann. Sci. For.*, 37(3), 249, 1980a. With permission.)

Table 7.21. Influence of varnish on Sitka spruce wood for piano: average values

| Parameters | Units | Specimen | | Differences (%) |
		Varnished	Uncoated	
Thickness	mm	2.159	2.056	+5
Density	kg/m³	398	337	+18
Frequency (L)	Hz	870.3	918.2	−5
Frequency (R)	Hz	346.3	205.6	+54
E_L	GPa	7.34	7.22	−3
E_R	GPa	1.15	0.426	+169
E_L/ρ	(m/s)²	18.3	21.6	+18
Velocity (L)	m/s	4277	4647	−9
E_R/ρ	(m/s)²	2.90	1.27	+128
Velocity (R)	m/s	1703	1127	+34
Q_L^{-1}	—	0.0118	0.0079	+49
Q_R^{-1}	—	0.0359	0.0178	+101
E_R/E_L	—	0.159	0.057	+179
Q_R^{-1}/Q_L^{-1}	—	3.04	2.25	+35
Po	dB	−39.8	−41.5	−4

Note: Varnish, 65 μm thick, was applied on both sides of the plate. The principal chemical constituents of varnish for piano were copal resin and toluene for solvent (50–60%), with acetone (10–20%) and ethyl acetate (10–20%).

From Ono, T., *J. Acoust. Soc. Jpn. (E).* 14(6), 397, 1993. With permission.

Table 7.22. Variations of physical properties of Sitka spruce at 20°C and 65% relative humidity induced by different chemical treatments

Direction	Density (kg/m³)	$E_{dynamic}$ (Pa)	tan δ (10^{-3})	E_{static} (Pa)	σ_R (Pa)
		Initial Parameter Values			
L	405	102.9	7.35	113.9	0.849
R	424	7.99	1.99	7.53	0.108
		Ratio of Parameters after Treatment			
		Formaldehyde with SO₂ at 120°C; 12 h; 4 × 10³ mol/dm³			
L	1.04	1.05	0.66	1.08	0.96
R		1.22	0.48	1.35	0.87
		Formaldehyde with SO₂ at 120°C; 24 h; 4 × 10³ mol/dm³			
L	1.05	1.03	0.63	1.06	0.82
R	1.05	1.24	0.48	1.40	0.95

Data from Yano et al., *Mokuzai Gakkaishi*, 36(11), 923, 1990. With permission.

Several authors (Hutchins, 1975, 1978; Haines, 1979) related high resonance wood anisotropy to the ratio of Young's moduli ($E_L:E_R \geq 8$ to 12) or to the ratio of Young's modulus and shear modulus ($E_R:G_{RT} \geq 23$) (Okano, 1991). As noted by C. M. Hutchins (1992), the high ratio of $E_L:E_R$ "makes possible to tune the plate modes so that mode #2 can be kept high enough to be an octave below mode #5, which all my plate tuning shows is the most desirable frequency relationship of these two modes". The ratio of elastic constants also can be related to the shape of the arch, "a smaller ratio for flat arches and a larger ratio for high arches" (Hutchins, 1992).

Bearing in mind the increasing scarcity of resonance spruce for the industrial production of musical instruments, efforts were made toward assessing suitable substituting materials (Haines et al., 1975; Haines, 1980; Holz, 1979b; Firth and Bell, 1988; Beldie and Ghelmeziu, 1968; Sobue, 1981; Decker, 1991; Schumacher, 1988; Vaidelich and Besnainou, 1989). It was reasonable to hypothesize that wood-based composites could be used for the mass-produced instruments. Stress was put on reinforced composites and efforts were made to reproduce the properties of resonance wood in these materials.

Ghelmeziu and Beldie (1968), Holz (1979), and Haines (1980) discussed the possible combinations of veneer sheets in plywood (Tables 7.23 and 7.24). The set of measurements performed with the frequency resonance method show that the velocities on plywood plates are in the range of the values measured on solid spruce. The main difference between solid wood and wood-based composites is induced by the presence of the adhesive layer, which considerably increases the density (650 kg/m³). Consequently, the Young's moduli are increased as is the damping constant. Holz (1979) analyzed the damping constant of triple-layered plywood in only one principal direction in the frequency range 80 to 8000 Hz. A very slight increase in logarithmic decrement is observed up to 200 Hz, which is followed by a dramatic increase in this value for high frequencies.

Table 7.23. Elastic parameters of plywood used in classical guitar manufacture

Density (kg/m³)	Velocity (m/s)		Moduli (10^8 N/m²)				2π tan δ (10^{-2})	
	//	⊥	$E_{//}$	$E_\perp$	$G_{//}$	$G_\perp$	//	⊥
			Spruce Plywood					
520	5100	1700	130	16	2.3	10	26 (995 Hz)	45 (343 Hz)

From Haines, D., *Catgut Acoust. Soc. Newslett.* No. 33, 19, 1980. With permission.

Table 7.24. Possible combinations of veneer sheets in plywood for musical instruments

| Species | Young's moduli (GPa) | | Sound velocity, (m/s) |
	//	⊥	//
Spruce // Spruce ⊥ Spruce //	7.6–11.20	9.4	4590
Spruce // Mahogany ⊥ Spruce //	9.4–12.0	10.8	4320
Spruce // Alder ⊥ Spruce //	11.0–13.1	11.9	4850
Alder // Alder ⊥ Alder //	11.6–14.0	12.9	—
Spruce ⊥ Mahogany // Spruce ⊥	0.8–0.9	0.8	4320
Alder ⊥ Alder // Alder ⊥	1.2–1.7	1.4	—

From Holz, D., *Holztechnologie.* 20(4), 201, 1979a. With permission.

In substitute materials for musical instruments the main difficulty (Haines et al., 1975) arose in finding a core material with sufficiently low damping to match spruce. Balsa and other low density species for core material could be explored further, as one always wants to add damping to the "sandwich structure" of wood-based composites. The utilization of reinforced composites could be considered an important step for a complete revitalization of the musical instrument industry. Further efforts are needed to match the properties of such composites with the very high mechanical and acoustical anisotropy (Bucur, 1990) of resonance wood, and to develop nondestructive techniques for the simultaneous measurement of elastic constants on full-size plates, as suggested by McIntyre and Woodhouse (1984 to 1986, 1988), Caldersmith and Freeman (1990), Tonosaki et al. (1985), and Sobue and Katoh (1990).

APPENDIX

RESONANCE WOOD OF VARIOUS ORIGINS

Velocity measured with ultrasonic technique at 1 MHz frequency

| Species | Density (kg/m³) | Ultrasonic velocities (m/s) | | | | | |
		V_{11}	V_{22}	V_{33}	V_{44}	V_{55}	V_{66}
Resonance Spruce (*Picea* spp.)							
P. abies	400	5050	2000	1500	300	1425	1375
P. rubens	485	6000	2150	1600	330	1240	1320
P. sitchensis	370	5600	2150	1450	300	1340	1400
Fiddleback Maple (*Acer* spp.)							
A. pseudoplatanus	670	4600	2500	1870	925	1529	1835
A. platanoides	740	4940	2491	1942	937	1350	1698
A. macrophyllum	600	4500	2340	1550	900	1340	1720
A. saccharum	700	4785	2376	1786	653	1352	1736
A. rubrum	560	3800	2510	1850	740	1450	1750

From Bucur, V., *J. Catgut Acoust. Soc.,* Ser. 1, 47, 42, 1987. With permission.

Diagonal terms of stiffness matrix measured with ultrasonic technique at 1 MHz frequency

Species	Density (kg/m³)	Diagonal terms of stiffness matrix (10⁸ N/m²)					
		C_{11}	C_{22}	C_{33}	C_{44}	C_{55}	C_{66}
Resonance Spruce (*Picea* spp.)							
P. abies	400	102.01	16.00	10.24	0.36	8.12	7.56
P. rubens	485	174.60	22.44	12.42	0.53	7.45	8.46
P. sitchensis	370	116.03	17.10	7.78	0.33	6.64	7.25
Fiddleback Maple (*Acer* spp.)							
A. pseudoplatanus	670	141.34	41.87	23.43	5.73	15.68	22.56
A. platanoides	740	180.59	45.91	27.90	7.20	13.68	21.34
A. macrophyllum	600	121.50	32.85	14.42	4.86	10.77	17.75
A. saccharum	700	160.27	39.52	22.33	2.82	12.79	21.09
A. rubrum	560	80.86	35.28	19.16	3.06	11.77	17.15

From Bucur, V., *J. Catgut Acoust. Soc.,* Ser. 1, 47, 42, 1987. With permission.

Off-diagonal terms of stiffness matrix measured with ultrasonic technique at 1 MHz frequency

Species	Density (kg/m³)	Off-diagonal terms (10⁸ N/m²)		
		C_{12}	C_{13}	C_{23}
Resonance Spruce (*Picea* spp.)				
P. abies	400	17.53	13.20	12.14
P. rubens	485	22.98	16.43	15.46
P. sitchensis	370	14.97	10.26	7.57
Fiddleback Maple (*Acer* spp.)				
A. pseudoplatanus	670	32.49	30.72	18.53
A. platanoides	740	54.17	43.70	16.26
A. macrophyllum	600	12.65	11.39	8.39
A. saccharum	700	43.47	32.19	20.38
A. rubrum	560	11.50	14.49	12.78

From Bucur, V., *J. Catgut Acoust. Soc.,* Ser. 1, 47, 42, 1987. With permission.

Technical terms calculated from ultrasonic technique data at 1 MHz frequency

Species	Density (kg/m³)	Diagonal terms of stiffness matrix (10⁸ N/m²)					
		E_1	E_2	E_3	G_{23}	G_{13}	G_{12}
Resonance Spruce (*Picea* spp.)							
P. abies	400	82.79	1.56	1.03	0.36	8.12	7.56
P. rubens	485	150.86	3.13	1.75	0.53	7.45	8.46
P. sitchensis	370	99.95	9.49	4.30	0.33	6.64	7.25
Fiddleback Maple (*Acer* spp.)							
A. pseudoplatanus	670	98.59	26.55	12.93	5.73	15.68	22.56
A. platanoides	740	89.53	29.08	16.99	7.20	13.68	21.34
A. macrophyllum	600	111.20	27.66	11.71	4.86	10.77	17.75
A. saccharum	700	104.37	19.17	10.97	2.82	12.79	21.09
A. rubrum	560	69.77	19.17	10.97	3.06	11.77	17.15

From Bucur, V., *J. Catgut Acoust. Soc.,* Ser. 1, 47, 42, 1987. With permission.

REFERENCES

Abbott, J. (1987). Mountain Mahogany, a domestic alternate for ebony. *J. Catgut Acoust. Soc.* Ser. 1, No. 48, 32.

Arbogast, M. (1992). L'érable à fibres ondulées et la lutherie. *Rev. For. Fr.* 44, 176–186.

Bamber, R. K. (1964). Musical instruments from native timbers. *Aust. Timber J.* 30, 33–37.

Bamber, R. K. (1988). The properties of wood and musical instruments. *J. Acoust. Assoc. Musical Instr. Makers.* 1(3), 3–9.

Barducci, I. and Pasqualini, G. (1948). Misura dell'attrito interno e delle costanti elastiche del legno. *Nuovo Cimento.* I(5), 416–466.

Bariska, M. (1978): Klangholz, Holzinstrument, Musik. *Naturwiss. Rundsch.* 31, 245–52.

Bariska, M. (1984). Tams-tams et violons. in *Des Forêts pour les hommes.* Payot, Paris, 49–80.

Barlow, C. Y., Edwards, P. P., Millward, R. G., Raphael, A. R., and Rubio, D. J. (1988) Wood treatment used in Cremonese instruments. *Nature.* 332, 313.

Barlow, C. Y. and Woodhouse, J. (1990). Bordered pits in spruce from old Italian violins. *J. Microsc.* 160 (Part 2), 203–211.

Bazhenov, V. A. (1961). *Piezoelectric Properties of Wood.* Consultants Bureau, New York.

Beldie, I. P. (1968). The determination of the modulus of shear of spruce wood. *Holz Roh Werkst.* 26(7), 261–266.

Beldie, I. P. (1975). Vibration and Sound Radiation of the Violin at Low Frequencies. Ph.D. thesis, Berlin (in German).

Benade, A. H. (1976). *Fundamentals of Musical Acoustics.* Oxford University Press, New York.

Bissinger, G. and Hutchins, C. M. (1980). "Q" measurements of mode 2 and mode 5 in a number of violin and viola top and back plates. *Catgut Acoust. Soc. Newslett.* No. 34, 5–7.

Bonamini, G., Chiesa, V., and Uzielli, L. (1991). Anatomical features and anisotropy in spruce wood with indented rings. *Catgut Acoust. Soc. J.* Ser. 3, 1(8), 12–16.

Bond, C. W. (1976). Wood anatomy in relation to violin tone. *J. Int. Wood Sci.* 7(3), 22–26.

Boutillon, X. (1986). The Piano Hammer Action. Paper presented at Catgut Acoust. Soc.-Hartford Int. Symp., July 20 to 23, Hartford, CT.

Bucur, V. (1975). Ageing, a modifying factor of resonance wood properties. *Ind. Lemnului.* 26(4), 169–174 (in Romanian).

Bucur, V. and Ghelmeziu, N. (1977). The influence of long term loading on acoustic properties of resonance woods. *Ind. Lemnului.* 28(4), 171–180 (in Romanian).

Bucur, V. (1979). Wood failure testing in ultrasonic method. In *Proc. 4th Nondestructive Testing of Wood Symp.* Vancouver, WA, August 28 to 30, 223–226.

Bucur, V. (1980a). Modifications des propriétés acoustiques du bois de résonance sous l'effet de sollicitations de longue durée. *Ann. Sci. For.* 37(3), 249–264.

Bucur, V. (1980b). Anatomical structure and some acoustical properties of resonance wood. *Catgut Acoust. Soc. Newslett.* No. 33, 24–29.

Bucur, V. (1983). Vers une appréciation objective des propriétés des bois du violon. *Rev. For. Fr.* 35, 130–137.

Bucur, V. (1984). Le profil densitométrique du bois de résonance et les corrélations des composantes de la densité avec la vitesse des ultrasons. Unpublished data.

Bucur, V. (1987). Varieties of resonance wood and their elastic constants. *J. Catgut Acoust. Soc.* Ser. 1, No. 47, 42–48.

Bucur, V. (1990). About the anisotropy of resonance wood for violins. in *Fortschr. Akust. DAGA '90 (Deutsche Arbeitsgemeinschft für Akustik).* Teil A. Technische Universtät Wien, Vienna, April 9 to 12, 553–557.

Bucur, V. (1992). La structure anatomique du bois d'érable ondé. *Rev. For. Fr.* 44, 3–8.

Bucur, V. and Chivers, B. (1991). Acoustic properties and anisotropy of some Australian wood species. *Acustica.* 74, 69–74.

Caldersmith, G. and Freeman, E. (1990). Wood properties from sample plate measurements. I. *Catgut Acoust. Soc. J.* Ser. 2 1(5) 8–12.

Caussé, R. (1992). Mise en point des archets numériques. Utilisation pour l'e'tude de la corde frottée. Thèse Doctorat d'Etates Sci. Physiques, Université du Maine, Le Mans, France.

Chaigne, A. and Askenfelt, A. (1994). Numerical simulation of piano strings. I. A physical model for struck string using finite difference methods. *J. Acoust. Soc. America*, 95(2), 1112–1118.

Chiesa, V. (1987). Influenza dell'intensita delle introflessioni sui parametri elastici del legno di abete rosso "di risonanza". Tesi di laurea in SCIENZE FORESTALI, UNIVERSITA di FIRENZE.

Corona, E. (1980). Richerche dendrochronologiche su due violini del XVIII secolo. *Ital. For. Mont.* 35, 112–115.

Corona, E. (1987). La dendrocronologia nella datazione degli strumenti musicali. in Per una carta Europea del restauro. Conservazione, restauro e riuso degli strumenti musicali antichi. Atti del Convegno Internazionale, Venice, October 16–19, 1985. Eds., E. Ferrari Barassi and M. Laini, p. 159–169.

Corona, E. (1988). Dendrocronologia. in *100 Ani di Ricerche Botaniche in Italia, 1888–1988*. Societa Botanica Italiana, Firenze, 891–901.

Corona, E. (1990a). Anomalie anulari nell'abete rosso di risonanza. *Ital. For. Mont.* 45(5), 393–397.

Corona, E. (1990b). Note dendrocronologiche sugli strumenti dell'istituto della PIETA di VENEZIA. STRUMENTI MUSICALI DELL'ISTITUTO DELLA PIETA DI VENEZIA. 83–87.

Cousins, J. (1974). Effects of strain-rate on the transverse strength of *Pinus radiata* wood. *Wood Sci. Tech.* 8(4), 307–321.

Cremer, L. (1983). *Physik der Geige*. S. Hirzel, Stuttgart; *The Physics of the Violin*. Massachusetts Institute of Technology Press; Cambridge, (1984).

Decker, J. A. (1991). Commercial Graphite Acoustic Guitars. CASA '91, 9th Int. Symp. Musical Acoustics. Annapolis, MD, May 3 to 5.

Delgado, O. A., Vazquez Montano, V., Carmona Valdovinos, T. F., and Robles de Benito, M. (1983). Fabricacion y Restauracion de Instrumentos Musicales Clasicos de Cuerda. Part I. Arcos. Instituto Nacional de Investigaciones sobre Recursos Bioticos, Xalapa, Veracruz, Mexico.

Delune, L. (1977). Le bois dans les industries de la musique. *Rev. For. Fr.* 29(2), 143–149.

Dettloff, J. A. (1985). Statistical relationships between acoustic parameters of violin tonewoods. *J. Catgut Acoust. Soc.* Ser. 1, No. 43, 13–15.

Deschamps, R. (1973). Note préliminaire concernant l'identification anatomique des espéces de bois utilisées dans la fabrication des xylophones de l'Afrique Centrale. *Africa-Tervuren.* 20(3), 61–66.

Dinwoodie, J. M. (1968). Failure in timber. *J. Int. Wood Sci.* 21, 37–53.

Douau, D. (1986). Evaluation des Propriétés Acoustiques, Mécaniques et Structurelles des Bois de Tables d'Harmonie de Guitare; leur Influence sur le Timbre de l'Instrument. Thesis, University of Maine, France.

Dunlop, J. I. (1988). Acoustic properties of timber. *J. Assoc. Aust. Musical Instr. Makers.* 1(3), 11–12.

Dunlop, J. I. (1989). The acoustic properties of wood in relation to stringed musical instruments. *Acoust. Aust.* 17(2), 37–40.

Dunlop, J. and Shaw, M. (1991). Acoustical properties of some Australian woods. *Catgut Acoust. Soc. J.* Ser. 2, No. 1(7), 17–20.

Dünnwald, H. (1990). An extended method of objectively determining the sound quality of violins. *Acustica.* 71, 269–276.

Eban, G. (1981). Musical instrument wood—a luthier's view. *Catgut Acoust. Soc. Newslett.* No. 36, 8–10.

Eban, G. (1991). Wood for Guitar Backs, Sides and Bridges. CASA '91, 9th Int. Symp. Musical Acoustics. Annapolis, MD, May 3 to 5.

Ek, L. and Jansson, E. (1985). Vibration and electroacoustical methods applied to determine vibration properties of wooden "blanks" for violin plates. *J. Catgut Acoust. Soc.* Ser. 1, No. 44, 16–22.

Fengel, D. and Wegener, G. (1989). *Wood. Chemistry, Ultrastructure, Reactions*. Walter de Gruyter, Berlin.

Firth, I. M. and Bell, A. J. (1988). Acoustic effects on wood veneering on the soundboards of harps. *Acustica.* 66(2), 113–116.

Firth, I. M. (1988). Temperature of the top plate of a guitar in playing position. *J. Catgut Acoust. Soc.* Ser. 2, No. 1(1) 30.

Fisher, P. (1986). Materials and acoustics. A guitar maker's view. *Proc. I.O.A.* 18(1), 113–115.

Fletcher, N. H. and Rossing, T. D. (1991). *The Physics of Musical Instruments*. Springer-Verlag, New York.

Fryxell, R. E. (1964). Influence of moisture on the behaviour of stringed instruments. *Catgut Acoust. Soc. Newslett.* No. 2, 8–9.

Fryxell, R. E. (1981). Further observations on moisture breathing of wood. *Catgut Acoust. Soc. Newslett.* No. 35, 7–11.

Fukada, E., Yasuda, S., Kohara, J., and Okamoto, H. (1956). Dynamic Young's modulus and piezoelectric constants of old timbers. *Bull. Kabayasi Inst. Phys. Res.* 6, 104–107.

Fukada, E. (1965). Piezoelectric effect in wood and other crystalline polymers. in *Proc. 2nd Symp. NDT of Wood.* Washington State University, Pullman, 143–170.

Fukazawa, K. and Ohtani, J. (1984). Indented rings in Sitka spruce. *Proc. Pac. Reg. Wood Anat. Conf.* Sodo, S. (Ed.). October 1 to 7, Tsukuba, Ibaraki, 25–27.

Fulton, W. (1991). The bent violin top and back plate. CASA '91, 9th Int. Symp. Musical Acoustics, Annapolis, MD, May 3 to 5.

Gadd, C. (1983). Stress relaxation of wood *J. Catgut Acoust. Soc.* No. 41, 17–18.

Ghelmeziu, N. and Beldie, I. P. (1972). On the characteristics of resonance spruce wood. *Catgut Acoust. Soc. Newslett.* No. 17, 10–16.

Ghelmeziu, N. and Beldie, I. P. (1969). Wood based composites for guitars. *Bull. Inst. Polytech. Brasov.* Ser. B, Vol. 11, 359–368 (in Romanian).

Gough, C. E. (1980). The resonant response of a violin G-string and the excitation of the Wolf tone. *Acustica.* 44(3), 113–123.

Gough, C. E. (1981). The theory of string resonance on musical instruments. *Acustica.* 49(2), 124–141.

Gough, C. E. (1987). Microcomputers for acoustic measurement and violin assessment. *J. Catgut Acoust. Soc.* Ser. 1, No. 48, 4–9.

Haines, D. and Chang, N. (1975). Application of graphite composites in musical instruments. *Catgut Acoust. Soc. Newslett.* No. 23, 13–25.

Haines, D., Hutchins, C. M., Hutchins, M. A., and Thompson, D. A. (1975). A violin and a guitar with graphite-epoxy composite soundboards. *Catgut Acoust. Soc. Newslett.* No. 24, 25–28.

Haines, D. (1979). On musical instrument wood. I. *Catgut Acoust. Soc. Newslett.* No. 31, 23–32.

Haines, D. (1980). On musical instrument wood. II. Surface finishes, plywood, light and water exposure. *Catgut Acoust. Soc. Newslett.* No. 33, 19–23.

Hall, D. E. (1992). Piano string excitation. *J. Acoust. Soc. Am.* 92(1), 95–105.

Hancock, M. (1989). Dynamics of musical strings. *J. Catgut Acoust. Soc.* Ser. 2, 1(3), 33–45.

Hancock, M. (1994). Personal communication.

Harajda, H. and Poliszko, S. (1971). Essai de détermination de l'influence de quelques caractéristiques techniques de la table de résonance d'épicéa sur certains paramétres des sons amplifiés. *Folia For. Pol. Ser. B:.* No. 10, 19–34.

Harris, J. M. (1988). *Spiral Grain and Wave Phenomena in Wood Formation.* Springer-Verlag, Berlin.

Hase, N. and Okuyama, T. (1986). Stress grading of piano action parts. *Wood Ind. (Japan).* 41(5), 15–17.

Hase, N. and Okuyama, T. (1987). Quality control of piano action parts. II. The grading of parts based on modulus of elasticity and weight. *Mokuzai Gakkaishi.* 33(2), 108–114.

Hase, N. (1987). A comparison between acoustic physical factors of Honduras rosewood for marimbas and xylophones and sensory evaluation of these instruments. *Mokuzai Gakkaishi.* 33(10), 762–768.

Hase, N., Sasaki, Y., and Suzuki, S. (1979). Strength properties of laminated wood hammershanks. III. *Mokuzai Gakkaishi.* 35(8), 703–709.

Heinrich, J. M. (1991). "Arundo donax" das Holz mit dem wir leben müssen. Kleiner Traktat über Zusammenhänge zwischen Holzstruktur und Spieleigenschaften der Mundstücke von Oboe, Klarinette und Fagott. in *Tibia.* Vol. 4. Moeck Verlag, Celle, Germany, 610–621.

Hix, P. (1991). Personal communication.

Holz, D. (1966). Untersuchungen an Resonanzhölzern. I. Beurteilung von Fichtenresonanzhölzern auf der Grundlage der Rohdichteverteilung und Jahrringbreite. *Arch. Forstwes.* 15(11/12), 1287–1300.

Holz, D. (1967a). Untersuchungen an Resonanzhölzern. II. Beurteilung von Resonanzholz der Oregon pine (*Pseudotsuga menziessii*) auf der Grundlage der Rohdichteverteilung über dem Stammquerschnitt sowie des Harzgehaltes. *Arch. Forstwes.* 16(1), 37–50.

Holz, D. (1967b). Untersuchungen an Resonanzholz. III. Über die gleichzeitige Bestimmung des dynamischen Elastizitätsmoduls und der Dämpfung an Holzstäben im hörbaren Frequenzbereich. *Holztechnologie.* 8(4), 221–224.

Holz, D. and Schmidt, J. (1968). Untersuchungen an Resonanzholz. IV. Über den Zusammenhang zwischen statisch und dynamisch bestimmten Elastizitätsdmodulen und die Beziehung zur Rohdichte bei Fichtenholz. *Holztechnologie.* 9(4), 225–229.

Holz, D. (1973). Untersuchungen an Resonanzholz. V. Über bedeutsame Eigenschaften nativer Nadel— und Laubhölzer im Hinblick auf mechanische und akustische Parameter von Piano— Resonanzböden. *Holztechnologie*. 14(4), 195–202.

Holz, D. (1974). On some important properties of non-modified coniferous and broad leaved woods in view of mechanical and acoustical data in piano soundboards. *Arch. Akustyki*. 9(1), 37–57.

Holz, D. (1979a). Untersuchungen zum Einfluss von Klebfugen und schichten auf die akustisch wichtigen Eigenschaften von Resonanzplatten aus Voll- und Lagenholz. *Holztechnologie*. 20(4), 201–206.

Holz, D. (1979b). Investigations on a possible substitution of resonant wood in plates of musical instruments by synthetic materials. *Arch. Acoust*. 4(4), 305–316.

Holz, D. (1979c). Über einige Zusammenhänge zwischen E-Modul, Dämpfung und strukturellem Aufbau bei GUP-Laminaten mit orientierten Verstärkungsmaterialen. *Plaste Kautsch*. 26(4), 206–210.

Holz, D. (1981). Zum Alterungsverhalten des Werkstoffes Holzeinige Ansichten, Untersuchungen, Ergebnisse. *Holztechnologie*. 22(2) 80–85.

Holz, D. (1984). On some relations between anatomic properties and acoustical qualities of resonance wood. *Holztechnologie*. 25(1), 31–36.

Hutchins, C. M. (1962). The physics of violins. *Sci. Am.* November, 79–93.

Hutchins, C. M. (Ed.) (1975). *Benchmark Papers in Acoustics. Musical Acoustics. Part I. Violin Family Components. Part II. Violin Family Functions.* Dowden, Hutchinson & Ross, Stroudsburg, PA.

Hutchins, C. M. (1978). Wood for violins. *Catgut Acoust. Soc. Newslett.* No. 29, 14–18.

Hutchins, C. M. (1981). The acoustics of violin plates. *Sci. Am.* 245, 171–186.

Hutchins, C. M. (1982). Problems of moisture changes when tuning violin plates. *Catgut Acoust. Soc. Newslett.* No. 37, 25.

Hutchins, C. M. (1983). A history of violin research. *J. Acoust. Soc. Am.* 73(5), 1421–1440.

Hutchins, C. M. (1987). Effects of five years of filler and varnish seasonings on the eigenmodes in four pairs of viola plates. *J. Catgut Acoust. Soc.* Ser. 1, No. 48, 25–26.

Hutchins, C. M. (1991). Body length found to be primary factor controlling tone quality between violins, violas, cellos and basses. *Catgut Acoust. Soc. J.* Ser. 2, No. 1(7), 36–38.

Hutchins, C. M. (1992). Personal communication.

Hutchins, C. M. and Rogers, O. E. (1992). Methods of changing the frequency spacing between the A1 and B1 modes of the violin. *Catgut Acoust. Soc. J.* Ser. 2, No. 2(1), 13–19.

Hutchins, M. A. (1981). Acoustical parameters for violin and viola-back wood. *Catgut Acoust. Soc. Newslett.* No. 36, 29–31.

Hutchins, M. A. (1983). Physical measurements on sampling of European spruce and maple for violin top and back plates. *Catgut Acoust. Soc. Newslett.* No. 40, 28–30.

Hutchins, M. A. (1984). Study of the physical parameters of 7 species of spruce and other softwoods with relation to their suitability for violin top plates (abstract). *Catgut Acoust. Soc. Newslett.* Ser. 1, No 41, 13.

Hutchins, M. A. (1984). Physical parameters of Chinese and Australian woods in relation to their use in violin making. Unpublished data.

Hutchins, M. A. (1991). Effects on spruce test strips of four-year application on four different sealers plus oil varnish. *Catgut Acoust. Soc. J.* Ser. 2, No. 1(7), 11–16.

Ille, R. (1975). Properties and processing of resonance spruce for making master violins. *Holztechnologie*. 16(2), 95–101 (in German).

Jahnel, F. (1981). *Manual of Guitar Technology.* Verlag Das Musikinstrument, Frankfurt/Main.

Janson, E. V., Molin, N. E., and Sundin, H. (1970). Resonances of violin body and acoustical methods. *Phys. Scr.* 2, 243–256.

Janson, E. V. (Ed.). (1983). Function, Construction and Quality of Guitar. Publ. No. 38. Royal Swedish Academy of Music, Stockholm.

Janson, E. V. and Niewczyk, B. (1989). Experiments with violin plates and different boundary conditions. *J. Acoust. Soc. Am.* 86(3), 895–901.

Janson, E. V., Molin, N. E., and Saldner, H. O. (1994). On eigenmodes of the violin. Electronic holography and admittance measurements. *J. Acoust. Soc. Am.* 95(2), 1100–1105.

Kahle, E. and Woodhouse, J. (1989). Modelling the microstructure of wood (abstract). *J. Catgut. Acoust. Soc.* Ser. 2, No. 1(4), 42.

Kataoka, A. and Ono, T. (1976). The dynamic mechanical properties of Sitka spruce used for sounding boards. *Mokuzai Gakkaishi*. 22(8), 436–443.

Keith, C. T. (1971). The anatomy of compression failure in relation to creep inducing stress. *Wood Sci.* 4(2), 71–82.

Klein, P., Mehringer, H., and Bauch, J. (1986). Dendrochronological and biological investigations on string instruments. *Holzforschung.* 40(4), 197–203.

Koberle, M. and Majek, M. (1978). Kiln drying of exotic wood species for musical instruments. *DREVO.* (33), 178–180.

Kondo, M., Hirano, N., Hirota, M., and Kubota, H. (1980). Principal Directions of Elasticity of a Bridge on Holographic Maps of the Soundbox Excited in Those Directions. 10th Cong. Acoustics, Sydney, K-7, 3.

Leipp, E. (1965). *Le Violon.* Editions Hermann, Paris.

Leipp, E. (1976). *Acoustique et Musique.* Masson, Paris.

Leonhardt, K. (1969). *Geigenbau und Klangfrage.* Verlag Das Musikinstrument, Frankfurt/Main.

Manasevici, A. D. (1962). Etude sur la résistance du bois de pin par sollicitation de résonance. *Lesnoi J.* 5(6), 106–113.

Marchal, K. D. (1985). Modal analysis of a violin. *J. Acoust. Soc. Am.* 77(2), 695–709.

Mark, R. (1967). *Cell Wall Mechanics of Tracheids.* Yale University Press, New Haven, CT.

Marok, M. (1990). Characteristic acoustic impedance of wood as a stochastic medium. *Acustica.* 72(2), 148–150.

McIntyre, M. E. and Woodhouse, J. (1988). Overview 66. On measuring the elastic and damping constants of orthotropic sheet materials. *Acta Metall.* 36(6), 1397–1416.

McIntyre, M. E. and Woodhouse, J. (1984–1986). On measuring wood properties. *J. Catgut Acoust. Soc.* Part I, 1984, No. 42, 11–15; Part II, 1985, No. 43, 18–24; Part III, 1986, No. 45, 14–24.

McLennan, J. E. (1980). The violin as a structure, a consideration of the static force in the instrument. *Catgut Acoust. Soc. Newslett.* No. 34, 15–20.

Minato, K. and Yano, H. (1990). Improvement of dimensional stability and acoustic properties of wood for musical instruments by sulfur dioxide catalyzed formalization. *Mokuzai Gakkaishi.* 36(5), 362–367.

Minato, K., Yasuda, R., and Yano, H. (1990). Improvement of dimensional stability and acoustic properties of wood for musical instruments with cyclic oxymethylenes. I. Formalization with trioxane. *Mokuzai Gakkaishi.* 36(10), 860–866; Part II. Formalization with tetraoxane. *Mokuzai Gakkaishi.* 36(11), 990–996.

Meyer, J. (1983). Acoustics and Performance of Music. Verlag E. Bochinsky, Das Musikinstrument, Frankfurt-am-Main.

Meyer, J. (1984). Tonal quality of violins (abstract). *Catgut Acoust. Soc. Newslett.* No. 41, 10.

Molin, N. E. and Jansson, E. V. (1989). Transient wave propagation in wooden plates for musical instruments. *J. Acoust. Soc. Am.* 85(5), 2179–2184.

Muller, H. A. and Geissler, P. (1984). Modal analysis applied to instruments of the violin family (abstract). *Catgut Acoust. Soc. Newslett.* No. 41, 12.

Muller, H. A. (1986). How the violin makers choose the wood and what the procedure means from a physical point of view (abstract). *J. Catgut. Acoust. Soc.* Ser. 1, No. 46, 41, and also the manuscriptum of the talk at Catgut Acoustical Symposium, July 20–23, 1986, Hartford, CT.

Nakamura, I. (1993). Vibrational and acoustic characteristics of soundboard. (Acoustical research on the piano. Part III). *J. Acoust. Soc. Jpn. (E).* 16(6), 429–439.

Niedzielska, B. (1972). Investigations on the interaction of anatomical elements of spruce and its acoustical properties. *Acta Agrar. Silvestria.* 12, 85–100.

Norimoto, M., Ono, T., and Watanabe, Y. (1984). Selection of wood used for soundboards. *J. Soc. Rheol. Jpn.* 12(3), 115–119.

Norimoto, M., Tanaka, F., Ohogama, T., and Ikimune, R. (1986). Specific dynamic Young's modulus and internal friction of wood in longitudinal direction. *Wood Res. Tech. Notes, Kyoto.* 2, 53–63.

Norimoto, M., Gril, J., and Rowell, R. M. (1992). Rheological properties of chemically modified wood: relationships between dimensional and creep stability. *Wood Fiber Sci.* 24(1), 25–35.

Ohta, M., Okano, T., and Asano, I. (1984). Vibrational properties of marimba soundboards related with its grain orientation and density variation. *Proc. Pac. Reg. Wood Anat. Conf.* Tsukuba, October 1 to 7, 25–27.

Okano, T. (1991). Acoustic properties of wood. *Mokuzai Gakkaishi.* 37(11), 991–998.

Ono, T. (1981). Relationship of the selection of wood used for piano soundboards to the dynamic mechanical properties. *J. Soc. Mater. Sci. Jpn.* 30(334), 719–724.

Ono, T. (1983). On dynamic mechanical properties in the trunks of woods for musical instruments. *Holzforschung* 37, 245–250.

Ono, T. (1993). Effects of varnishing on acoustical characteristics of wood used for musical instrument soundboards. *J. Acoust. Soc. Jpn.* (E) 14(6), 397–407.

Ono, T. and Kataoka, A. (1979). The frequency dependence of the dynamic modulus and internal friction of wood used for the soundboards of musical instruments. I. Effect of rotary inertia and shear on the flexural vibration of free-free beams. *Mokuzai Gakkaishi* 25(7), 461–468; II. The dependence of Young's modulus and internal friction on frequency. *Mokuzai Gakkaishi* 25(8), 535–542.

Ono, T. and Norimoto, M. (1983). Study on Young's modulus and internal friction of wood in relation to the evaluation of wood for musical instruments. *Japan. J. Appl. Phys.* 22(4), 611–614.

Ono, T. and Norimoto, M. (1984). On physical criteria for the selection of wood for soundboards of musical instruments. *Rheologica Acta* 23, 652–656.

O'Toole, B. and Gilet, G. (1983). Making music with Australian trees. *Forest, Timber.* 19, 25–27.

Parmantier, J. L. (1981). Le grand livre international du bois. (Adaptation française). Paris, Fernand Nathan, 276–417.

Pearson, F. G. O. and Weraiter, C. (1967). Timber Used in the Musical Instrument Industry. FOR. PROD. LAB. Ministry of Technology, London.

Pishik, I. I., Fefilon, V. V., and Burkovskaya, V. I. (1971). Chemical composition and chemical properties of new and old wood. *Lesnoi J.* 14(6), 89–93 (in Russian).

Reinicke, W. and Cremer, L. (1970). Application of holographic interferometry to the body of stringed instruments. *J. Acoust. Soc. Am.* 48, 988–992.

Richardson, B. E. (1986). Wood for the guitar. *Proceedings of the Institute of Acoustics U.K.,* 8(1), 107–112.

Richardson, B. E., Roberts, G. W., and Walker, G. P. (1987). Numerical modelling of two violin plates. *J. Catgut Acoust. Soc.* No. 47, 12–16.

Richardson, B. E. (1988). Vibrations of stringed musical instruments. *Univ. Wales Rev.* No. 3, 13–20.

Richter, H. G. (1988). *Holz als Rohstoff für den Musikinstrumentenbau.* No. 4043. Moeck Verlag, Celle, Germany.

Rocaboy, F. and Bucur, V. (1990). About the physical properties of wood of twentieth century violins. *Catgut Acoust. Soc. J.* Ser. 2 No. 1(6), 21–28.

Rodgers, O. E. (1986). Initial results on finite element analysis of violin backs. *J. Catgut Acoust. Soc.* Ser. 1, No. 46, 18–23.

Rodgers, O. E. (1991). Influence of local thickness changes on violin top plate frequencies. *Catgut Acoust. Soc. J.* 1(7), 6–10.

Rossing, T. D. (1981). Physics of guitars, an introduction. *J. Guitar Acoust.* 4, 45–67.

Rubin, C. and Farrar, D. F., Jr. (1987). Finite element modelling of violin plate vibrational characteristics. *J. Catgut Acoust. Soc.* Ser. 1, No. 47, 8–11.

Schelleng, J. C. (1963). The violin as a circuit. *J. Acoust. Soc. Am.* 35(3), 326–338.

Schelleng, J. C. (1968). Acoustical effects of violin varnish. *J. Acoust. Soc. Am.* 44, 1175–1183.

Schelleng, J. C. (1982). Wood for violins. *Catgut Acoust. Soc. Newslett.* No. 37, 8–19.

Schleske, M. (1990). Speed of sound and damping of spruce in relation to the direction of grains and rays. *Catgut Acoust. Soc. J.* 1(6), 16–20.

Schmidt-Vogt, H. (1986). Die Fichte. (The Spruce). Vol. 1 and 2, Parey, Hamburg, Germany.

Scholes, P. (1972). *The Concise Oxford Dictionary of Music.* Oxford University Press, London.

Schumacher, R. T. (1988). Compliances of wood for violin top plates. *J. Acoust. Soc. Am.* 84(4), 1223–1235.

Schweingruber, F. W. (1983). *Der Jahrring: Standort, Methodik, Zeit und Klima in der Dendrochronologie.* Verlag Paul Haupt, Bern, Switzerland.

Shigo, L. A. and Roy, K. (1983). *Violin Woods, A New Look.* University of New Hampshire Press, Durham.

Sobue, N., Nakano, H., and Asano, I. (1984). Vibrational properties of spruce plywood for musical instruments. *Mokuzai Gakkaishi.* 31(1), 93–97.

Sobue, N. and Katoh, A. (1990). Simultaneous measurements of anisotropic elastic constants of standard full-size plywood by vibration techniques. *Proc. Int. Timber Eng. Conf.* October 23 to 25, Tokyo.

Sobue, N. and Okayasu, S. (1992). Effects of continuous vibration on dynamic viscoelasticity of wood. *J. Soc. Mater. Sci. Jpn.,* 41(461), 164–169.

de Souza, M. R. (1983). Classification of Wood for Musical Instruments. Ser. Tech. No. 6. Departamento de Pesquisa, Instituto Brasileiro de Desenvolvimento Florestal, Brazil.

Suzuki, H. (1986). Vibration and sound radiation of a piano soundboard. *J. Acoust. Soc. Am.* 80(6), 1573–1582.

Suzuki, H. and Nakamura, I. (1990). Acoustics of piano. *Appl. Acoust.* 30, 147–205.

Thompson, R. (1979). The effect of variations in relative humidity on the frequency response of free violin plates. *Catgut Acoust. Soc. Newslett.* No. 32, 25–27.

Tonosaki, M., Okano, T., and Asano, I. (1985). Measurement of plate vibration as a testing method of wood for musical instruments. *Mokuzai Gakkaishi.* 31(3), 152–156.

Vaidelich, S. and Besnainou, C. (1989). About the mechanical properties of wood used in instrument making and their replacement by carbon fibers composites (abstract). Catgut Symp. Musical Acoustics, Mittenwald, Germany, August 19 to 22.

Vintoniv, I. S. (1973a). Effect of the growth conditions of *Acer pseudoplatanus* on the acoustic properties of its wood. *Lesnoi J.* 16(2), 103–105.

Vintoniv, I. S. (1973b). Effect of the growth conditions on the physical and mechanical properties of the wood of sycamore (*Acer pseudoplatanus*) growing in the Ukrainian Carpathians. *Lesnoi J.* 16(5), 154–155.

Vintoniv, I. S. (1973c). Physical and mechanical properties of *Acer pseudoplatanus* growing in Carpathians. *Derevoob. Prom.* 11, 12.

Weinreich, G. and Caussé, R. (1991). Elementary stability consideration for bowed-string motion. *J. Acoust. Soc. Am.,* 89(2), 887–895.

Woodhouse, J. (1986). Spruce for soundboards. Elastic constants and microstructure. *Proc. Inst. Acoust.* 8(1), 99–105.

Woodhouse, J. (1993a). On the playability of violins. I. Reflexion function. *Acustica.* 78, 125–135, II. Minimum bow force and transients. *Acustica.* 78, 137–153.

Woodhouse, J. (1993b). Idealised models of a bowed string. *Acustica.* 79, 233–250.

Woodhouse, J. (1994). On the stability of bowed string motion. *Acustica.* in press.

Yankovski, B. A. (1967). Dissimilarity of the acoustic parameters of unseasoned and aged wood. *Sov. Phys. Acoust.* 13(1), 3.

Yano, H. and Yamada, T. (1985). Study on the timbre of wood. I. Sound spectrum of wood in radial direction. *Mokuzai Gakkaishi.* 31(9), 719–724.

Yano, H., Mukudai, J., and Norimoto, M. (1988). Improvements in the piano pin-block. *Mokuzai Gakkaishi.* 34(2), 94–99.

Yano, H. and Mukudai, J. (1989). Acoustic properties in the radial direction in Sitka spruce used for piano soundboards. *Mokuzai Gakkaishi.* 35(10), 882–885.

Yano, H., Kanou, N., and Mukudai, J. (1990a). Changes in acoustic properties of Sitka spruce due to saligenin treatment. *Mokuzai Gakkaishi.* 36(11), 923–929.

Yano, H., Oonishi, K., and Mukudai, J. (1990b). Acoustic properties of wood for the top plate of guitar. *J. Soc. Mater. Sci. Jpn.* 39(444), 1207–1212.

Yano, H. and Minato, K. (1992). Improvement of the acoustic and hygroscopic properties of wood by a chemical treatment and applications to the violin parts. *J. Acoust. Soc. Am.* 92(3), 1222–1227.

Yano, H., Norimoto, M., and Rowell, R. M. (1993). Stabilization of acoustical properties of wooden musical instruments by acetylation. *Wood Fiber Sci.* 25(4), 395–403.

Yano, H., Kajita, H., Minato, K. (1994). Chemical treatment of wood for musical instruments. *J. Acoust. Soc. Am.,* 96(6), 3380–3391.

Zimmermann, M. H. (1978). Letter to C. M. Hutchins, concerning wood seasoning. *Catgut Acoust. Soc. Newslett.* No. 30, 7.

Chapter 8

Ultrasound as a Complementary Nondestructive Tool for Wood Quality Assessment

8.1. ULTRASOUND AND WOOD QUALITY

The methods employed to evaluate the quality of wood material are based on the assumption that some simple physical properties can be used to give a reasonably good indication of the characteristics which determine it. Relevant to the evaluation of wood products is the definition of a general concept of "defect" or "flaw" as a discontinuity of the structural unity (in solid wood: knots, cracks, decay, etc., and in wood composites: blister, blow, delamination, etc.)

A number of ways of classifying defects that occur in solid wood or in wood-based materials are possible. Practically different processes, from the original growth of the tree down to the operation finishing on the wood material, can and does induce discontinuities into the structure that nondestructive testings could find. One broad grouping which is useful in ultrasonic testing is based on location, whether on the surface (i.e., in solid wood, curly grain, excessive slope of grain, etc.) or the body of the test specimen (i.e., knots, decay, cracks, etc.). Another possible system is to classify defects via the biological process (i.e., reaction wood, juvenile wood, decay, etc.) or the technological process that produced them (i.e., blisters, delaminations in wood composites, etc.).

As defined in the literature (Bond and Saffari, 1984; Wedgewood, 1987), the various stages that are usually taken into consideration during ultrasonic inspection are detection, localization, characterization, and decision to act on it, if the defect is important enough. Ultrasonic techniques can be divided into two families:

- The scattering-based techniques that use travel time, frequency, amplitude ratio, waveform shape, etc. Most of these methods are performed using waves in a frequency range between 50 kHz and 2 MHz (see below).
- The imaging techniques which seek to provide a picture of the discontinuity. Ultrasonic tomography is performed with both velocity and attenuation as contrast producing parameters. These techniques are not very popular for wood products. Tomikawa et al. (1990) developed a system for the detection of heartwood and rotted zones in poles, using transducers of 78 kHz. The equations derived to locate a defect in three-dimensional space using two probes, a transmitter and a receiver, were presented by Mak (1987).

The success of ultrasonic nondestructive methods is related primarily to understanding the phenomenon of ultrasonic wave propagation in testing material, and ultimately to defining how to use the results of the basic research to improve the technology.

8.2. ULTRASONIC METHODS FOR TREES, TIMBER, AND WOOD COMPOSITES

8.2.1. ULTRASOUND AND TREES

8.2.1.1. Detection of the Slope of the Grain

The slope of grain can be produced from twisted grain in trees and bowed logs, or by poorly sawed lumber. The slope of grain is usually expressed as "the angle between the direction of fibers and the edge of the piece" (ASTM D245-81). Also, Bechtel and Allen (1987) noted that "the grain angle refers to the angle between the longitudinal wood elements and the axis of the stem". Harris (1989) cited such important effects of grain angle on wood properties as shrinkage, rate of moisture movement, strength, machining properties of sawn timber, plywood, veneer, etc. Different methods were proposed to measure the overall grain angle on trees, roundwood, sawn timber, veneer, and normal and small sized samples as increment cores. These methods can be classified following the physical principles:

- *Mechanical methods*, based on the geometrical angle estimation, as noted by Ferrand (1982) and Harris (1989)

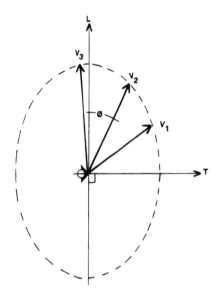

Figure 8.1. Considering wood's elliptical pattern of ultrasonic wave propagation, three measurements of longitudinal velocities (V_1, V_2, V_3) can precisely determine the slope of grain (angle $\varnothing$), bearing in mind that along the fibers the velocity value is at maximum. (Bucur and Perrin, 1990).

- *Dielectrical methods*, based on the dielectric constants measurements, as reported by Baker and Carlson (1978), McLauchlan and Kusec (1978), Bechtel and Allen (1987), Chazelas (1990), James et al. (1985), King (1985), McDonald and Bendtsen (1985, 1986), Martin et al. (1987), Samson (1988), and Portala and Cicotelli (1989)
- *Radiation methods*, as reported by Keller et al. (1974) and Szymani and McDonald (1981)
- *Ultrasonic methods*, as reported by Lee (1958), Foulger (1969), Bucur (1984), Suzuki and Sasaki (1990), and Bucur and Perrin (1989, 1990)

Lee (1958) is believed to be the first to draw attention to the use of the ultrasonic velocity method for the nondestructive systematic estimation of the grain angle in wood. He noted that "ultrasonic pulse measurements may give a rapid guide to the conditions of grain orientation detection such as spiral and interlocking grain which may not be visible from the outer face of a specimen. This may even be possible on standing trees, as some exploratory tests demonstrated with an ultrasonic flow-detector . . ." (Waid and Woodman, 1957). Foulger (1969) proposed an empirical procedure that employed the ultrasonic velocity method to detect the grain angle. He marked reading points for the receiver position on the surface of the tree at 5°, 10°, 15°, 20°, 30° vs. the stem axis. Bearing in mind that the grain direction corresponds to the maximum value of velocity, the slope of grain was estimated with an accuracy of 5°.

Precise determination of the slope of the grain was possible through understanding the mechanism of ultrasonic wave propagation in wood material. The development of the transmission technique for precise measurement of the general grain angle in wood was based on the assessment that the propagation of ultrasonic waves in wood is governed by Christoffel's equations for orthotropic solids.

Simulation of the effect of grain orientation on specimens in three main orthotropic planes was achieved using suitable specimens at 0°, 15°, 30°, 45°, 60°, 75°, and 90° (Bucur, 1984). The characteristic acoustic anisotropy of wood material determines the parameters of the ellipsoidal propagation pattern. Moreover, three velocities (Figure 8.1) can very precisely define the grain angle on this pattern (Bucur and Perrin, 1990), bearing in mind that along the fibers the velocity value is maximum. An accurate prediction of the grain angle depends on the accuracy of the velocity measurements ($<1\%$). For practical purposes an array of transducers could be set on the tested tree or specimen, as in Figure 8.2. The measurements can be performed in green or dry conditions. The results of the grain angle measurements on 50 trees, using the ultrasonic velocity method are given in Figure 8.3. Good agreement with optical measurements is achieved.

For the local variation of the slope of the grain around knots the same theoretical considerations are valid, and we can refer to Chazelas et al. (1988) for the results for softwoods (Figure 8.4). Stiffness (C_{LL}, C_{RR}, C_{TT}) mapping shows the changes in the elastic behavior of wood induced by the presence of

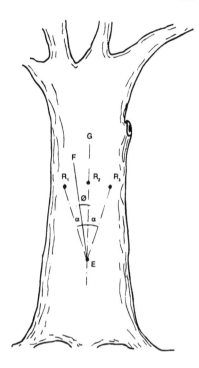

Figure 8.2. Array of transducers for grain angle measurements on trees or samples. (E) Emission point, (∅) grain angle, (F) fibers longitudinal axis, (G) geometric axis of tree, (α) measurement angle, (R) receiving points (Bucur and Perrin, 1990).

tight or loose knots. As for the visual pattern of grain, the stiffness variations reveal the nature of the knot: tight or loose.

8.2.1.2. Detection of Reaction Wood

The *Multilingual Glossary of Terms Used in Wood Anatomy* (Anonymous, 1964) refers to reaction wood observed on softwood as compression wood; the reaction wood observed is on hardwood referred to as tension wood. The presence of reaction wood is strongly associated with juvenile wood. The core of juvenile wood size has 5 to 25 annual rings, depending on species. Reaction wood (compression and tension) was extensively studied by Archer (1986), Timell, (1986), and Zobel and van Buijtenen (1989). The reasons for reaction wood formation have been largely discussed and it appears to be principally a gravitropic response of the tree, completed by hormonal stimuli. If we briefly review some of the more important characteristics of compression wood we can say that in this tissue the tracheids are shorter than in normal wood and the fibril angle is usualy 30° to 50°, which is much flatter than in normal wood. A localized increase in specific gravity associated with a higher lignin content is also observed. Tension wood fibers are longer than those in normal wood, the vessels are fewer and smaller, the cellulose content is higher, the gelatinous layers are present in cellular wall, and the specific gravity is greater than that of normal wood. Juvenile wood also has some unique characteristics: very low specific gravity, shorter cells with larger lumen diameter, thinner cellular walls, higher lignin content, and lower strength than that of mature wood. Attempts to use ultrasound for the characterization of reaction wood were made by Bucur (1987), Bucur et al. (1991), Feeney (1987), Hamm and Lam (1989), and Janin et al. (1990).

The presence of compression wood in lumber produces warp. Hamm and Lam (1989) detected the presence of compression wood on green western hemlock, with slowness plot in polar coordinates, as picks of highest slowness (Figure 8.5). The authors stated that commercial-sized plank specimens with compression wood and normal wood were correctly separated with velocity measurements "based on mean values of the 4 to 8 measurements taken per board". Nondestructive specimens of the increment core type (5 mm diameter), bored from Sitka spruce, were used by Feeney (1987) to detect the presence of compression wood, as well as the boundary between juvenile and adult wood. Precise detection was related to the selection of an appropriate measurement frequency (2 MHz, for a wavelength of 0.5 to 2.5 mm) and corresponding probes. Longitudinal waves propagating along the longitudinal and tangential

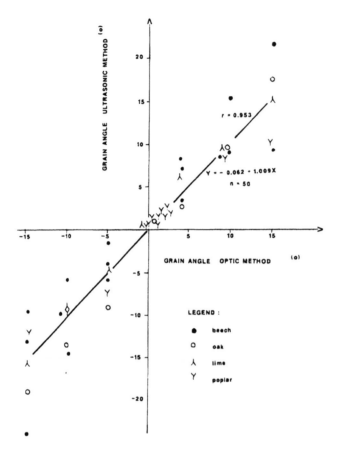

Figure 8.3a.

axes of wood were analyzed. Figure 8.6 shows the variation of longitudinal velocity in the longitudinal anisotropic direction of wood at several high levels on the tree. The continuous increase in velocity was observed in the 1st year and from years 14 to 18 at all height levels. This region is followed by a zone of constant velocity, corresponding probably to the mature wood. Comparison and intersection of the regression lines of the two zones can determine the mature zone.

The pattern of variation displayed in Figure 8.6 can be compared to Tsoumis' results (1968) concerning the increase in tracheid length with annual ring development. It would appear from this statement that the increase in the tracheid length with age is linked to the increase of ultrasonic velocity. This could indicate that the length of the tracheids plays an important role in the ultrasonic path in wood.

Velocity measurements in the tangential direction show a generally decreasing trend. The difference between juvenile and mature wood is less important in the tangential anisotropic direction.

At the fine anatomic scale of earlywood and latewood, the capability of the ultrasonic velocity method to detect the presence of reaction wood in beech, Douglas fir, and pine was proved by using longitudinal ultrasonic waves (Bucur et al., 1991) and narrow band piezoelectric transducers of 45 kHz, which had a very small contact area with the specimen (<1 mm^2). The variation of ultrasonic velocity, in both early- and latewood, from the pith to the bark of Douglas fir, is plotted in Figure 8.7. A detailed examination of the curves shows that the compression wood exhibited slightly low velocities when compared to normal wood. If we take into account the fact that the fiber is the basic anatomic element for ultrasonic wave propagation, the previous statement agrees with the fact that in compression wood the tracheids are shorter than in normal wood, and have a higher lignin content and a flatter microfibril angle. Measurements of Douglas fir green material revealed that ultrasonic velocity values diminished about 20% when compared to air-dried material. However, the pattern of the variation of the ultrasonic

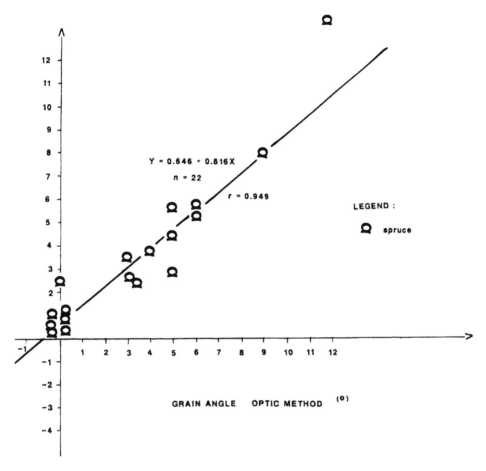

Figure 8.3b. Grain angle on trees by ultrasonic and optic methods (a) on various species and (b) on spruce (Bucur and Perrin, 1991).

velocity from the pith to the bark of green wood is similar to that measured in air-dried material (Bucur et al., 1991).

The polar envelope of velocities in beech with tension wood is homomorphous with those of fiber lengths (Bucur et al., 1991). In the tension wood zone, in which long fibers are located and contain an important amount of cellulose, the highest value of velocity was measured.

The juvenile wood zone can be recognized in Figure 8.8 as a zone in which a continuous increase in the longitudinal velocity V_{LL} is observed (from the ring no. 1 to 9). Measurements of Sitka spruce (Table 8.1) confirmed the observations for pine. The values of V_{LL} and the shear velocities (V_{LR}, V_{LT}, V_{TR}) are higher in adult wood than in juvenile wood.

8.2.1.3. Detection of Pruning Trees

A considerable body of knowledge has been amassed over the past 20 years on the influence of pruning on the quality of wood (Zobel and van Buijtenen, 1989; Flammarion et al., 1990). While it is important to increase this knowledge, it is also important to use it properly to detect the effects of this silvicultural treatment on standing trees. The principal effects of pruning on wood quality are the disappearance of knots, the improvement of the cylindrical shape of the stem of the tree, the reduction of the influence of the juvenile wood, and the improvement of such physical properties as density and shrinkage (Polge et al., 1973). The ultrasonic technique is able to complete the list of physical parameters related to the

182

"STIFFNESS" IMAGE OF KNOTS

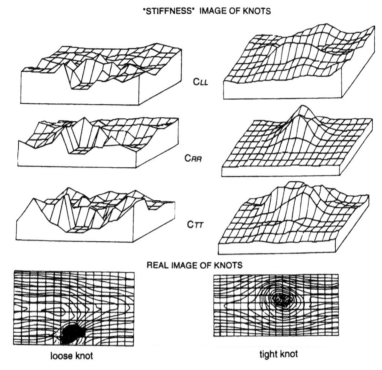

C_{LL}

C_{RR}

C_{TT}

REAL IMAGE OF KNOTS

loose knot   tight knot

Figure 8.4. Mapping of stiffnessess (C_{LL}, C_{RR}, C_{TT}) to show the location, size, and nature of knots (Left) loose knot; (right) tight knot (Chazelas, Vergne, and Bucur 1988).

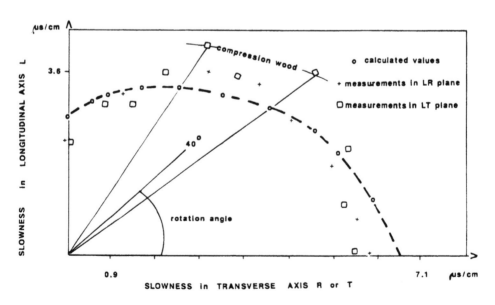

Figure 8.5. Slowness curves in polar coordinates on green western hemlock. Theoretical curves, and measured data in *LT* and *LR* planes. The compression wood was identified in specimens at 35° and 55°. 1989.) (From Hamm, E. A. and Lam, F., *G. Prove Nondestruct.*, 1, 40, 1989. With permission.)

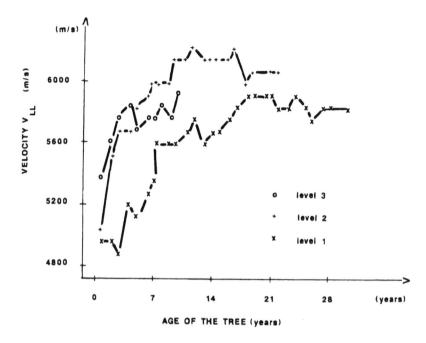

Figure 8.6. Sitka spruce. Variation of longitudinal velocity vs. annual ring, measured on increment core samples. Level 1, defined at "breast height", 1.30 m above the ground; level 2, defined as "mid-point", between breast height and 15 cm, top diameter of the log; level 3, defined as the "top", where the diameter of the trunk has narrowed to 15 cm (Feeny, 1987).

quality of wood with some terms of the stiffness matrix (C_{LL}, C_{RR}, C_{TT}, etc.) measured on increment cores collected from pruned trees and with data about the velocity of propagation of surface waves measured on the stem.

Table 8.2 gives the values of V_{RR} measured along the diameter of the trunk and the surface velocities measured at the periphery of the trunk at 1.30 m, with 20-cm distance between the ultrasonic emitter and receiver. The increasing velocities values measured on pruned trees could be interpreted as the result improving the wood quality. It is also interesting to observe the variation of surface velocity related to the height of the tree (Figure 8.9). The dispersion of values of surface velocity is higher in control trees than in pruned trees. This means that the variability of pruned tree wood quality is smaller than that of the control tree.

Stiffness constants and radiographic density are tabulated (Table 8.3) for pruned and control trees. Note that for wood having almost the same average density the stiffness C_{LL} is 8% greater for pruned trees than for control trees. A possible explanation is based on several structural factors such as the increase in the length of sapwood tracheids, the width of the annual ring and the corresponding latewood zone, or the thickness of the cell wall.

Bearing in mind the results of the research described above, it is possible to suggest that the actual technology of the detection of pruning treatment on trees may be modernized with ultrasound. The acoustic methods developed for trees can also be used for testing poles (Dunlop, 1981), on which internal decay needs to be localized. Jartti (1978), Kaiserlik (1978), and Abbott and Elcock (1987) analyzed nondestructive acoustic methods that were able to predict the strength in degraded wood or in electrical supply poles.

8.2.1.4. Detection of Curly Figures in Trees

The figures in wood originate from grain patterns that give material pleasing decorative characteristics. The pattern known as "wavy" or "curly" figures in maple trees are also named "fiddleback figures" because this wood is utilized in the manufacture of violin backs. This pattern results from an undulation

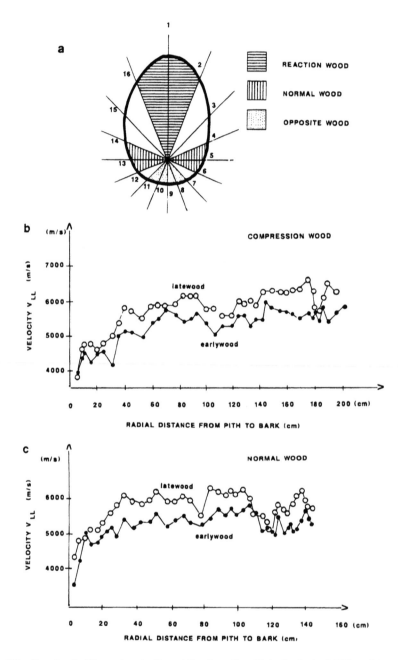

Figure 8.7. Douglas fir. Ultrasonic velocity variation from the pith to the bark at 10% moisture content. (a) Array of 16 points for ultrasonic measurements on the transverse section of a tree; (b) measurements of compression wood; (c) measurements of normal wood (Bucur et al., 1991).

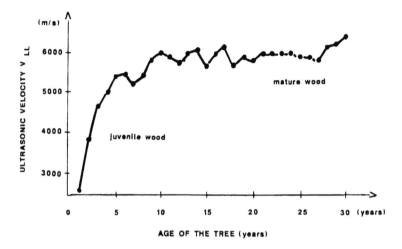

Figure 8.8. Pine. Ultrasonic velocity on juvenile and mature wood at 10% moisture content (Bucur et al., 1991).

Table 8.1. Ultrasonic velocities on juvenile and adult wood of Sitka spruce (Bucur, 1987)

Structure	Ultrasonic velocites (m/s)				
	V_{LL}	V_{TT}	V_{LR}	V_{LT}	V_{TR}
Adult wood	4928	1677	1605	1450	592
Juvenile wood	4894	1758	1535	1448	521

Table 8.2. Ultrasonic velocities on Douglas fir logs, without bark, at 1.30 m over the ground

Parameters	Symbol	Ultrasonic velocity (m/s)			
		Pruned tree		Control tree	
		Mean	Range	Mean	Range
Radial velocity	V_{RR}	1589	1545–1607	1272	1240–1304
Surface velocity in L direction	V_y	6006	5720–6424	5528	4515–5699

From Bucur, V., *Ultrasonics.* 23(6), 269, 1985. By permission of the publishers, Butterworth Heinemann Ltd.©

of fibers and abundant medullary rays. The environmental and the genetic conditions that produce curly figures are discussed at length by Harris (1989).

Detection of curly structure in maple using ultrasound was attempted on trees and on increment cores of 10 mm diameter. The surface velocity measured at the periphery of the trunks of standing trees show values (3200 to 3800 m/s) inferior to those measured in ordinary trees (4800 to 5100 m/s) growing under the same conditions.

The shear velocity V_{RL} seems to give pertinent indications about the presence of wavy figures (Figure 8.10) in increment cores. As expected, this velocity is higher in curly maple (1800 m/s) than in normal wood (1400 m/s). The explanation for this difference may be provided by the presence in curly maple of very abundant medullary ray tissues disposed in the radial anisotropic direction.

8.2.2. ULTRASOUND AND LUMBER GRADING

Ultrasound for lumber grading was discussed by Sandoz (1989, 1991) and Steiger (1991). Basically, the interaction of propagating ultrasonic waves with the structure of the material gives parameters for the characterization of material properties. It is important to point out that the presence of defects in testing specimens modifies the parameters of the ultrasonic wave. Furthermore, to link ultrasonic velocity

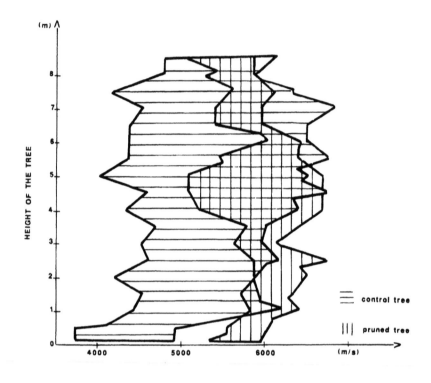

SURFACE VELOCITY (m/s)

Figure 8.9. Influence of pruning on the variation of surface velocity versus the height of the tree on Douglas fir. (From Bucur, V., Ultrasonics, 23(6), 269, 1985. With permission.)

Table 8.3. Physical characteristics of Douglas fir sapwood measured on increment cores at 12% moisture content

	Ultrasonic data: stiffnesses (10^8 N/m^2)					X-ray data: density (kg/m^3)		
	C_{LL}	C_{TT}	G_{LR}	G_{LT}	G_{TR}	D	D_{min}	D_{max}
Pruned tree	121.8	16.4	11.6	10.6	1.5	547	291	936
Control tree	112.8	16.4	12.2	10.9	1.3	548	273	877

From Bucur, V., *Ultrasonics.* 23(6), 269, 1985. By permission of the publishers, Butterworth Heinemann Ltd.©

measurements with the physics of propagation phenomena provides a basis for the interpretation of the results of nondestructive measurements of lumber. An early contribution to the development of the ultrasonic method for grading structural lumber was made by establishing relationships between the ultrasonic velocity and the overall mechanical parameters of lumber (Elvery and Nwokoye, 1970).

Ultrasonic pulse velocity testing was also considered to be a safety measure for timber structure, for checking nondestructively the residual strength in members of roof structures, to replace expensive proof-loading tests (Lee, 1965), or for specifying the strength properties of materials shipped to consumers. An on-line nondestructive measurement of these properties alleviates the problem of testing only random portions of materials using destructive methods.

8.2.2.1. Ultrasonic Velocity Method for Grading Timber and Round Wood
Tests for defect detection (knots, decay, slope of grain, internal voids) in timber were conducted for the most part using two techniques: immersion and through transmission (McDonald, 1978; Szymani

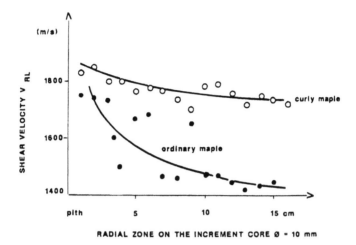

Figure 8.10. Shear velocity V_{RL} on increment cores of curly maple and of maple of ordinary structure (*Acer pseudoplatanus* L.) (Bucur, 1987).

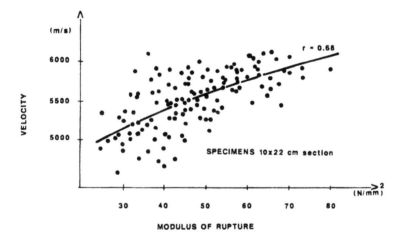

Figure 8.11. Relationships between the ultrasonic longitudinal velocity and modulus of rupture in static bending on spruce beams of 10 × 22 cm cross-section and 4.40 m length. (From Sandoz, J. L., *Wood Sci. Technol.,* 23(2), 95, 1989. With permission.)

and McDonald, 1981; Okyere and Cousin, 1980). The immersion technique is able to scan lumber or logs in water. The scanning time is 4 min for a 1′ × 8′ plank. No commercial device has been designed for in-line quality control of logs or lumber, but improving inspection velocity may assist in the implementation of this technique in sawmills and could help computerized log-sawing. For practical purposes it seems that the through transmission technique, which uses a portable apparatus, is very convenient. The results obtained by Sandoz (1989, 1991) and Steiger (1991) are discussed later.

The relationships between the ultrasonic velocity of waves and some mechanical parameters of timber such as the modulus of rupture or the modulus of elasticity in static bending (Figure 8.11) permits the comparison with classes that were established in agreement with standard visual grading rules (Figure 8.12). Corresponding to these classes three velocity levels were established: for the first class, $V > 5600$ m/s; for the second class, 5250 m/s $< V > 5600$ m/s, and for the third class $V < 5250$ m/s. A comparison between the ultrasonic method and visual grading shows that the maximum of error for visual examination was observed to be 45% for the first class and only 6% for the second

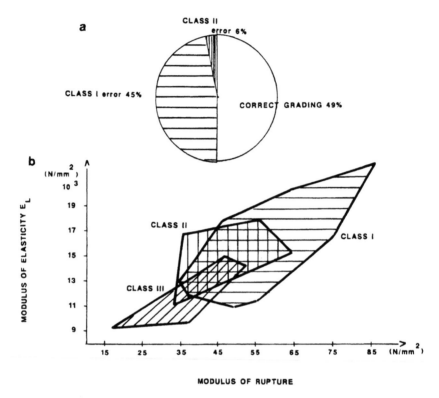

Figure 8.12. Visual grading rules (Swiss standard SIA 553164) and relations with moduli of elasticity and rupture on spruce beams. (a) Error diagram for visual grading, (b) correlation between the moduli of elasticity and rupture for different classes. (From Sandoz, J. L., *Wood Sci Technol.*, 23(2), 95, 1989. With permission.)

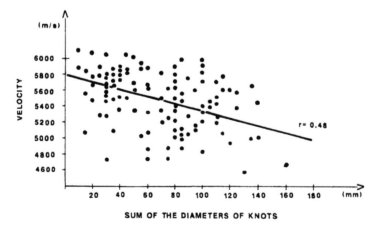

Figure 8.13. Relationship between the knots area on the beam and the longitudinal ultrasonic velocity. (From Sandoz, J. L., *Wood Sci. Technol.*, 23(2), 95, 1989. With permission.)

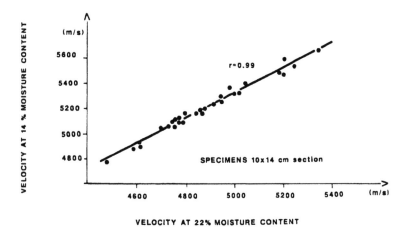

Figure 8.14. Relationships between velocity values of structural lumber measured at 22 and 14% moisture content. (From Sandoz, J. L., *Wood Sci. Technol.*, 23(2), 95, 1989. With permission.)

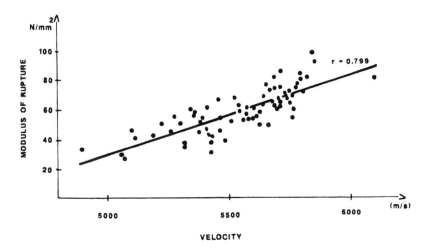

Figure 8.15. Relationship between the modulus of rupture and ultrasonic velocity on spruce roundwood without bark. (From Sandoz, J. L., *Wood Sci. Technol.*, 25(3), 203, 1991. With permission.)

Table 8.4. Ultrasonic velocities and physical properties of logs and squared timbers under green and dry conditions

Species	Velocities (m/s)			Density of timber (kg/m³)		Moisture content (%)			
	Logs	Timber				Green		Dry	
		(1)	(2)	(1)	(2)	(1)	(2)	(1)	(2)
Sugi	4232	4141	3039	380	420	145	59	11	11
Hinoki	4668	4983	3945	440	540	156	47	11	12

Note: (1) sapwood; (2) heartwood.

From Kodama, Y., *Mokuzai Gakkaishi.* 36(11), 997, 1990. With permission.

Table 8.5. Correlation coefficient between the modulus of elasticity E_L in static and in stress-wave modes reported in different research programs in the U.S., 1954–1982, for various wood materials (Ross and Pellerin, 1991a).

Material	Static loading mode	Correlation coefficient (r)
Clear wood	Compression	0.98
	Bending	0.98
Lumber	Bending	0.96
Veneer	Tension	0.99

class. Sandoz concluded that "the visual grading is unreliable and because of this one must use considerable factors of safety . . . , i.e. 2.25".

Several influences on the ultrasonic velocity are the knots area (Figure 8.13), moisture content (Figure 8.14), and slope of grain. The dynamic experimental approach proposed by Sandoz integersall these factors, givingan overall estimation of the quality of the tested specimen. Furthermore, this method can be adapted to the mechanical or acoustic classification for round timber and logs before milling. Figure 8.15 and Table 8.4 support this idea.

Finally, it is important to note again that the ultrasonic transmission technique is a powerful tool for grading structural lumber and logs in industry, where elements of large size are handled. It is hoped that in the near future this practical and cost-effective technique would be considered a valuable alternative to actual visual grading.

8.2.2.2. Stress-Wave Grading Technique for Testing Timber and Wood Composites

The nondestructive stress-wave technique for wood quality assessment is based on the measurement of the velocity of propagation of a stress wave generated by a shock. This technique was developed at Washington State University to determine the dynamic elastic modulus of small clear specimens, to assess strength properties of wood species, or to nondestructively test logs, poles, green and dry lumber, veneer sheets, plywood, glued laminated timber, structural members, particleboards, fiberboards, and mineral bonded wood composites (Pellerin, 1965; Pellerin and Ross, 1989; Hoyle and Pellerin, 1978; Kennedy, 1978; Galligan and Courteau, 1965; Ross and Pellerin, 1988, 1991a,b; Ross, 1985; Ross and Vogt, 1985; Gerhards, 1982a,b; Jung, 1979; Kunesh, 1978, Bender et al., 1990; Smulski, 1991; Bodig and Anthony, 1994).

Ross and Pellerin (1991a) summarized the results of different research reports from the U.S. from 1954 to 1982 related to the relationships between the modulus of elasticity in the static loading and dynamic modes using the stress-wave technique (Table 8.5). The experimental correlation coefficient was strong, between 0.87 and 0.99, and enabled the authors to conclude that the stress-wave method is a valid, nondestructive method for wood. (Instrumentation used in this technique was developed by Metriguard, Inc., Pullman, WA). The parameter measured was the time of propagation of the stress-wave signal. Furthermore, the velocity of propagation and the elastic modulus can be calculated.

Sobue (1986) employed a small hammer to generate the stress-wave signal and introduced a very elegant method of calculation of Young's modulus using fast Fourier transform of the power spectrum of the vibrating specimen to provide instantaneous vibration analysis. This method was successful for both small, clear specimens and heavy beams ($20 \times 20 \times 600$ cm size and 114 kg weight). The stress-wave technique was recently applied by Nanami et al. (1992) for the quality assessment of logs and standing trees.

8.3. ULTRASONIC CONTROL OF WOOD-BASED COMPOSITES

Different acoustic techniques have been used with varying degrees of success in predicting the quality of wood-based composites. Here, attention is focused upon the defect detection in wood-based composites, for which scattering-based ultrasonic methods were employed.

Defect detection and location in waferboard is shown in Figure 8.16. To verify the initial quality of the panel, the transit time was measured. The visual delamination was also mapped. The effectiveness of the ultrasonic method was verified by comparing the results with visual prediction. The visual assessment showed only half of the defect area that was detected ultrasonically. The technique reported

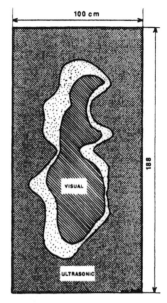

VISUAL...35 μs
ULTRASONIC ...20 μs
VELOCITY RANGE 93...1111m/s

Figure 8.16. Ultrasonic and visual detection of delamination in a 11-mm thickness waferboard panel (Szabo, 1978).

has the potential to be successfully implemented in industry, if a peculiar pattern recognition technique is developed.

Industrial continuous control of particleboard using the ultrasonic velocity method is possible, according to Roux et al. (1980). The ultrasonic device was implemented after the press. Measurements were performed through the thickness of the board at the current speed of fabrication. Conical transducers were in direct contact with the surface of the board, without coupling medium, on cool boards. The contact area was <2 mm^2.

Another important aspect in board utilization is their behavior under aggressive conditions. The influence of aging conditions under treatments V100 and V313, as imposed by French Standards NF 51-262 and NF 51-263, on structural flakeboards was studied by Petit et al. (1991). The first treatment required the immersion of specimens in hot water (100°C) for 90 min, while the second treatment was six cycles of immersion in water at 20°C for 72 h, followed by 24 h at −12°C and finally 72 h at 70°C in dry medium. After treatment the specimens were conditioned at normal temperature and moisture content. The influence of the V313 treatment on ultrasonic velocities and mechanical parameters of structural flakeboards is shown in Table 8.6. After six cycles of treatment the velocity diminished 16%, the density diminished 5%, and σ_R diminished 49%.

For both treatments Petit et al. (1991) observed that the velocities and moduli of rupture are correlated (strongly, $r = 0.949$, for V100 treatment and significantly, $r = 0.629$, for V313 treatment). The relationships established between the ultrasonic velocity and Young's modulus are also very significant

Table 8.6. Influence of V313 treatment on ultrasonic velocities and on mechanical parameters of structural flakeboard, measured on specimens cut in the perpendicular direction of flake alignment (Petit et al., 1991)

Parameters	Treatment						
	Initial	Cycle 1	Cycle 2	Cycle 3	Cycle 4	Cycle 5	Cycle 6
Velocity (m/s)	2500	2433	2492	2465	2168	1987	2098
Density (kg/m^3)	710	718	717	700	711	688	671
σ_R (MPa)	22.4	20.9	20.0	6.7	13.6	6.1	11.3

Table 8.7. Manufacturing characteristics of flakeboard (19 mm) and acoustic parameters

Characteristics	Ratio of ultrasonic parameters	Values	
		Minimum	Maximum
Flake alignment			
Longitudinal velocities	V_{11}/V_{22}	1.11	1.13
Surface velocities	(on 1/on 2)	1.03	1.10
Acoustic invariants	I_{12}/I_{13}	1.32	1.46
Skin damage			
Acoustic emission at 5 cm (axis 1/axis 2)			
Rise time		1.07	1.10
Peak amplitude		1.04	1.08
Duration		1.18	0.80
Count		0.15	0.11
Energy		1.72	1.68
Interlaminar heterogeneity			
Shear velocity			
In planes 12 and 13	V_{66}/V_{55}	1.86	1.91
In planes 12 and 23	V_{66}/V_{44}	1.87	1.99
Global board anisotropy			
Acoustic invariants	I_{ratio}	0.72	0.82

Note: The symmetry directions are (1) parallel to the flake alignment; (2) perpendicular to the flake alignment; (3) the thickness of the board. Velocities of $V_{11} \ldots V_{66}$ were subsequently measured.

From Bucur, V., *Wood Fiber Sci.* 24(3), 337, 1992. With permission.

($r = 0.877$ for V100 and $r = 0.99$ for V313). The relationships deduced from regression analysis could be useful for the development of a routine technique for the nondestructive evaluation of mechanical properties of boards. An important role could be played by frequency domain analysis (Gibson, 1985).

It is well known that wood-based composites were designed and fabricated with the aim of obtaining boards having specific mechanical properties for particular uses. The estimation of the anisotropy of these composites is one of the pertinent questions to be answered. Thus, we concentrate our attention on structural flakeboard. The aspects related to the implementation of ultrasonic testing may be analyzed in terms of the solution of an inverse acoustic problem, because with ultrasonic measurements one seeks to recover from signals the characteristics of the material. The anisotropy of commercial-sized boards or of specimens cut from boards can be expressed as the ratios of velocities of longitudinal, shear, or surface waves or as the ratio of acoustic invariants (Table 8.7). Moreover, the anisotropy induced by the flake orientation can also be expressed by some acoustic emission parameters, stimulated by breaking a 0.5-mm pencil lead on the surface of the specimen. The internal bond and the shear behavior of layers related to the thickness interlaminar heterogeneity may be linked to the ratio of measured shear velocities or to acoustic invariants. The global board anisotropy may also be expressed by the ratio of acoustic invariants in the three anisotropic planes.

With the recent advances in transducers, the production and inspection configuration for any type of product requires a good fundamental understanding of wave propagation phenomena into the member under test. The influence of frequency range, geometric beam spreading, pulse-receiver contact pressure, acoustic coupling, etc., can be controlled in practice, and simultaneous measurements of several transient waveform parameters and velocity can be performed.

Criteria need to be established, which involve changes in the measurements and which reliably indicate the position of the defects. A multiparameter technique such as mode conversion can provide a convenient means for inspection of large areas using relatively simple probes. This subject is obviously worthy of detailed investigation and a great deal of work remains to be done to provide a reliable indication of defect location in specimens under test.

8.4. OTHER NONDESTRUCTIVE TECHNIQUES FOR THE DETECTION OF DEFECTS IN WOOD

Other technologies were developed for detecting internal defects in wood material. Computed tomography (Gilboy and Foster, 1982; Reimers et al., 1984; Habermehl et al., 1986; Funt and Bryant, 1987; Davis et al., 1989) and magnetic resonance imaging (Wang and Chang, 1986; Kucera, 1989) can detect defects in logs and observe the moisture gradient in wood at the laboratory scale (Menon et al., 1989); thermal conditions monitoring (Bond et al., 1991) can evaluate the composites wood blades for wind turbine; infrared spectroscopy (von Wienhaus et al., 1988; Niemz et al., 1989, 1990) is used for laboratory and industrial purposes such as detection of needle and bark in chip mixture for particleboards, etc.) and for measuring small, clear specimens related to the microfibril orientation or to the presence of decay (Kuo et al., 1988). The development of noncontact laser technology (Soest, 1987; Jouaneh et al., 1987) opens the very broad field of small defect detection of specimens under stress, measurements of roughness, measurements of moisture content repartition, three-dimensional mapping of the slope of grain, detection of the incipient decay, detection of reaction wood, etc. However, these techniques must be adapted for log scanning at production speed and software must be developed to delineate scanned defects.

REFERENCES

Abbott, A. R. and Elcock, G. (1987). Pole testing in European context. Proc. 6th Symp. Nondestructive Testing of Wood. Washington State University, Pullman, 277–302.

Anon. (1964). *Multilingual Glossary of Terms Used in Wood Anatomy.* ETH, Zürich.

Archer, R. R. (1986). *Growth Stresses and Strains in Trees.* Springer-Verlag, Berlin.

ASTM D-245-81. Standard Methods for Establishing Structural Grades and Related Allowable Properties for Visual Graded Lumber. American Society for Testing and Materials, Philadelphia.

Baker, D. E. and Carlson, D. C. (1978). On line product inspection by non-contact ultrasonics. Proc. 4th Symp. Nondestructive Testing of Wood. Washington State University, Pullman, 233–238.

Bechtel, F. K. and Allen, J. R. (1987). Methods of implementing grain angle measurements in the machine stress rating process. Proc. 6th Symp. Nondestructive Testing of Wood. Washington State University, Pullman, 303–353.

Bender, D. A., Burk, A. G., and Hooper, J. A. (1990). Predicting localized MOE and tensile strength in solid and finger-jointed laminated lumber using longitudinal stress waves. *For. Prod. J.* 40(3), 45–47.

Bodig, J. and Anthony, R. W. (1994). Non-destructive strength classification of new-wood utility poles. *Proc. Canadian Electric Assoc.,* Engineering and Operations Division Meeting, March 1994, Toronto, ON.

Bond, L. J. and Saffari, M. (1984). Mode conversion ultrasonic testing. in *Nondestructive Testing.* Vol. 7. Sharpe, R. S. (Ed.). Academic Press, London, 145–189.

Bond, L. J., Aftab, N., and Clayton, B. R. (1991). Condition monitoring techniques for composite wind turbine blade. In press.

Bucur, V. (1984). Relationship between grain angle of wood specimens and ultrasonic velocity. *Catgut Acoust. Soc. Newslett.* 41, 30–35.

Bucur, V. (1985). Ultrasonic hardness and X-ray densitometric analysis of wood. *Ultrasonics.* 23(6), 269–275.

Bucur, V. (1987). Wood characterization through ultrasonic waves. in *Ultrasonic Methods in the Evaluation of Inhomogeneous Materials.* Alippi, A. and Mayer, W. G. (Eds.). Martinus Nijhoff, Dordrecht, 323–342.

Bucur, V. and Perrin, J. R. (1989). Slope of grain ultrasonic measurements in living trees and timber. *Holz Roh Werkst.* 47(2), 75.

Bucur, V. and Perrin, J. R. (1990). Procédé Ultrasonore pour l'Estimation de l'Angle du Fil du Bois. Brevet Français No. 90-06589. INRA, Paris.

Bucur, V., Janin, G., Herbe, C., and Ory, J. M. (1991). Ultrasonic Detection of Reaction Wood in European Species. Paper presented at Xème Congrès Forestier Mondial. September 17 to 26, Paris.

Bucur, V. (1992). Anisotropy characterization of structural flakeboards with ultrasonic methods. *Wood Fiber Sci.* 24(3), 337–346.

Chazelas, J. L., Vergne, A., and Bucur, V. (1988). Analyse de la variation des propriétés physiques et mécaniques locales du bois autour des noeuds. Actes du Colloque Comportement Mecanique du Bois. Bordeaux, June 8 to 9, 376–386.

Chazelas, J. L. (1990). Caractéristiques Physiques et Mécaniques Locales du Bois dans la Zone des Noeuds. Ph.D. thesis, Université "Blaise Pascal", Clermont-Ferrand, France.

Davis, J. R., Wells, P., Morgan, M., and Shadbolt, P. (1989). Wood research applications of computerized tomography. Proc. 7th Symp. Nondestructive Testing of Wood. Washington State University, Pullman, 71–75.

Dunlop, J. I. (1981). Testing of poles by using acoustic pulse method. *Wood Sci. Tech.* 15(4), 301–310.

Elvery, R. H. and Nwokoye D. N. (1970). Strength assessment of timber for glued laminated beams. Symp. Nondestructive Testing of Concrete and Timber. The Institute for Civil Engineering and the British Commission for NDT, London, June 11 and 12, 105–110.

Feeney, F. (1987). The Adaptation of an Ultrasonic Pulse Velocity Technique for Use on Small Samples of Sitka Spruce for Evaluation of Structural Wood Quality. M.Sc. thesis, St. Patrick College, Maynooth, Ireland.

Ferrand, J. C. (1982). Mesure de la variation de l'angle de la fibre torse avec l'âge. *Ann. Sci. For.* 39(1), 99–104.

Flammarion, J. P., Keller, R., and Mosnier, J. C. (1990). Production économique de bois de qualité par l'élagage des plantations de resineux. Actes du 3-éme Colloque Science et Industrie du Bois. Vol. 2, Bordeaux, May 14 and 15, 437–446.

Foulger, A. N. (1969). Through-bark measurements of grain direction: preliminary results. *For. Sci.* 15(1), 92–94.

Funt, B. V. and Bryant, B. V. (1987). Detection of internal log defects by automatic interpretation of computer tomography images. *For. Prod.J.* 37(1), 56–62.

Galligan, W. L. and Courteau, R. W. (1965). Measurements of the elasticity of lumber with longitudinal stress waves and the piezoelectric effect of wood. Proc. 2nd Symp. Nondestructive Testing of Wood. Washington State University, Pullman, 223–244.

Gerhards, C. C. (1982a). Effects of Knots on Stress Waves in Lumber. Res. Pap. No. 384. USDA Forest Products Laboratory, Madison, WI.

Gerhards, C. C. (1982b). Longitudinal stress waves for lumber stress grading. Factors affecting applications: state of the art. *For. Prod. J.* 32(2), 20–25.

Gibson, R. F. (1985). Frequency domain testing of materials. Proc. 5th Symp. Nondestructive Testing of Wood. Washington State University, Pullman, 385–406.

Gilboy, W. B. and Foster, J. (1982). Industrial applications of computerized tomography with X-and gamma radiation. in *Research Techniques in Nondestructive Testing*. Vol. 6. Sharpe, R. S. (Ed.). Academic Press, London, 255–287.

Habermehl, A., Pramann, F. W., and Ridder, H. W. (1986). Untersuchung von Alleebäumen mit einem neuen Computer-Tomographie Gerät. *Neue Mitt. Landwirtsch.* 31, 806–812.

Hamm, E. A. and Lam, F. (1989). Compression wood detection using ultrasonics. *G. Prove Nondestruct.* 1, 40–47.

Harris, J. M. (1989). *Spiral Grain and Wave Phenomena in Wood Formation.* Springer-Verlag, Berlin.

Hoyle, R. J. and Pellerin, R. F. (1978). Stress wave inspection on wood structures. Proc. 4th Symp. Nondestructive Testing of Wood. Washington State University, Pullman, 33–45.

James, W. L., Yen, Y., and King, R. J. (1985). A Microwave Method for measuring moisture content, density and grain angle in wood. Res. Pap. No. 0250. USDA Forest Products Laboratory, Madison, WI.

Janin, G., Ory, J. M., and Bucur, V. (1990). Les fibres des bois de reaction. *Rev. A.T.I.P.* 44(8/9), 268–275.

Jartti, P. (1978). Summary on the measurement of internal decay in living trees. *Silva Fenn.* 12(2), 140–148.

Jouaneh, M., Lemaster, R. L., and Dornfeld, D. A. (1987). Measuring work piece dimensions using a non-contact laser detector system. *Int.J.Adv. Manuf. Technol.* 2(1), 59–74.

Jung, L. (1979). Stress Grading Techniques on Veneer Sheets. Gen. Tech. Rep. FPL 27. USDA Forest Products Laboratory, Madison, WI.

Kaiserlik, J. H. (1978). Selected methods for quantifying strength in degraded wood. Proc. 4th Symp. Nondestructive Testing of Wood. Washington State University, Pullman, 95–117.

Keller, R., Azoeuf, P., and Hoslin, R. (1974). Détermination de l'angle de la fibre torse d'arbres sur pied à l'aide d'un traceur radioactif. *Ann. Sci. For.* 31(3), 161–169.

Kennedy, J. (1978). Ultrasonic testing of wood and wood products. Proc. 4th Symp. Nondestructive Testing of Wood. Washington State University, Pullman, 185.

King, R. J. (1985). Microwave diagnostics for stress-rating of dimension lumber. Proc. 5th Symp. Nondestructive Testing of Wood. Washington State University, Pullman, 445–464.

Kodama, Y. (1990). A method of estimating the elastic modulus of wood with variable cross-section forms by sound velocity. I. Application for logs. *Mokuzai Gakkaishi.* 36(11), 997–1003.

Kucera, L. (1989). Current use of the NMR tomography on wood at the Swiss Federal Institute of Technology. Proc. 7th Symp. Nondestructive Testing of Wood. Washington State University, Pullman, 71–72.

Kunesh, R. H. (1978). Using ultrasonic energy to grade veneer. Proc. 4th Symp. Nondestructive Testing of Wood. Washington State University, Pullman, 275–278.

Kuo, M., McLelland, J. F., Luo, S., Chien, P., Walker, R. D., and Hse, C. (1988). Application of infrared photoacoustic spectroscopy for wood samples. *Wood Fiber Sci.* 20(1), 132–145.

Lee, I. D. G. (1958). A nondestructive method for measuring the elastic anisotropy of wood using an ultrasonic pulse technique. *J. Int. Wood Sci.* 1(1), 43–57.

Lee, I. D. G. (1965). Ultrasonic pulse velocity testing considered as a safety measure for timber structures. Proc. 2nd Symp. Nondestructive Testing of Wood. Washington State University, Pullman, 185–205.

Mak, D. K. (1987). Locating flaws in three dimensional space using a transmitter-receiver system. *NDT Int.* 20(2), 117–120.

Martin, P., Collet, R., Barthelemy, P., and Roussy, G. (1987). Evaluation of wood characteristics: internal scanning of the material by microwaves. *Wood Sci. Tech.* 21(4), 361–371.

McDonald, K. A. (1978). Lumber Defect Detection by Ultrasonics. Res. Pap. No. FPL 311. USDA Forest Products Laboratory, Madison, WI.

McDonald, K. A. and Bendtsen, B. A. (1985). Localized slope of grain—its importance and measurement. Proc. 5th Symp. Nondestructive Testing of Wood. Washington State University, Pullman, 477–489.

McDonald, K. A. and Bendtsen, B. A. (1986). Measuring localized slope of grain by electrical capacitance. For. *Prod. J.* 36(10), 75–78.

McLauchlan, T. A. and Kusec, D. J. (1978). Continuous noncontact slope of grain detection. Proc. 4th Symp. Nondestructive Testing of Wood. Washington State University, Pullman, 67–76.

Menon, R. S., Mackay, A. L., Flibotte, S., and Hailet, J. R. T. (1989). Quantitative separation of NMR images of water in wood on the basis of T_2. *J. Magnetic resonance.* 82, 205–210.

Metriguard, Inc. (1986). Portable stress wave timer. For. *Prod. J.* 36(11/12), 1.

Nanami, N., Nakamura, N., Arima, T., and Okuma, M. (1992). Measuring the properties of standing trees with stress waves. I. The method of measurement and the propagation path of the waves. *Mokuzai Gakkaishi.* 38(8), 739–746; II. Application of the method to standing trees. *Mokuzai Gakkaishi.* 38(8), 747–752; III. Evaluating the properties of standing trees for some forest stands. *Mokuzai Gakkaishi.* 39(8), 903–909.

NF B 51 262. Panneaux de Particules. Epreuve d'Immersion dans l'eau Bouillante. Méthode dite "V100" (French standard).

NF B 51 263. Panneaux de particules. Epeuve de Vieillissement accéléré. Méthode dite "V313" (French standard).

Niemz, P., Wienhaus, O., Schaarschmidt, K., and Ramin, R. (1989). Untersuchungen zur Holzartendifferenzierung mit Hilfe der Infrarotspektroskopie. II. *Holzforsc. Holzverwert.* 41(2), 22–26.

Niemz, P., Wienhaus, O., Korner, S., and Szaniawski, M. (1990). Untersuchungen zur Holzartendifferenzierung mit Hilfe der Infrarotspektroskopie. III. *Holzforsch. Holzverwert.* 42(2), 25–28.

Okyere, J. G. and Cousin, A. J. (1980). On flaw detection in live wood. *Mater. Eval.* March, 43–47.

Pellerin, R. F. and Ross, R. J. (1989). Nondestructive evaluation of wood and wood products. in *Concise Encyclopedia of Wood and Wood-Based Materials.* Schnipwind, A. P. (Ed.). Pergamon Press, Oxford, 203–206.

Pellerin, R. F. (1965). The contribution of transverse vibration grading to design and evaluation of 55-foot laminated beams. Proc. 2nd Symp. Nondestructive Testing of Wood. Washington State University, Pullman, 337–348.

Pellerin, R. F., DeGroot, R. C., and Esenther, G. R. (1985). Nondestructive stress wave measurements of decay and termite attack in experimental wood units. Proc. 5th Symp. Nondestructive Testing of Wood. Washington State University, Pullman, 319–352.

Petit, M. H., Bucur, V., and Viriot, C. (1991). Ageing monitoring of structural flakeboards by ultrasound. Proc. 8th Symp. Nondestructive Testing of Wood. Washington State University, Pullman.

Polge, H., Keller, R., and Thiercelin, F. (1973). Influence de l'élagage de branches vivantes sur la structure des accroissements annuels et sur quelques caractéristiques du bois de douglas. *Ann. Sci. For.* 30(2), 127–140.

Portala, J. F. and Cicotelli, J. (1989). Nondestructive testing for evaluation of wood characteristics. Proc. 7th Symp. Nondestructive Testing of Wood. Washington State University, Pullman, 97–126.

Reimers, P., Gilboy, W. B., and Goebbels, J. (1984). Recent developments in the industrial application of computerized tomography with ionizing radiation. *NDT Int.* 17(4), 197–207.

Ross, R. J. (1985). Stress wave propagation in wood products. Proc. 5th Symp. Nondestructive Testing of Wood. Washington State University, Pullman, 291–318.

Ross, R. J. and Vogt, J. J. (1985). Nondestructive evaluation of wood-based particle and fiber composites with longitudinal stress waves. Proc. 5th Symp. Nondestructive Testing of Wood. Washington State University, Pullman, 121–157.

Ross, R. J. and Pellerin, R. F. (1988). NDE of wood-based composites with longitudinal stress waves. *For. Prod. J.* 38(5), 39–45.

Ross, R. J. and Pellerin, R. F. (1991a). Nondestructive Testing for Assessing Wood Members in Structures. A Review. FPL-GTR-70. USDA Forest Products Laboratory, Madison, WI.

Ross, R. J. and Pellerin, R. F. (1991b). NDE of green material with stress waves: preliminary results using dimension lumber. *For. Prod. J.* 41(6), 57–59.

Roux, J. E., Amede, P. E., and Duchier, H. R. (1980). Dispositif Ultrasonore d'Aquisition de Paramètres Mécaniques de Panneaux de Particules et Application au Controle Continu Non-Destructif. Brevet Français no. 80 18231.

Samson, M. (1988). Transverse scanning for automatic detection of general slope of grain in lumber. *For. Prod. J.* 38 (7/8), 33–38.

Sandoz, J. L. (1989). Grading of construction timber by ultrasound. *Wood Sci. Technol.* 23(2), 95–108.

Sandoz, J. L. (1991). Form and treatment effects on conical roundwood tested in bending. *Wood. Sci. Technol.* 25(3), 203–214.

Smulski, S. (1991). Relationship of stress-wave and static bending determined properties of four North-eastern hardwoods. *Wood Fiber Sci.* 23(1), 44–57.

Sobue, N. (1986). Measurements of Young's modulus by the transient longitudinal vibration of wooden beams using a fast Fourier transformation on spectrum analyser. *Mokuzai Gakkaishi.* 32(9), 744–747.

Soest, J. F. (1987). Potential of future technologies in lasers. Proc. 6th Symp. Nondestructive Testing of Wood. Washington State University, Pullman, 357–368.

Steiger, R. (1991). Festigkeitssortierung von Kantholz mittels Ultraschall. *Holz Zentralbi.* 117(59), 985–989.

Suzuki, H. and Sasaki, E. (1990). Effect of grain angle on the ultrasonic velocity of wood. *Mokuzai Gakkaishi.* 36(2), 103–107.

Szabo, T. (1978). Use of ultrasonics to evaluate or characterize wood composites. Proc. 4th Symp. Nondestructive Testing of Wood. Washington State University, Pullman, 239–260.

Szymani, R. and McDonald, K. A. (1981). Defect detection in lumber: state of the art. *For. PROD. J.* 31(11), 34–44.

Timell, T. E. (1986). *Compression Wood in Gymnosperms.* Springer-Verlag, Berlin.

Tomikawa, Y., Iwase, Y., Arita, K., and Yamada, H. (1990). Nondestructive inspection of a wooden pole using ultrasonic computed tomography. *Ultrasonic Technol.* 2(2), 26–37 (in Japanese); *IEEE TRANS. UFFC.* 33(4), 354–358 (in English).

Tsoumis, G. (1986). *Wood as Raw Material.* Pergamon Press, Oxford.

Waid, J. S. and Woodman, M. S. (1957). A nondestructive method for detecting diseases in wood. *Nature.* 180(4575), 47, cited by Lee (1958).

Wang, P. C. and Chang, S. J. (1986). Nuclear magnetic resonance imaging of wood. *Wood Fiber.* 18(2), 308–314.

Wedgewood, A. (1987). Data processing in ultrasonic NDT. Proc. Ultrasonic Int. 87, London, 381–386.

von Wienhaus, O., Niemz, P. and Fabian, J. (1988). Untersuchungen zur Holzartendifferenzierung mit Hilfe der Infrarotspektroskopie. I. *Holzforsch. Holzverwert.* 40(6), 120–124.

Zobel, B. J. and van Buijtenen, J. P. (1989). *Wood Variation: Its Causes and Control.* Springer-Verlag, Berlin.

Chapter 9

Environmental Modifiers of the Structural Parameters of Wood Detected with Ultrasonic Waves

The scientific literature lacks sufficient data on the dynamic, nondestructive characterization of wood material, wood-based composites, or wooden structures operating in hostile environmental conditions (Meyer and Kellog, 1982). Little basic data are available on the relationship between the microstructural features induced by adverse environmental conditions and physical parameters able to be deduced by nondestructive techniques. Within this context the full potential of ultrasonic techniques is yet to be realized.

Earlier in the volume, it was noted that the mechanical properties of solid wood or of wood composites are strongly affected by the fluctuation of relative humidity and temperature that control the level of moisture absorbed and desorbed by the material. In addition, the activity of microorganisms that attack wood is controlled by ambient conditions of temperature and humidity. The effects of those parameters are not always easily separated.

In order to understand the interaction of wood material with environmental conditions it is necessary to consider different levels of temperature (below freezing, from freezing to the temperature at which the thermal decomposition begins, and up to the temperature of combustion) and relative humidity which induce moisture content fluctuation, below or up to the fiber saturation point.

The purpose of this chapter is to identify the dependence of ultrasonic velocities and related mechanical parameters of wood or wood-based composites on moisture content and temperature. An attempt is made also to identify, via the ultrasonic technique, the biological deterioration of wood by bacterial attack, fungi, insects, molluscs, crustaceans, etc.

9.1. DEPENDENCE OF ULTRASONIC VELOCITIES AND RELATED MECHANICAL PARAMETERS OF WOOD ON MOISTURE CONTENT AND TEMPERATURE

9.1.1. INFLUENCE OF MOISTURE CONTENT ON SOLID WOOD

In his excellent book Skaar (1988) gives an extensive treatment on wood-water relations. The approach outlined below contains some novel elements and serves to complement Skaar's results, emphasizing those aspects related to ultrasound. In contrast to the procedure adopted for other materials, in which the moisture content is expressed in terms of the wet weight of the material, it is customary to express the moisture content of wood in terms of its oven-dry weight (USDA Forest Products Laboratory, 1987). The determination of moisture content in small, clear specimens is usually carried out using the gravimetric technique. Other methods for moisture content measurement in solid wood, lumber, and wood composites are described by Kollmann and Höckele (1962) and more recently by Siau (1984). The equilibrium moisture content of wood and wood products in various environments can be seen in Figure 9.1. This equilibrium is considered to be reached when for any combination of the relative humidity and temperature of the environment, no inward or outward diffusion of water will occur.

In living trees, depending on the season of the year, species of tree, or location in the tree, the moisture content of green wood varies from about 60 to 200% (Zimmermann, 1983). Green wood generally contains water in three forms: liquid water partially or completely filling the cell cavities, water vapor in the empty cell cavity spaces, and water in the cell wall. The liquid water in the cell cavities is also called "free water" to distinguish it from the cell wall water which may be called "bound water". The transition from "bound water" to "free water" occurs in the range of the fiber saturation point, which corresponds to the moisture content of a wood specimen placed in a relative humidity near 100%, or more exactly at 98%, as indicated by Siau (1984). In this case the cell cavities contain no liquid water, but the cell walls are saturated with moisture. For wood species originating from the temperate zone, the equilibrium moisture content corresponding to the fiber saturation point is about 28 to 30% (Kollmann, 1951). Elevating the temperature reduces the equilibrium moisture content at a given relative humidity, and at the same time reduces the hygroscopicity. At temperatures below 0°C the equilibrium moisture content of wood decreases with decreasing temperature. All these environmental conditions affect wood mechanical properties.

198

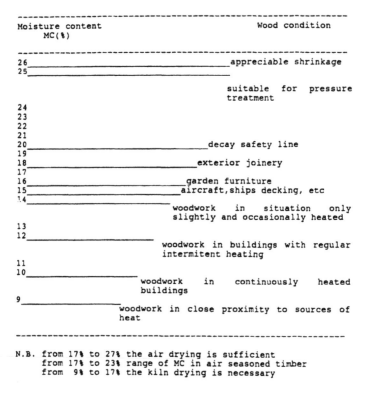

```
---------------------------------------------------------------
Moisture content                              Wood condition
     MC(%)

---------------------------------------------------------------
26_____appreciable shrinkage
25_____

                                          suitable   for   pressure
                                          treatment
24
23
22
21
20_____decay safety line
19
18_____exterior joinery
17
16_____garden furniture
15_____aircraft,ships decking, etc
14_____
                                          woodwork    in    situation    only
                                          slightly and occasionally heated
13
12_____
                                          woodwork in buildings with regular
                                          intermitent heating
11
10_____
                                          woodwork    in    continuously    heated
                                          buildings
 9_____
                                          woodwork in close proximity to sources of
                                          heat

---------------------------------------------------------------
  N.B. from 17% to 27% the air drying is sufficient
       from 17% to 23% range of MC in air seasoned timber
       from  9% to 17% the kiln drying is necessary
```

Figure 9.1. Equilibrium moisture content of timber. (Adapted from Dinwoodie, J. M., *Timber: Its Nature and Behaviour,* Van Nostrand Reinhold, New York, 1981.)

Before addressing the aspects of the influence of temperature and moisture content on the ultrasonic velocities, let us consider the evaluation of those influences on the mechanical parameters (strengths and moduli of elasticity) to which the ultrasonic velocities may be related.

Tiemann (1906) also cited by Siau (1984) was the first to note that the strength properties of wood are not affected by the "free water" because only the cell wall is effective for strength. Gerhards (1982) summarized the studies in the literature on the effects that moisture content and temperature had on several mechanical properties of clear wood specimens. Confirmation of the reduction in strength and elastic moduli with increasing moisture content to about 30% is seen in Figure 9.2. Further increases in moisture content have no influence on mechanical parameters. In order to compare the mechanical parameters of different wood species under various environmental conditions, the measured properties are often adjusted to expected values at 12% moisture content (Skaar, 1988).

The use of ultrasound for monitoring the moisture content in wood is relatively new. Ultrasonic longitudinal waves in large specimens were reported by Facaoaru and Bucur (1974), James et al. (1982), and Bucur and Sarem (1992). Table 9.1 shows some ultrasonic longitudinal velocity values, impedances, and attenuation (expressed by the value of maximum of amplitude from fast Fourier transform, FFT, spectrum) measured on all anisotropic directions of wood. As expected in dry conditions the velocity values are higher than in saturated conditions. The attenuation illustrates that the higher internal friction is observed in the T direction.

The variation of longitudinal velocity (V_{LL}) and of corresponding attenuation vs. moisture content is shown in Figure 9.3. The ultrasonic velocity decreases with moisture content, whereas the attenuation increases with increasing moisture content. The maximum velocity and the minimum attenuation were measured in dry conditions. On this graph it is interesting to note that the variation of velocity vs. moisture content has a critical point, U1, at 38%, corresponding to the fiber saturation point, whereas

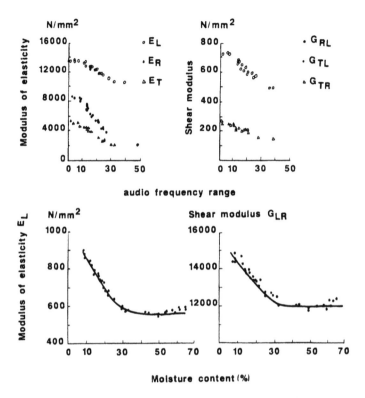

Figure 9.2. Effect of moisture content on the Young's and shear moduli of Sitka spruce. (a) Measured statically; (b) measured in audio-frequency range. (Adapted from Dinwoodie, J. M., *Timber: Its Nature and Behaviour,* Van Nostrand Reinhold, New York, 1981.)

Table 9.1. Velocity and attenuation of ultrasonic longitudinal waves on small, clear specimens of spruce (24 × 20 × 20 mm) measured on saturated and air-dry conditions (Bucur and Sarem, 1992)

Statistic parameters Anisotropic directions	Velocities (m/s)			Impedances (10^6 kg/sm²)			Amplitude at 1 MHz (dB)		
	L	R	T	L	R	T	L	R	T
Saturated Condition									
Mean	4576	1453	1153	4.5	1.48	1.17	−31.55	−31.03	−37.62
CV[a] (%)	6	10	6	—	—	—	23	19	19
Air-Dried Condition (12%)									
Mean	5203	1958	1057	1.94	0.73	0.39	−5.34	−13.87	−18.20
CV (%)	7	8	7	—	—	—	48	40	29

[a]CV, coefficient of variation.

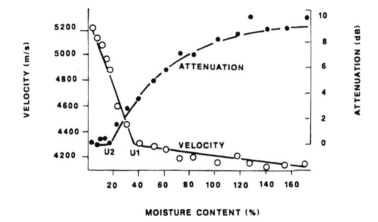

Figure 9.3. Ultrasonic velocity (V_{LL}) and corresponding attenuation vs. the moisture content in small, clear specimens of metasequoia. (From Sakai, H. et al., *Ultrasonics*, 28(6), 382, 1990. With permission of the publishers, Butterworth Heinemann Ltd.)

the variation of attenuation vs. moisture content has a different critical point, U2, at 18%. An attempt was made by Sakai and co-work (1990) to explain the differences:

- the velocity decreases dramatically with moisture content up to the fiber saturation point, and thereafter, the variation is very small
- the attenuation is quite constant in the low moisture content region and increases after critical point U2

At low moisture content (U <18%), when the water is present in the cell walls as bound water, the ultrasonic pulse is scattered by the wood cells and by cell boundaries. The side units of OH or other radicals of the cellulosic material may reorient their position under the ultrasonic stress. In this case the attenuation mechanism related to the cellulosic cell wall material is probably the most important. At higher moisture content but under the fiber saturation point, the scattering at cell boundaries could be the most important loss mechanism. After the fiber saturation point, when free water is present in cellular cavities, the porosity of the material intervenes as a predominant factor in ultrasonic scattering.

Because the aim of this research was to obtain an indication of the sensitivity of the ultrasonic technique to the moisture content of wood, and to determine whether the reliability of the test is likely to be affected by the structural wood parameters, it may be noted that the velocity is related to the presence of bound water, whereas the attenuation is related to the presence of free water. The critical point U1 corresponds to the fiber saturation point and the critical point U2 corresponds to the point at which the wood cells begin to retain free water. The difference between critical points U1 and U2 is related to the accuracy of velocity and attenuation measurements. It is well known that ultrasonic velocity is a very accurate parameter (1% or less), whereas ultrasonic attenuation, expressed as the received amplitude, is less accurate (10 to 20%).

Because this chapter shows the relationships between ultrasonic and mechanical parameters and the moisture content of wood, it is useful to note the variation of the stiffness C_{LL} vs. the moisture content for different species (Figure 9.4). As in the case of the variation of velocity vs. moisture content, the variation of C_{LL} vs. moisture content shows a critical point, corresponding to the fiber saturation point. Below this point stiffness decreases with moisture content. Thereafter, stiffness increases with moisture content because of the increasing mass density of the specimen and of the presence of free water related to the porosity of the material. The ultrasonic technique reaches a new field that could be developed further, and may be related to the nondestructive measurement of wood porosity. The theory for porous materials was developed by Biot (1956) and Plona and co-workers (1980, 1987). Adapted to wood characteristics this theory can reveal further ultrasonic methods for the nondestructive control of impregnation processes.

The anisotropy of wood, expressed by ultrasonic parameters in various hygroscopic conditions, also may be used as an indicator of the dynamics of kiln drying. The anisotropy can be expressed as a ratio of velocities, impedances, and attenuation in different axes, as can be seen from Table 9.2.

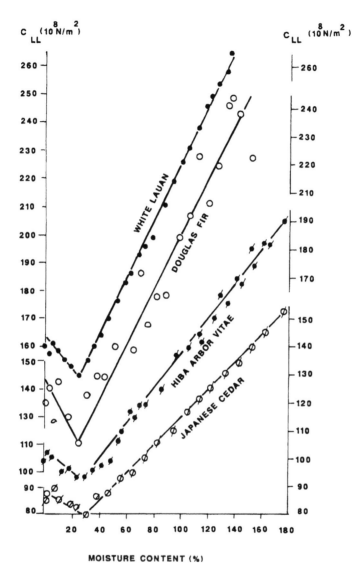

Figure 9.4. Variation of longitudinal stiffness vs. the moisture content in small clear specimens of different species. (From Sakai, H. et al. *Ultrasonics*, 28(6), 382, 1990. With permission of the publishers, Butterworth Heinemann Ltd.)

9.1.2. INFLUENCE OF TEMPERATURE ON SOLID WOOD

Launay and Gilleta (1988) studied the simultaneous influence of temperature (20° to 80°C) and moisture content (6, 12, 18%) on ultrasonic velocities V_{LL}, V_{RR}, V_{TT}, V_{RT}, V_{LT}, and V_{LR}. In this relatively low moisture content domain temperature-inalterations of ultrasonic velocities are less important than those due to variations in moisture content. The data reported by DINWOODIE (1981) on *Pinus radiata* revealed simultaneous influences of moisture content and temperature on the relative modulus of elasticity. However, at 0% moisture content the reduction of the modulus between −20°C and +60°C is about 6%, whereas at 20% moisture content the reduction is 40%, when the same temperature range is considered.

Strong decreasing values of moduli E_L and E_R, the result of increasing temperatures between 100 and 300 K (−173° and +25°C), were also reported by Polisko (1986). The range in the moisture content

Table 9.2. Anisotropy of spruce (small clear specimens) expressed by the ratios of different acoustic parameters in saturated and air-dry conditions (Bucur and Sarem, 1992)

	Hygroscopic conditions	
	Saturated	Air-dried (12%)
Velocity ratios in axes		
V_{LL}/V_{RR}	3.15	2.65
V_{LL}/V_{TT}	4.12	4.92
V_{RR}/V_{TT}	1.26	1.85
Ratio of impedances in different axes		
L/R	3.04	2.65
L/T	3.84	4.97
R/T	1.26	1.87
Velocity/mass density in different axes		
V_{LL}/mass density	4.50	13.90
V_{RR}/mass density	1.43	5.24
V_{TT}/mass density	1.13	2.83
Attenuation ratios in different axes		
A_{LL}/A_{RR}	0.81	1.06
A_{LL}/A_{TT}	0.94	1.17
A_{RR}/A_{TT}	1.16	1.10

was limited from 0 to 11% because in this interval only the bound water affects the properties of the wood. For higher moisture content we could expect phenomena induced by water phase transition. Phase transition effects reported by Kollmann (1951) were related to the variation of mechanical properties (strengths) determined from static tests between 0° and −80°C.

The influence of temperature on elastic parameters is a key question for full-scale structural lumber testing. Barrett et al. (1989) proposed temperature adjustments for field testing of American and Canadian lumber. Adjustments were proposed in the range of −40° to 100°F for the percentage of property change for green and dry material at 12% moisture content.

Discussion of the influence of temperature and moisture content on the velocities of ultrasonic waves propagating in wood is conveniently approached using three ranges of temperature:

- Low temperature, below freezing
- Intermediate temperature, from freezing to the temperature at which thermal decomposition begins (180°C)
- High temperature, from thermal decomposition up to the temperature of combustion (1000°C)

As for the results concerning the influence of low temperature (−30°C to 0°C) and intermediate temperature (0°C to +50°C) on the velocities of propagation of longitudinal waves in spruce (Figure 9.5), in air-dried and green moisture content conditions, note that the ultrasonic velocity generally increases linearly with decreasing temperature (+50° to −30°C) and moisture content below the fiber saturation point.

When green wood is considered, the question arises of ultrasonic wave propagation through a frozen porous medium, which exhibits a phase transition at 0°C, similar to other porous materials [kaolinite, boom clay, etc. (Deschatres et al., 1989). At this temperature abrupt changes in velocity, probably induced by ice segregation in cells, is associated with capillary transition (Sellevold et al., 1975) and gradual solidification of the free water in the lumen. When the liquid component changes from a fluid to a rigid state, the porous solid may become more rigid and may exhibit a higher velocity (Auge et al., 1989; Sandoz, 1990). When the moisture condition of wood is below the saturation point, the cellular lumen is empty and the interfibrillar adsorbed water can nucleate in ice and expand easily. The desorption process of cellular wall could explain the linearly increasing ultrasonic velocity with decreasing temperature.

Similar phase transitions were observed for other basic properties of wood such as piezoelectricity. Fukada (1968) and Hirai (1974) considered the influence of temperature and moisture content on the piezoelectric moduli of wood (Figure 9.6). Three points of discontinuity can be observed in this figure −80°C, 0°C, and 30°C, which seem to be related to different phase transitions in various components

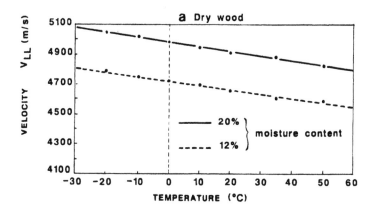

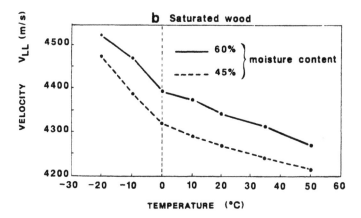

Figure 9.5. Spruce. The influence of temperature ranging from −30° to +50°C on velocity V_{LL}, for solid wood. (a) Specimens in air-dry conditions, at 12 and 20% moisture content; (b) specimens in green conditions, at 45 and 60% moisture content (Auge, 1990).

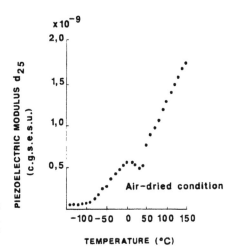

Figure 9.6. Relationship between the piezoelectric modulus d_{25} and temperature in sugi. (From Hirai, N., *Bull. Shizuoka Univ. For.* No. 3, 11, 1974. With permission.)

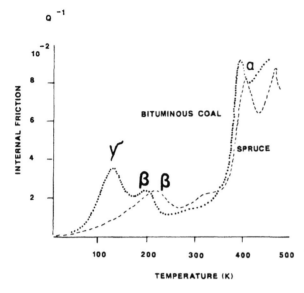

Figure 9.7. Internal friction (1/Q) vs. temperature for different materials. (From Wert, C. A. et al., *J. Appl. Phys.*, 56(9), 2453, 1984. With permission.)

of wood. Fukada suggested that in the low temperature region, piezoelectric activity is governed by a "mechanical relaxation in cellulose molecules within the amorphous regions. This relaxation seems to be associated with torsional vibration of cellulose molecules." The increase of temperature is associated with the increase in piezoelectric modulus, probably generated by modifications in the crystal lattice of cellulose. From the similar evolution of piezoelectric moduli and ultrasonic parameters, it seems reasonable to assume that such properties may be used as a quantitative measure of the internal structural modifications of wood. Therefore, a substantial research effort in wood physics relating the ultrasonic and piezoelectric parameters will be justified for many years to come.

If all elastic constants must be determined in the low temperature range, attention must be paid to the measurements of density and dimensions of specimens, bearing in mind the important "coldness shrinkage" (Kubler, 1983). On the other hand, the evaluation of technical properties of lumber and wood-based composites requires methods for adjusting the measurements with temperature and moisture conditions. Using velocity as a reference parameter, as determined from the ultrasonic or frequency resonance methods, several statistical, empirical models were established for corresponding practical applications (Barrett et al., 1989; Sandoz, 1990; Tsuzuki et al., 1976, 1983). Studies on the relationships among ultrasonic velocities, moisture content, and low temperatures may help in developing an apparatus for *in situ* testing of poles (Sandoz, 1990), or for an automatic control procedure for predrying. [Actually, drying hardwoods, from green conditions to 25% moisture content, at low temperatures, is a technique used for furniture-grade 1-in. oak lumber (Quarles and Wengert, 1989)].

We now turn to the influence of high temperatures on ultrasonic velocities (bear in mind that several characteristic stages of temperature increase induce phase transition in wood structure). In examining the behavior of wood and other polymers (Wert et al., 1984) over a wide range of temperatures (100 to 500 K) it is generally accepted that three peaks (α, β, Γ) can be seen in the internal friction spectrum (1/Q) (Figure 9.7). The α peak (475 K or 200°C) is sharp and probably induced by the segmental motion of the main cellulosic chain in crystalline and amorphous layers. The β peak is related to water that is chemically connected to the macromolecules. The third peak, γ (130 K), is not present in wood.

The elevated temperatures that can be used in the many kinds of wood processing (drying, pulping, size stabilization, production of particle and fiber-boards) are

- From 80° to 190°C, when water is absorbed from all constituents. This domain is interesting for the kiln drying process. It is well known (Quarles and Wengert, 1989) that drying lumber at temperatures over 100°C continues to gain popularity for softwoods.
- From 190° to 380°C, when degradation of chemical components is induced.
- From 380° to 600°C, when pyrolysis is observed.
- From 600° to 1000°C, when combustion is produced.

Table 9.3. Influence of thermal treatment of beech, poplar, spruce, and pine on ultrasonic velocities of longitudinal and shear waves (Böhnke, 1989)

Ultrasonic velocities (m/s)						Mechanical constants		
V_{LL}	V_{RR}	V_{TT}	V_{RT}	V_{LT}	V_{LR}	σ_R (daN/cm^2)	E_L (10^8 N/m^2)	R_{invar}
Hardwoods								
Beech, Natural Wood, 760 kg/m^3, 10% Moisture Content								
4345	1996	1666	534	1354	1549	88–118	81–102	0.526
Beech, Treated Wood, 280°C, 30 min 680 kg/m^3, 3% Moisture Content								
3940	2216	1441	526	1324	1411	42–79	73–85	0.576
Poplar, Natural Wood, 431 kg/m^3, 9% Moisture Content								
5335	2071	1150	480	1100	1882	58–73	52–85	0.411
Poplar, Treated Wood, 280°C, 10 min 390 kg/m^3, 3% Moisture								
5272	2101	1115	584	1054	1829	42–69	69–79	0.426
Softwoods								
Spruce, Natural Wood, 430 kg/m^3, 10% Moisture Content								
4560	2106	926	822	1259	1217	61–63	53–60	0.501
Spruce, Treated Wood, 260°C 15 min 440 kg/m^3, 4% Moisture Content								
5361	2060	1084	648	1320	1523	34–39	54–73	0.420
Pine, Natural Wood, 560 kg/m^3, 10% Moisture Content								
3710	2174	1459	891	1422	1347	65–75	40–70	0.634
Pine, Treated Wood, 260°C, 15 min 530 kg/m^3, 4% Moisture Content								
4091	2558	1447	682	1515	1598	48–56	70–75	0.608

For our purposes, the focus is solely on the temperature region 190° to 380°C, because it was demonstrated by Bourgois and Guyonnet (1988) that a semirefined material, an intermediate between wood and charcoal, could be produced under an inert atmosphere (nitrogen). This roasted wood was called "torrefied" or "retified" wood. The interest in this thermal treatment is in reducing the hygroscopicity of solid wood, at 3 to 4% moisture content at room temperature and normal relative humidity, without doing too much damage to its mechanical properties.

Tables 9.3 and 9.4 reproduces the results of the influence of thermal treatment (280°C for 30 min under nitrogen atmosphere) on hardwoods and softwoods. The values of longitudinal and shear velocities as well as those of the acoustic invariants were employed to study the influence of the thermal treatment on the structural modifications of the treated material. From the tables, one immediately notices that for hardwoods the rule is that the treatment induces a decrease in nearly all acoustical parameters, density, and stiffnesses. Also, the acoustical invariant ratios show a small tendency to reduce anisotropy by 3 to 9%.

An interesting aspect revealed by Bohnke and Guyonnet (1991) is related to the existence of peaks on all curves of velocity vs. the loss of mass induced by thermal treatment. These peaks (Figure 9.8) can be observed in all anisotropic directions for both longitudinal and shear waves, for the same mass loss, induced by the same temperature (230°C). This effect was supposedly due to the chemical changes in the amorphous matrix of wood which contained lignin and corresponded to a point of phase transition of the treated material.

Because improvement in the hygroscopicity of thermally treated wood under a nitrogen atmosphere is expected, the central dilemmas to be solved are related to a compromise between the loss of mechanical capacities and the reduction by several percent of the hygroscopicity of the material.

Table 9.4. Influence of thermal treatment on diagonal terms of stiffness matrix (Böhnke, 1989)

Stiffnesses (10^8 N/m^2)									
C_{11}	C_{22}	C_{33}	C_{44}	C_{55}	C_{66}	C_{12}	C_{13}	C_{23}
Beech, Natural Wood								
143.38	30.28	21.09	2.17	13.93	18.23	28.39	15.08	20.90
Beech, Treated Wood								
105.56	33.39	14.12	1.88	11.92	13.54	29.75	5.07	17.85
Differences (%)								
26.3	−10.3	33.0	13.4	14.4	25.7	−0.1	66.4	14.6
Poplar, Natural Wood								
122.39	18.44	5.69	0.99	5.20	15.23	17.88	15.31	8.06
Poplar, Treated Wood								
108.39	17.21	4.85	1.33	4.33	13.05	23.29	16.87	6.84
Differences (%)								
11.4	6.7	14.8	−34.3	16.7	14.3	−30.3	−10.2	15.1

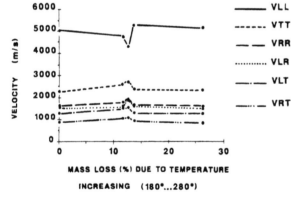

MASS LOSS (%) DUE TO TEMPERATURE
INCREASING (180°...280°)

Figure 9.8. Beech. Variation of velocity vs. the mass loss (%) induced by thermal treatment (180° to 280°C) in a nitrogen atmosphere. (From Bohnke, I. and Guyonnet, R. *Proc. Ultrasonics Int. '91.*, Le Touquet, France, July 1–4, 1991. With permission of the publishers, Butterworth Heinemann Ltd.)

For the domain corresponding to the pyrolysis and combustion of wood no data on ultrasonic velocity have been reported.

9.1.3. INFLUENCE OF HYGROTHERMAL TREATMENT ON THE QUALITY OF WOOD COMPOSITES

The durability of wood-based composites can be tested by accelerated aging treatment, usually an aggressive hygrothermal treatment. Such treatment is characterized by very important variations of temperature and humidity in the environment. This section focuses on monitoring the aging of structural flakeboards via the ultrasonic velocity method. The aging treatment for particleboards is recommended by French standard NF B 51-263, also called the V313 treatment. This treatment is cyclic, beginning with 72 h of immersion in water of specimens at 20°C, followed by 24 h at −12°C, and ending with 72 h at 70°C in a dry atmosphere. The boards used are industrially made three-layer oriented flakeboards, 19 mm thick. The pine flakes are glued using melamine urea formaldehyde resin. On the exterior layers of the board the flakes are oriented parallel to the length of the board. In the middle layer the flakes are oriented parallel to the width of the board. Ultrasonic velocity measurements are performed before and after treatment on standard-size specimens. This test is completed with static measurements of Young's modulus and the modulus of rupture (Table 9.5). After treatment, differences of 16% in velocity measured on specimens cut perpendicular to the length of the board and of 50% in the modulus of rupture indicate that modification of the internal structure of the specimens as induced by hygrothermal treatment.

Table 9.5. Influence of the aging treatment V313, recommended by NF B 51-263 on ultrasonic velocity and mechanical characteristics of structural flakeboard (Petit et al., 1991)

Parameters	Cycle	0	1	2	3	4	5	6
		\multicolumn{7}{c}{**Specimens Cut Parallel to Board Length**}						
Velocity	(m/s)	3315	3071	3367	3517	3409	3021	3135
Density	(kg/m³)	731	726	743	691	692	679	723
E	(MPa)	6376	6129	5608	4430	4711	4635	4736
$R_{rupture}$	(MPa)	53.5	56.3	46.8	43.2	42.8	39.4	32.1
		\multicolumn{7}{c}{**Specimens Cut Perpendicular to Board Length**}						
Velocity	(m/s)	2500	2433	2492	2465	2168	1987	2098
Density	(kg/m³)	710	718	717	700	711	688	671
E	(MPa)	2269	1754	1531	1460	1196	1230	1086
$R_{rupture}$	(MPa)	22.4	20.9	20.0	16.7	13.6	16.1	11.3

9.2. ULTRASONIC PARAMETERS AND BIOLOGICAL DETERIORATION

Various methods, using X- or γ-radiations, electrical resistance, vibrations, etc. (Kaiserlic, 1978) were employed to detect the biological deterioration of wood. This section discusses in particular the ultrasonic velocity method, developed for the detection of attacks by bacteria, fungi and insects. This nondestructive technique allows measurements to be performed in the field with minimal disturbance to the structure. The ease of measurement permits a large number of tests to be conducted, and this allows overall conclusions to be drawn with a reasonable amount of statistical confidence. The resulting database could be large enough to examine limiting values. Because of the statistical procedures required, the necessity for an objective parameter to monitor the degradation of wood is obvious in order to avoid falling into the trap of a stochastic rather than into a deterministic analysis. Therefore, ultrasonic velocities would be useful for improving confidence in reliably based evaluations of existing wood structutres or trees affected by decay or other injuries.

9.2.1. BACTERIAL ATTACK

Bacterial attack on sound wood facilitates the permeability of sapwood with preservative liquids (Liese, 1975). However, some undesirable effects on strength as a result of this treatment may be expected. If better liquid flow in tracheids can be achieved, the tori of pits would be destroyed using a specific bacterial treatment as a long-term water storage. To this end, Ehransjah et al. (1989) reported the utilization of *Bacillus subtilis* to improve the permeability of spruce sapwood.

Ultrastructural modifications induced by bacterial attack during water storage and the impact of this treatment on acoustical and mechanical properties of wood are analyzed in Tables 9.6 and 9.7. The examination of ultrasonic velocity values measured in sound and treated wood leads to the following comments. A slight difference exists between longitudinal velocities (V_{LL}, V_{RR}, V_{TT}) on sound and treated

Table 9.6. Spruce: ultrasonic velocities and stiffnesses on sound and treated wood[a]

\multicolumn{6}{c}{**Ultrasonic velocities (m/s)**}						\multicolumn{6}{c}{**Stiffnesses (10^8 N/m²)**}					
V_{LL}	V_{RR}	V_{TT}	V_{RT}	V_{LT}	V_{LR}	C_{11}	C_{22}	C_{33}	C_{44}	C_{55}	C_{66}
\multicolumn{12}{c}{**Sound Wood, Density 407 kg/m³**}											
4619	1994	1351	665	1276	1395	86.9	16.0	6.38	1.8	6.6	7.9
\multicolumn{12}{c}{**Treated Wood, Density 383 kg/m³**}											
4740	2115	1309	567	1222	1480	86.1	17.1	6.56	1.2	5.8	8.4
\multicolumn{12}{c}{**Differences (%)**}											
−2.5	−5.7	2.6	17	4.2	−6	1	−6.9	−2.8	31	13	−6

[a]Water storage, 5 months, under *Bacillus subtilis* attack.
Data from Efransjah et al., *Wood Sci. Tech.*, 23(1), 35, 1989.

Table 9.7. Strength properties of sound and attacked spruce wood

| | Young's moduli (10^8 N/m²) | | | | Rupture bending (daN/cm²) | Shock energy (kg/cm²) |
| | Ultrasonic | | Static | | | |
	E_L	E_R	E_T	E_L	σ Rupture	W_L
Sound wood	86	13	5	82	638	430
Attacked wood	78	11	4	78	529	380
Difference (%)	10	15	20	5	17	12

Data from Efransjah et al., *Wood Sci. Tech.*, 23(1), 35, 1989. With permission.

wood, but the shear wave velocity values (V_{TR}, V_{LT}, V_{LR}) seem to be significantly different than the previously mentioned velocities for structural modifications in planes that include the R axis (e.g., planes *LR* and *RT*).

Ultrastructural observations in scanning electronic microscopy (Figure 9.9) show the modified structure in which the area of aspirated pits was affected by bacteria after 4 weeks of treatment in a pure culture. On the other hand, the shear moduli are strongly changed. The modulus G_{RT} shows a difference of 31%. A possible explanation is that the treatment affects the continuity of the tracheid wall and, implicitly, a modification of the ultrasonic pathway is induced. Liese (1975) noted modification of the properties of decayed wood expressed mostly in Young's modulus in bending or in toughness strength.

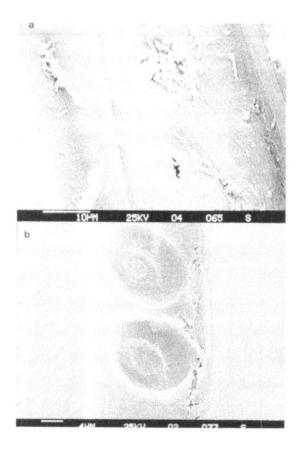

Figure 9.9. Ultrastructure of spruce attacked by bacteria. (a) SEM micrograph of intact tori; (b) SEM micrograph of bacteria colonies around the pits after treatment. (From Efransjah, F. et al., *Wood Sci. Technol.*, 23(1), 35, 1989. With permission.)

Compare the ultrasonic and static values of E_L in Table 9.7. This modulus is not affected by water storage treatments despite E_R and E_T being distinctly changed (15 to 20%). These diminished values were previously confirmed in the G_{RT} measurement by shear waves. It was further estimated that the cross-section of wood is affected by bacterial attack.

Other techniques are capable of identifying bacterial attack in wood. Ross et al. (1992) and Verkasalo et al. (1993) reported on the stress wave technique for the detection of wetwood before kiln drying red and white oak lumber. Notice that wetwood (also called "bacterial oak") is produced by the anaerobic bacterial infection of trees. This defect in lumber produces a decrease in the mechanical properties of wood.

9.2.2. FUNGAL ATTACK

Lee (1965) reported data were on the deterioration caused by the decay of structural members of old buildings, which was detected via ultrasonic velocity measurements. The data were connected to the loss of strength caused by fungi attack and deduced from measurements on specimens cut from the purlins. The presence of a decayed zone was illustrated by a chart relating ultrasonic velocities and strength. Some results were reported by Konarski and Wazny (1974), relating bending strength, modulus of elasticity, and density to ultrasonic velocity on pine attacked by *Coniophora cerebella*, during 1, 3, 4, 5, and 6 months.

Suyima (1965), using the direct transmission technique by employing longitudinal ultrasonic waves of 20, 50, and 100 kHz, was able to detect the presence of decay produced by *Coriollus palustris* on small clear specimens. When ultrasonic velocities (V_{RR} and V_{TT}) were regressed on the weight loss and bending strength (Figure 9.10), the presence of decay on the inner or outer parts of the specimen can be detected.

Detection of the early stages of decay produced by *Gloeophyllum trabeum* and *Poria placenta* in Douglas fir by using the ultrasonic pulse velocity method was noted by Wilcox (1988). He used the through transmission technique, with longitudinal waves of 35, 54, 150, and 500 kHz. The measurements were taken (approximately 15 × 36 cm section and 3 to 4 m in length) along the grain and in a transverse section of beam decayed by brown rot in service. Wilcox also reported laboratory tests on decayed wafers of Douglas and white fir. Pulse velocities measured on the transverse section of the wafers were regressed vs. the weight loss; the regression coefficient r^2 was reported to be 0.67 to 0.91. Relatively strong relationships between weight loss measured on attacked specimens and pulse velocity allowed Wilcox to deduce that the early stages of decay could be detected using this approach.

Decaying of beech and pine by white and brown rot induced by *Coriolus versicolor* and *Gloeophyllum trabeum*, estimated through longitudinal and shear ultrasonic waves, was reported by Bauer et al. (1990). They stated that fungi attack induces a more sensitive decrease in velocities than weight loss of the specimen. The results (Table 9.8) confirmed that ultrasonic velocities detect the degrading wood from the earliest stage. White rot on beech produced a 42% decrease in V_{TT} and a 34% decrease in V_{TR}. More dramatic was the modification of velocities produced by brown rot on pine (31% on V_{RR} and 50% on V_{TR}), when weight loss is only 10% (Figure 9.11).

When the regression equations were calculated between ultrasonic velocities and weight loss it was observed that

- For white rot on beech, the decrease in V_{TT} is from 29 to 36%, when the weight loss is from 8 to 16%.
- For brown rot on pine, the decrease in V_{TT}, V_{RR}, and V_{LL} is 52, 34, and 20%, respectively, when the weight loss is from 12 to 16%.

These numbers confirm that physical interpretation of the modifications of material properties induced by fungi attack is meaningful when the ultrasonic velocities are measured.

A more refined analysis, which gives a complete picture of this modification of the wood structure, permits the calculation of acoustical invariant, considering simultaneously the propagation along all symmetry axes of wood (Table 9.9). The invariant values in the *RT* plane are of considerable interest when examining the structural deterioration produced in this plane by white and brown rot. Accordingly, the deterioration process probably starts in this plane. Moreover, computing invariants leads to the statement that the voluminous experimental data are compactly reduced and, consequently, easy to handle. The synthetic treatment of invariants allows the observation that the transverse anisotropic plane *RT* is the most affected by fungi, having different capacities to decay wood.

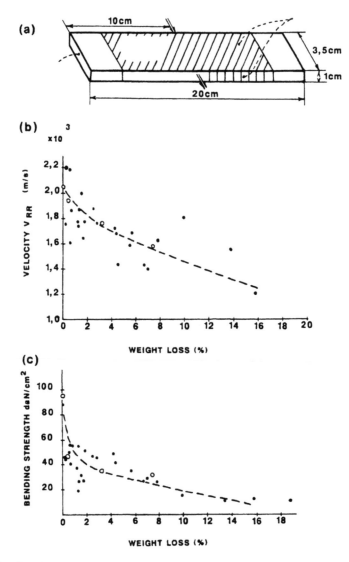

Figure 9.10. *Coriollus palustris* attack on *Cryptomeria japonica*. (a) Type of specimen used for ultrasonic test; (b) velocity V_{RR} vs. weight loss; (c) effect of decay on bending strength. (From Sumiya, K., *Wood Res.* 34, 22, 1965. With permission.

Detection of the presence of brown rot decay fungi on southern pine specimens of $^3/_4 \times {}^3/_4 \times 12$ in. or $19 \times 19 \times 300$ mm (Pellerin et al., 1985; Pellerin, 1989) also may be made using the stress wave velocity measurement. The samples were tested after incubation periods of 2, 4, 6, 9, and 12 weeks, and the following variables were measured: weight loss, stress wave time (used for velocity and modulus of elasticity calculation), and compression strength parallel to the grain. Relationships between stress wave time or stress wave modulus of elasticity and ultimate compression stress were established. As expected, long-term exposure to decay reduces the ultimate compression stress and increases stress wave propagation time in the decayed zone.

Patton-Mallory and DeGroot (1989) developed an interesting and promising acousto-ultrasonic technique, similar to those proposed by Vary (1980) for composites, for the evaluation of brown rot decay

Table 9.8. Beech and pine: ultrasonic velocities measured on sound and decayed wood
when weight loss is 10%

	Ultrasonic velocities (m/s)					
	V_{LL}	V_{RR}	V_{TT}	V_{RT}	V_{LT}	V_{LR}
Beech						
Sound wood	5074	2200	1580	960	1270	1510
Decayed wood	4235	1920	912	630	1100	1463
Differences (%)	16	13	42	34	13	3
Pine						
Sound wood	5000	2100	1200	600	1030	1050
Decayed wood	4348	1444	1123	300	1119	917
Differences (%)	13	31	6	50	—	13

From Bauer, C. et al., *Holzforschung,* 45(1), 41, 1991. With permission.

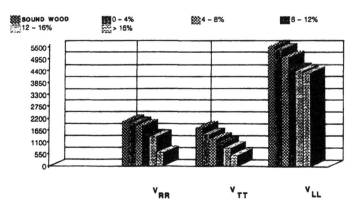

Figure 9.11. Histograms of ultrasonic velocities and weight loss. (From Bauer, C. et al., *Holzforschung,*
45(1), 41, 1991. With permission.)

Table 9.9. Beech and pine: values of acoustic invariants measured on sound and decayed wood when weight loss is 10%

	Acoustic invariants (m/s)		
	LR plane I_{12}	LT plane I_{13}	RT plane I_{23}
Beech			
Sound wood	2960	2800	1509
Decayed wood	2545	2301	1152
Differences (%)	14	18	24
Pine			
Sound wood	2811	2672	1282
Decayed wood	2381	2381	939
Differences (%)	15	11	27

From Bauer, C. et al., *Holzforschung*, 45(1), 41, 1991. With permission.

in Southern yellow pine wood. The ultrasonic pulse is treated the same as in acoustic emission technique. Several parameters of the ultrasonic pulse (waveform pattern, start or central time for velocity measurement, amplitude, frequency, peak voltage, average signal level, or root mean square voltage) which give indications of the energy contained in the waveform, were analyzed at the receiving point of the signal. Combinations of waveform parameters were used to define an acoustic signature of the specimen under test. The frequency spectrum on a control specimen and on a specimen after 5 days of fungus attack are shown in Figure 9.12. The loss of high-frequency components of the spectrum occurred on

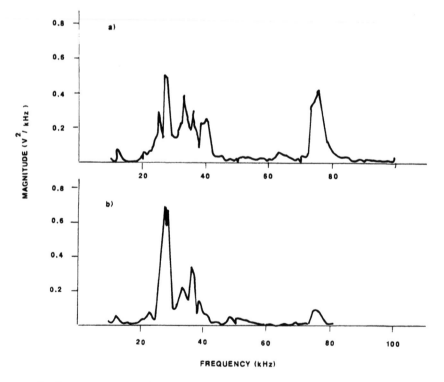

Figure 9.12. Frequency spectra of received wave on a specimen attacked by fungi and on a sound specimen. (a) Control specimen; (b) specimen exposed to fungi for 5 days. (From Patton-Mallory, M. and DeGroot, R. C., *Proc. 2nd Pacific Timber Eng. Conf.,* 2, 185, 1989. With permission.)

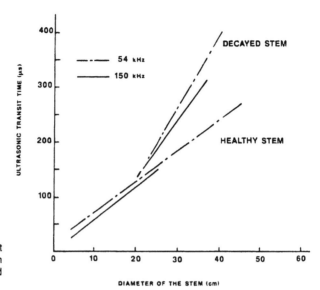

Figure 9.13. Predicted transit time (μs) of ultrasonic pulses in green ash; healthy and decayed stems, (McCracken, 1985).

the decayed specimen. In spite of the fact that spectral analysis on wood is in its infancy, attempts have been made to predict the strength as well as the rate of strength degradation of on-site poles (Anthoni and Bodig, 1989). Using this technique, a list of acousto-ultrasonic parameters which are very sensitive to the early stages of decay was established. The velocity measured from start time or central time decreases with increased decay degradation. Waveform amplitude, measured as root mean square, decreased with decay degradation. The high frequency components of the waveform were attenuated from very early stages of decay.

A subject of considerable interest is the detection of decay in trees. Significant annual economic losses result from the decay of the world's forests. McCracken (1985) proposed an ultrasonic test to detect and estimate decay in standing hardwood trees. This test would be a scattering-based method that used travel time of waves in a frequency range between 50 and 150 kHz. This relatively narrow band of frequency induces wavelengths that are of the same order of magnitude as the dimension of defects. It was proved that a direct relationship exists between the diameter of a healthy stem and the ultrasonic transit time (Figure 9.13). In decayed wood the transit time is increased when compared to a sound stem. The 150-kHz transducers were effective on stems of up to about 30 cm in diameter. Consistent results were obtained with 54 kHz transducers on 58 cm diamater stems.

To locate hollow rot in American beech in standing trees, Okyere and Cousin (1980) used an ultrasonic pulse echo method. For 250 kHz the maximum depth of penetration of echos is about 4.5 cm. The feasibility of the ultrasonic method for defect detection on trees was also demonstrated by Bethge et al. (1992). They compiled a catalog of defects in cross-sections of standing trees related to the radial stress wave velocity.

9.2.3. WOOD-BORING AGENTS

The category of "wood-boring agents" includes the marine borers (molluscs and crustaceans developed in coastal constructions structural members immersed in seawater), and the wood-boring insects (termites, insects causing power-post types of damage, and carpenters) responsible for important losses of living trees, timber that has a large portion of sapwood, subterranean structural elements, roof beams, etc. Figure 9.14 is an illustration of the deterioration produced by wood-boring agents, which induce a distinct reduction in ultrasonic velocity.

Sumiya (1965) simulated the biological boring of wood specimens using a sample strategy that allowed investigation of the effect on ultrasonic velocity and bending strength of the volume and size of the pores or of the cavities of wood (Figure 9.15). From all the figures analyzed, it can be understood that increasing the size and volume of the artificially produced defects induced a decrease in the strength and ultrasonic velocity values.

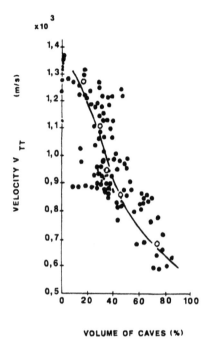

Figure 9.14. Relationship between ultrasonic velocity and the total volume of small caves on *Cryptomeria japonica* attacked by *Teredo japonica* (Ciessin) (After Sumiya, 1965.)

Prieto (1990) used the ultrasonic velocity method with narrow band longitudinal wave transducers (45 kHz) to detect the damage produced by *Hylotropes bajulus* L. larvae on Scotch pine. Larvae were inserted into predrilled holes on small, clear specimens (2 × 2 × 34 cm). The optimum conditions for the larvae development were 28°C and 80% relative humidity. The developing larvae produced caves in the specimens. The cave surfaces represent at least 40% of the initial section. The corresponding reduction of velocity could be 60% of the value in sound specimens. Using the same technique Prieto and Fernandez-Cancio (1990) proposed the circular scanning of round timber or poles to locate and evaluate internal holes. This approach completed the acoustic methods proposed by Miller et al. in 1965 for the detection of decay in poles.

Attack by marine borers was studied by Agi (1978). He suggested an ultrasonic inspection of a field marine piling by a diver to locate the damage and to evaluate the extent of cross-sectional loss of wooden elements.

Pellerin et al. (1985) reported on the relationship between the modulus of elasticity determined from longitudinal stress wave velocity and ultimate compression stress measured on wood specimens exposed to subterranean termite attack. The high experimental correlation coefficient (r) obtained, ($r_{\text{attacked specimens}}$ = 0.793), allowed the authors to state that this technique may be suitable for field testing of wood material exposed to various degrees of termite attack. Another point on which emphasis was placed was the utility of such nondestructive methods for field testing of wood preservatives.

9.2.4. ARCHAEOLOGICAL WOOD

Archaeological wood, as defined by Florian (1990), is "dead wood, used by an extinct human culture, that may or may not have been modified for or by use, and that was discarded into a specific natural environment". Ultrasonic testing of archaeological wood was reported by Schniewind (1990) and by Uzielli (1986) for structural members extracted from old churches and monuments. Characteristics of ten beams of oak and poplar (15 to 18 cm × 17 to 25 cm × 3.5 m), extracted from old buildings in Italy, were determined by Bonamini et al. (1990), as seen in Tables 9.10 and 9.11. The techniques used for mechanical characterization of the beams were static bending tests for rupture, shock tests with a hammer for inducing longitudinal and bending vibrations (analyzed via FFT), and ultrasonic tests with longitudinal and surface waves at 45 kHz and with pilodyn as a hardness tester. Different experimental relationships were established between parameters such as ultrasonic velocity, frequency, and penetration

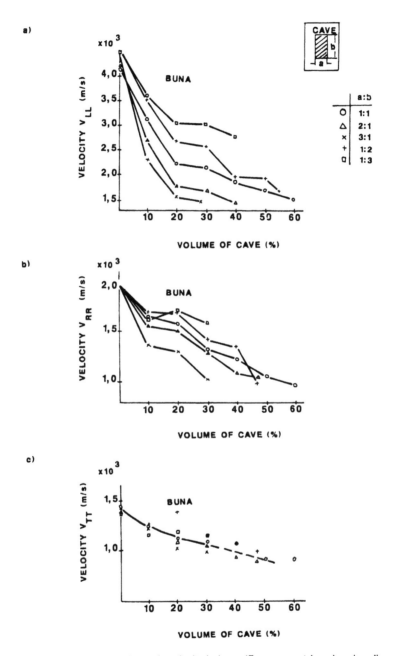

Figure 9.15. Effect of boring on ultrasonic velocity in buna (*Fagus crenata*) and on bending strength on sugi (*Cryptomeria japonica*). (a) velocity V_{LL} vs. volume percentage of caves; (b) velocity V_{RR} vs. volume percentage of caves; (c) velocity V_{TT} vs. volume percentage of caves. (From Sumiya, K., *Wood Res.* 34, 22, 1965. With permission.)

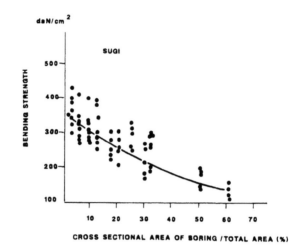

Figure 9.15. Effect of boring on ultrasonic velocity in buna (*Fagus crenata*) and on bending strength on sugi (*Cryptomeria japonica*). (d) bending strength vs. percentage of cross-sectional area of boring/ total area. (From Sumiya, K., *Wood Res.* 34, 22, 1965. With permission.)

Table 9.10. Physical characteristics of old beams from Italian monuments

Beam No.	Species	Moisture (%)	Density (kg/m³)	Ultrasonic velocities (m/s) 1	2	3	Pilodyn (mm)
0	Oak (1)	22	859	1828	2800	3754	11.16
1	Oak (1)	16	881	1869	3130	3697	11.10
3	Oak (1)	21	815	1047	2524	3776	7.90
4	Oak (2)	22	723	2012	—	3187	9.50
5	Oak (2)	20	887	1846	—	3781	7.00
2	Oak (2)	17	832	1970	4303	4555	7.00
6	Oak (2)	15	782	1989	4288	4566	7.40
7	Poplar	17	569	1785	4863	4766	10.40
8	Poplar	15	408	1685	4538	4538	16.80
9	Poplar	15	367	1391	3835	4316	16.10
10	Poplar	16	475	1776	4909	2499	13.30

Note: Oak (1), *Quercus pedunculata* Oak (2), *Q. cerris;* Poplar = *Populus* spp. 1, Velocity of longitudinal waves measured through the section; 2, velocity of longitudinal waves measured on the beam length; 3, velocity of surface waves on 40 cm. Beam 0 is a new beam; beams 1 to 10 were old beams.

From Bonamini, G. et al., in *Legno, Materiale per l'Ingegnaria Civile*, Racolta Monogr., University of Florence Italy, 69, 1990. With permission.

of pilodyn. The dispersion of measurements, expressed by the coefficient of variation, gives an idea about the quality of the beam. For the ultrasonic longitudinal wave velocity measured through the section of the beam, this coefficient was 9% for a new beam and 34% for the oldest defective beam. The coefficient of variation was between 13 and 45%, under the experimental conditions noted previously, for the velocities of surface waves.

It is possible to draw a map of the beam using ultrasonic velocity values. The experimental correlation coefficient established between the ultrasonic longitudinal velocity and the modulus of rupture was very high ($r = 0.80 - 0.98$). The equation between the elasticity modulus E and $R_{rupture}$, written as

$$E = 12,000 \ (R_{rupture})^{1/2} \ (MPa),$$

can give the value of the $R_{rupture}$ and a good estimation of the minimal load-carrying capacity of the beam *in situ*.

Table 9.11. Mechanical characteristics of old beams from Italian monuments

Beam	Vibrational methods				Ultrasound	Static methods	
	F_f (Hz)	E_f (MPa)	F_l (Hz)	E_l (MPa)	E_u (MPa)	E_s (MPa)	R (MPa)
0	32.25	8,313	438	8,240	6,872	6,378	32.5
1	24.25	7,839	496	10,840	8,809	9,159	35.4
3	22.60	6,297	443	7,997	5,298	6,141	11.7
4	18.00	3,543	362	4,736	2,008	2,859	13.3
5	21.00	6,356	425	8,013	6,224	7,760	28.8
2	28.85	14,499	578	13,898	15,720	11,248	46.0
6	28.25	13,077	570	12,714	14,685	11,583	42.6
7	37.25	9,871	624	11,079	13,732	13,546	26.9
8	37.60	7,203	608	7,533	7,163	6,319	11.1
9	31.75	5,338	584	6,272	5,520	5,391	7.5
10	36.75	7,664	612	8,892	11,676	8,728	26.1

Note: F_f, resonance frequency from flexural excitation; E_f, modulus of elasticity from resonance flexural excitation; F_l, resonance frequency from longitudinal excitation; E_l, modulus of elasticity from resonance longitudinal excitation; E_u, modulus of elasticity from ultrasonic waves; E_s, modulus of elasticity from static bending test ISO 8375; $R_{rupture}$, modulus of rupture from bending test.

From Bonamini, G. et al., in *Legno, Materiale per l'Ingegnaria Civile*, Racolta Monogr., University of Florence, Italy, 69; 1990. With permission.

Routine maintenance, repair, and preservation of fine art objects or of old ships can be improved by using ultrasonic inspection. As an example, the results reported by Witherell et al. (1992) relate to the ultrasonic tests used in the restoration and repair of the "USS Constitution", launched on October 21, 1797, which is currently the oldest commissioned ship afloat.

Murray et al. (1991) developed a combination of nondestructive tests, an ultrasonic air coupled system, and a radiographic technique for the detection of voids and cracks in wooden panel paintings. Studies are in progress at the Smithsonian Institution (Washington, D.C.), the Canadian Conservation Institute (Ottawa), The National Gallery of Art (Washington, D.C.), and the Tate Gallery (London).

REFERENCES

Agi, J. J. (1978). Nondestructive testing and structural analysis of in-place wood marine piling. Proc. 4th Nondestructive Testing of Wood, Washington State University, Pullman, 83–93.

Anthoni, R. W. and Bodig, J. (1989). Nondestructive evaluation of timber structures for reliable performance. Proc. 2nd Pacific Timber Engineering Conference. Vol. 2. 197–200.

Auge, F., Bucur, V., and Sandoz, J. L. (1989). Influence des Paramètres Technologiques et Physiques sur la Propagation des Ondes Ultrasonores dans les Bois des Structures. Internal report. Ecole Polytechnique Federale de Lausanne, Lausanne, Switzerland.

Auge, F. (1990). Influence de l'Humidité du Bois et de sa Temperature sur la Propagation des Ultrasons. Rep. D.E.A. Universite de Nancy I, Nancy, France.

Barrett, D., Green, D. W., and Evans, J. W. (1989). Temperature adjustments for the North American in-grade testing program. Proc. In-Grade Testing of Structural Lumber. Forest Product Research Society, Madison, WI, 27–38.

Bauer, C., Kilbertus, G., and Bucur, V. (1991). Technique ultrasonore de caractérisation du degré d'altération des bois de hétree et de pin soumis à l'attaque de differents champignons. *Holzforschung*, 45(1), 41–46.

Bethge, K., Mattheck, C., and Thun, G. (1993). A Catalogue of Stress Wave Velocities in Defect Containing Cross-Sections of Trees. Primärbericht 22.03.20P01B. Kernforschungszentrum, Karlsruhe, Germany.

Biot, M. A. (1956). Theory of propagation of elastic waves in a fluid-saturated porous solid. Low-frequency range. *JASA*. 28 (2), 168–178; II. Higher frequency range. *JASA*. 28 (2), 179–192.

Bohnke, I. and Guyonnet, R. (1991). Spectral analysis of ultrasonic waves for the characterization of thermally treated wood. *Proc. Ultrasonics Int. '91.* Le Touquet, France, July 1–4, 499–502.

Bohnke, I., Guyonnet, R., and Bucur, V. (1990). Ultrasonics for the Characterization of Wood Properties Before and After Thermal Treatment. Paper presented at Int. Cong. Recent Developments in Air and Structure Borne Sound and Vibration. Auburn University, Auburn, Al March 6 to 8.

Bohnke, I. (1989). Contribution à l'étude des Propriétes Physico-Chimiques et Mécaniques du Bois Rétifié. DEA Matériaux Macromoléculaires et Composites. Ecole des Mines de St. Etienne, St. Etienne, France.

Bonamini, G., Cecotti, A., and Montini, E. (1990). Indagini non distruttive per la verifica strutturale di travi di legno antico. in *Legno, Materiale per l'Ingegneria Civile*. Racolta Monogr. University of Firenze, Firenze, Italy, 69–118.

Bourgois, J. and Guyonnet, R. (1988). Characterization and analysis of torrefied wood. *Wood Sci. Tech.* 22(2), 143–145.

Bucur, V. and Sarem, M. (1992). An Experimental Study of Ultrasonic Wave Propagation in Dry and Saturated Solid Wood. Paper presented at 4th Encontro Brasileiro em Madeiras e em Estruturas de Madeira. University of São Paulo, São Paulo, Brazil.

Deschatres, M. H., Cohen-Tenoudji, J., Aguirre-Puente, J., and Thimus, J. F. (1989). Ultrasonic propagation through frozen porous media-liquid phase content determination. Proc. Ultrasonics Int. '89, Madrid.

Dinwoodie, J. M. (1981). *Timber, Its Nature and Behaviour*. Van Nostrand Reinhold, New York.

Efransjah, F., Kilbertus, G., and Bucur, V. (1989). Impact of water storage on mechanical properties of spruce as detected by ultrasonics. *Wood Sci. Technol.* 23(1), 35–42.

Facaoaru, I. and Bucur, V. (1974). Moisture content influence on pulse velocity and natural frequency on black poplar. 2nd Symp. Int. Comm. 7 NDT-Rilem Nouveaux Développements dans l'Essai Non Destructif des Matériaux Non Métaliques. Vol. 1. 48–53.

Florian, M. L. E. (1990). Scope and history of archaeological wood. in *Archaeological Wood*. ACS Ser. 225. Rowell, R. M. and Barbour, R. J. (Eds.). American Chemical Society, Washington, D.C., 3–34.

Fukada, E. (1968). Piezoelectricity as a fundamental property of wood. *Wood Sci. Tech.* 2, 299–307.

Gerhards, C. (1982). Effect of moisture content and temperature on the mechanical properties of wood. An analysis of immediate effects. *Wood Fiber.* 14(1), 4–36.

Hirai, N. (1974). Studies on piezoelectric effect of wood. *Bull. Shizuoka Univ. For.* No. 3, 11–77.

James, W. L., Boone, R. S., and Galligan, W. L. (1982). Using speed of sound in wood to monitor drying in a kiln. *For. Prod. J.* 32(9), 27–34.

Jones, R. (1953). A Preliminary Test of the Use of an Ultrasonic Pulse Technique for Testing Deterioration in Telegraph Poles. Road. Research Laboratory, Harmonsworth, West Drayton, U.K.

Kaiserlik, J. H. (1978). Nondestructive Testing Methods to Predict Effect of Degradation on Wood: A Critical Assessment. Gen. Tech. Rep. F.P.L. 19, USDA Forest Products Laboratory, Madison, WI.

Kollmann, F. F. P. and Höckele, G. (1962). Kritischer Verleich einiger Bestimmungsverfahren der Holzfeuchtigkeit. *Holz Roh Werkst.* 20, 461–473.

Kollmann, F. (1951). *Technologie des Holzes und des Holzwerkstoffe*. Springer-Verlag, Berlin.

Konarski, B. and Wazny, J. (1974). Use of ultrasonic waves in testing of wood attacked by fungi. Proc. 2nd Symp. 7th NDT-Rilem. Vol. 1. Constanta, Romania, 11–18.

Kubler, H. (1983). Mechanism of frost cracks formation in trees. A review and synthesis. *For. Sci.* 29, 559–568.

Launay, J. and Gilleta, F. (1988). L'influence de la température et du taux d'humidité sur les constantes élastiques du bois d'épicea de Sitka. Actes du Colloques Mechanical Behaviour of Wood. Bordeaux, June 8–9, 91–102.

Liese, W. (Ed.) (1975). Biological transformation of wood by micro-organisms. in *Proc. 2nd Int. Cong. Plant Pathology*, Springer-Verlag, Berlin.

Lee, I. D. G. (1965). Ultrasonic pulse velocity testing considered as a safety measure for timber structures. Proc. 2nd Symp. Nondestructive Testing of Wood, Washington State University, Pullman, 185–206.

McCracken, F. (1985). Using sound to detect decay in standing hardwood trees. Proc. 5th Symp. Nondestructive Testing of Wood, Washington State University, Pullman, 281–287.

Meyer, R. W. and Kellog, R. M. (Eds.). (1982). *Structural Use of Wood in Adverse Environments*. Van Nostrand Reinhold, New York.

Miller, B. D., Taylor, F. L., and Popeck, R. A. (1965). A sonic method for detecting decay in wood poles. *Proc. Am. Wood Preserv. Assoc.* 61, 109–115.

Murray, A., Green, R. E., Mecklenburg, M. F., and Fortunko, C. M. (1991). NDE applied to the conservation of wooden panel paintings. in *Nondestructive Characterization of Materials*. Vol. 6. Ruud, C. O., Bussiere, J. F., and Green, R. E. (Eds.) Plenum Press, New York, 73–80.

NF B 51–263. Panneaux de Particules. Epreuve de Vieillissement Accéléré par la Méthode dite "V313". AFNOR. May 1979.

Okyere, J. G. and Cousin, A. J. (1980). On flow detection in live wood. *Mater. Eval.* March, 43–47.

Patton-Mallory, M. and DeGroot, R. C. (1989). Acousto-ultrasonics for evaluating decayed wood products. Proc. 2nd Pacific Timber Engineering Conference. Vol. 2. 185–190.

Pellerin, R. F. DeGroot, R. C., and Esenther, G. E. (1985). Nondestructive stress wave measurements of decay and termite attack in experimental wood units. Proc. 5th Symp. Nondestructive Testing of Wood, Washington State University, Pullman, 319–352.

Pellerin, R. F. (1989). Inspection of wood structures for decay using stress waves. Proc. 2nd Pacific Timber Engineering Conference. Vol. 2. 191–195.

Petit, M. H., Bucur, V., and Viriot, C. (1991). Ageing monitoring of structural flakeboards by ultrasound. Proc. 8th Int. Symp. on Nondestructive Testing of Wood. Washington State University, Pullman.

Plona, T. J. (1980). Observation of a second bulk compressional wave in a porous medium at ultrasonic frequencies. *Appl. Phys. Lett.* 36(4), 259–261.

Plona, T., Winkler, K. W., and Schoenberg, M. (1987). Acoustic waves in alternating fluid/solid layers. *JASA.* 81(5), 1227–1234.

Polisko, S. (1986). Anisotropy of dynamic wood viscoelasticity. in *Rhéologie des Matériaux Anisotropes*. Huet, C., Bourgoin, D., and Richmond, S. (Eds.). Editions Cepadues, Toulouse, France, 453–460.

Prieto, G. (1990). Detection and Estimation of *Hylotrupes bajulus* L. Wood Damage by Ultrasound. Pap. No. 2350, 21st Annu. Meet. Int. Res. Group on Wood Preservation, Rotoura, New Zealand, May 13 to 15.

Prieto, G. and Fernandez-Cancio, A. (1990). Hole Delimitation Inside Round Timber via Ultrasonic Technique. Pap. No. 2358, 21st Annu. Meet. Int. Res. Group on Wood Preservation, Rotoura, New Zealand, May 13 to 15.

Quarles, S. L. and Wengert, E. M. (1989). Applied drying technology 1978 to 1988. *For. Prod. J.* 39(6), 25–38.

Ross, R. J., Ward, J. C., and Tenwolde, A. (1992). Identifying Bacterially Infected Oak by Stress Wave Nondestructive Evaluation. Rep. No. 512. USDA Forest Products Laboratory, Madison, WI.

Sakai, H., Minamisawa, A., and Takagi, K. (1990). Effect of moisture content on ultrasonic velocity and attenuation in woods. *Ultrasonics.* 28(6), 382–385.

Sandoz, J. L. (1990). Triage et Fiabilité des Bois de Construction. Thesis no 851/1990. Ecole Polytechnique Fédérale de Lausanne, Lausanne, Switzerland.

Schniewind, A. (1990). Physical and mechanical properties of archaeological wood. in *Archaeological Wood.* ACS Ser. 225. Rowell, R. M. and Barbour, R. J. (Eds.). American Chemical Society, Washington, D. C., 87–110.

Sellvold, E. J., Radji, F., Hoffmeyer P., and Bach, L. (1975). Low temperature, internal friction and dynamic modulus for beech wood. *Wood Fiber.* 7(3), 162–169.

Siau, J. F. (1984). *Transport Processes in Wood.* Springer-Verlag, Berlin.

Skaar, C. (1988). *Wood-Water Relations.* Springer-Verlag, Berlin.

Sumiya, K. (1965). Relations between defects in wood and velocity of ultrasonic waves. *Wood Res.* 34, 22–36.

Tiemann, H. D. (1906). Effect of Moisture Upon the Strength and Stiffness of Wood. USDA For. Serv. Bull. 70, cited by Siau (1984).

Tsuzuki, K., Takemura, T., and Asano, I. (1976). Physical properties of wood-based materials at low temperatures. *Mokuzai Gakkaishi.* 22(7), 381–386.

Tsuzuki, K. and Yamada, N. (1983). Bending strength of wood and wood-based materials at low temperature. *J. Soc. Mater. Sci. Jpn.* 32(359), 864–868.

USDA. (1974). *Wood Handbook: Wood as an Engineering Material.* Agric. Handb. No. 72. USDA (1974). Forest Products Laboratory, Madison, WI.

Uzielli, L. (1986). Personal communication.

Vary, A. (1980). Acousto-ultrasonic characterization of fiber reinforced composites. Concepts and techniques for ultrasonic evaluation of material mechanical properties. in *Mechanics of Nondestructive Testing.* Plenum Press, New York, 123–141.

Verkasalo, E., Ross, R. J., Tenwolde, A., and Youngs, R. L. (1993). Properties Related to Drying Defects in Red Oak Wetwood. Rep. No. 5. USDA Forest Products Laboratory, Madison, WI.

Wert, C. A., Weller, M., and Caulfield, D. (1984). Dynamic loss properties of wood. *J. Appl. Phys.* 56(9), 2453–2458.

Wilcox, W. W. (1988). Detection of early stages of wood decay with ultrasonic pulse velocity. *For. Prod. J.* 38(5), 68–73.

Witherell, P. W., Ross, R. J., and Faris, W. R. (1992). Using today's technology to help preserve USS Constitution. *Nav. Eng. J.* May, 124–134.

Zimmermann, M. H. (1983). *Xylem Structure and the Ascent of Sap.* Springer-Verlag, Berlin.

Chapter 10

Acoustic Emission

10.1. PRINCIPLES AND INSTRUMENTATION

10.1.1. PRINCIPLES

As defined by Liptai et al. (1972), Nichols (1976), Lord (1983); Lynnworth (1989); Stephens and Levinthall (1974), and Williams (1980), acoustic emission is a transient elastic wave generated by the rapid release of energy within a material. The aim of acoustic emission analysis is to obtain information about the source of this phenomenon from the detected ultrasonic signal. Acoustic emission in solid wood or in wood-based composites can be generated by stress level—(high or low rate) plastic deformation, crack propagation, drastic variation of temperature and moisture content, drying, sap cavitation in vessels, freezing, phase transformation, anisotropy and nonhomogeneity of the anatomic structure in adverse environmental conditions, rapid collective motion of a group of anatomic elements, modification of the orientation of crystallites in microfibrils, dislocations in cellulosic chains, etc.

Kaiser (1953) was the first to analyze acoustic emission from spruce specimens under tensile stress. Simultaneously studying wood and different metallic materials (zinc, copper, aluminum, lead, and steel), he stated that the effect, known today as the "Kaiser effect", is an irreversible characteristic of acoustic emission resulting from an applied stress. For a specimen subjected to repeated stress the Kaiser effect is observable, if no acoustic emission occurs until the previously applied stress levels are exceeded. This means that if the damage is only mild, the acoustic emission threshold is constant and the emissions occur at the same stress level. Above a specific level of damage, the threshold evolves and the acoustic emission occurs at a lower stress. This is known as the Felicity Effect.

Acoustic emission studies in Europe, North America, and Asia have increased since 1965, probably due to technological advances in testing equipment and to basic and applied studies both in the laboratory and in the field. Table 10.1 shows the frequency range of various types of acoustic emission studies for various media. For solid wood and wood-based materials the frequency range is between 100 kHz and 2 MHz.

The physical description of the acoustic emission phenomenon should proceed from kinetic theory, which leads to dynamic equations for parameters such as density, velocity, and stress field. Therefore, the exhaustive description of acoustic emission can be made by the solution of equations representing a stress field in unstable matter.

The material instabilities generated by inelastic deformations in a solid induce modifications of its properties. These modifications are reflected by nonlinearity introduced into dynamic equations. An exact description of the instability in the source represents a very complicated, specific problem. The complexity of the problem increases when the model of the solid sample is varied from unbounded to half space, unbounded plate, beam, etc. (Pao, 1978; Ceranoglu and Pao, 1981; Sachse and Kim, 1987; Castagnede et al., 1989).

Brindley et al. (1973), Grabec (1980), Kollmann (1983), Lord (1983), Pollock (1970, 1973), Stephens and Levinthall (1974), Swindlehurst (1973), Vary (1980), Williams (1980), and Niemz et al. (1983) tried to understand the properties of acoustic emission signals, essential for material characterization or for monitoring the integrity and safety of structures.

The technology developed to locate and characterize the source is also called acoustic emission. If the reviewed literature is any guide it seems that recent developments are directed at the instrumentation, signal processing, and practical applications of monitoring systems of different technological processes.

The parameters that characterize an acoustic emission signal, described in Figures 10.1 and 10.2 and ASTM STP 505 and ASTM STP 571 are:

- Mode of emission, continuous or burst
- Rate of emission
- The acoustic emission event, defined as a rapid physical change in a material that releases energy appearing as acoustic emission
- The accumulated activity—the total number of events observed during a specific period of time
- The threshold set at a selected discriminator level

Table 10.1. Frequency range of different types of acoustic emission studies for various media, completed with solid wood and wood-based materials

Frequency range	Acoustic emission studies on
10–100 Hz	Micro-earthquake studies
200–300 kHz	Small specimens, geological material
	Concrete and rocks
	Composites
100 kHz–2 MHz	Metallic material and structures
	Solid wood and wood-based composites

From Liptai, R.G. et al., An Introduction to Acoustic Emission. STP 505. American Society for Testing Materials, Philadelphia, 1972.

- The duration of the event—from initially crossing the threshold until alternating below it
- The ring-down count—the number of wave peaks above the threshold
- The amplitude of the highest peak—the maximum amplitude of each recorded event in arbitrary units
- The rise time—the time from the crossing of the set threshold to the apex of the highest peak
- Frequencies within the emitted wave
- Energy as the area under the envelope of the amplitude-time curve, measured for each burst
- Cumulative energy recorded progressively since the beginning of the test; the energy is an effective method of differentiating between acoustic emission signals that have different frequency and damping characteristics
- Energy rate—the sum of the energy emitted by all events observed per unit time
- "Take off" point—on the stress or strain vs. cumulative energy graph; this point corresponds to the stress level at which the cumulative energy of the microfractures increases dramatically (Figure 10.3)
- Mean-square voltage—a measure of the energy
- Root mean square (rms) voltage value or signal level—used for the measurement of the signal amplitude averaged over a period of time

One of the most popular acoustic emission techniques is ringdown counting. The principle of this technique is "to count the number of times a threshold voltage is exceeded by the oscillating transducer output considered by acoustic activity" (Brindley et al., 1973). The main advantages of this technique are given by the simplicity of the measurement of acoustic activity, the suitability for comparative testing of identical samples, and the automatic improvement of noise rejection. The greatest disadvantage

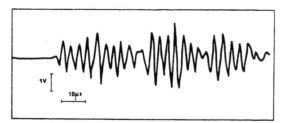

Figure 10.1. Typical acoustic emission signal bursts in plant tissue. The signal was produced by breaking a pencil on a birch dowel. The source was 25 cm from the transducer. (From Tyree, M. T. and Sperry, M. T., *Plant Cell Environ.*, 12(4), 371, 1989. With permission.)

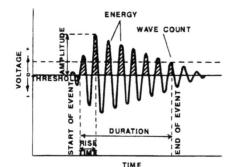

Figure 10.2. Parameters of an acoustic emission event, based on ASTM STP 571.

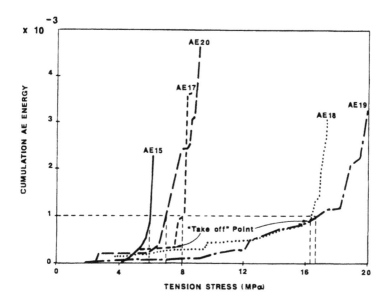

Figure 10.3. Acoustic emission cumulative energy vs. tension stress on different pine specimens labelled AE15 to AE20. (From Knuffel, W. E., *Holzforschung*, 42, 195, 1988. With permission.)

of this technique is related to the dependence of the sample geometry and signal amplitude on defect growth.

Emission parameters noted in Table 10.2 contribute to the identification of sources and allow different deformation mechanisms to be distinguished in terms of their characteristic "signature". This parameter is defined in ASTM STP 505 as "a set of identifiable characteristics of acoustic emission signals attributable to a particular type of source". The differences in the characteristic signature of wooden structures or living trees can be used for defect detection. However, a great deal of work remains to be done to provide reliable indications (between laboratory data on samples or large field-scale laboratory structures and on effective structures) of defect location and type, but the characteristics of acoustic signature from various types of discontinuities do indicate a potential way of differentiating between them.

10.1.2. INSTRUMENTATION
10.1.2.1. Systems

An acoustic emission system includes transducers, preamplifier, mean amplifier, signal processors, transient recorders, spectrum analyzer, microprocessor, and data storage system. System design depends on several factors such as the physical location of the equipment or the type of investigation to be fulfilled, for laboratory studies or for field structural integrity monitoring in adverse environmental conditions. The single-channel system is shown in Figure 10.4, and comprises the transducer, signal leads connected to the electronic signal processor, counter, and recorder. The test specimen configuration

Table 10.2. Parameters relating to acoustic emission signals

Emission parameter	Type of information carried
Waveform	Fine structure of source event
Frequency spectrum	Nature of source event; integrity of specimen
Amplitude	Energy of source event
Amplitude distribution	Type of damage occurring
Rate	Rate of damage occurring
Distribution in time	Type of damage
Relative arrival time at several transducers	Source location

From Stephens R. W. B. and Pollock, A. A., *J. Acoust. Soc. Am.*, 50(3, Part 2), 904, 1971. With permission.

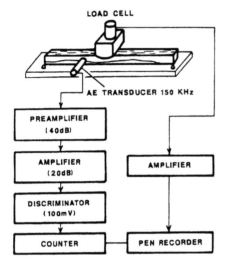

LOAD CELL

AE TRANSDUCER 150 KHz

PREAMPLIFIER
(40dB)

AMPLIFIER
(20dB)

AMPLIFIER

DISCRIMINATOR
(100mV)

COUNTER

PEN RECORDER

Figure 10.4. Acoustic emission system with one transducer. Stress waves were generated by bending static loading. The level of electronic noise was < 22 mV and the threshold on the counter was set at 100 mV for a gain of 60 dB. (From Noguchi, M. et al. *Acoust. Lett.*, 9(6), 79, 1985b. With permission.)

and the detection systems used in wood science are generally the same as the systems used for different materials. Starting with a small load, the specimen is strained at a uniform rate until the final stress is reached. The piezoelectric transducer placed on the specimen, with an acoustical impedance matching coupling medium, senses the acoustic activity. The signal is then amplified and processed or recorded and processed later. The arrangement of the acoustic emission sensor for different tests (tension, compression, bending) under static loading is presented by Ansell (1982a,b), Beall and Wilcox (1987), and Yoshimura et al. (1987).

Using an array of transducers spread over the structure in a prescribed geometrical pattern, and measuring the time of arrival of bursts to the transducers, the position of the acoustic emission source within the structure can be determined by a computer-controlled signal analysis system. This approach adds a new dimension to signal discrimination in monitoring different processes. For the interpretation of experimental data, the unwanted signals are avoided by appropriate arrangement of sensors, performed by a skilled operators having a good knowledge on the physics of wave propagation within the material under test. Moreover, a skilled operator is able to eliminate undesired sources of noise.

It is important to bear in mind that for an idealistic configuration, in order to reliably assess the experimental data, it is normally assumed that:

- Each acoustic emission event will be counted only once
- All damaging events will produce acoustic emission signals of sufficient amplitude to be counted and said signals are equaly damaging to the structure

Indeed, these conditions are not always fulfilled in practical applications. It is recognized that the mechanical configuration of the specimen and an inappropriate transducer in the frequency response could modify the acoustic wave considerably. Consequently, the operator may misinterpret the acoustic event.

The main requirements of an acoustic emission system are

- To distinguish between signals from pertinent and insignificant sources
- To exclude mechanical and electrical interference from the field
- To produce records suitable for comparison with past and future records

10.1.2.2. Material Conditioning

Acoustic emission testing requires that the material under test be conditioned in such a way that structural discontinuities emit pulses. Under specific conditions the generated stress waves are propagated into the elastic medium to the transducer where they excite an electric signal which is processed further by electronic instruments.

The central problem of the acoustic emission technique is to isolate the emissions that identify the quality of the specimen from those that are inconsequential to the test. Many times the emissions consist of distinct bursts of energy from one or more sources, and they can appear as continuous, low-level energy signals.

In a viscoelastic medium such as wood, the dispersion phenomena occur during acoustic emission signal propagation. This is caused by damping the vibration in the material, by the geometrical shape and size of the specimen, by multiple scattering induced by anatomical structure and inhomogeneities, or by reflections at the boundaries of a waveguide. Elsewhere, the effect of multiple scattering is less important if the dominant wavelength of source signals is longer than that of the characteristic dimension of the inhomogeneities.

Frequencies intrinsic to acoustic emission sources for wood materials range from audible to ultrasonic (often between 20 and 300 kHz; for higher frequencies, material attenuation substantially reduced the system sensibility) and the threshold used varies from 0.1 to 0.7 V (or less), depending on the nature of the studied phenomena.

In the majority of wood science studies, acoustic emission activity was stimulated by the application of an external load to the sample which is sufficient to generate stress waves (this aspect is discussed in more detail later). Stress waves can also be induced by ultrasonic pulses. This procedure was described by Vary and co-workers (1979, 1980). For solid wood, Beall (1987) used this technique for monitoring glueline curing in a lap joint in order to develop a quantitative method for cure time measurement.

10.1.2.3. Transducers

During acoustic emission part of the energy radiated from the source in the form of elastic waves can be detected at the surface of the material under test. Transducers are used as receivers of mechanical vibrations, converting acoustic energy into electrical energy. The design or the selection of a particular transducer is prescribed by its specific application—the level of acoustic activity of the material under test, the noise background, the signal attenuation, etc. Capacitive transducers are recommended for quantitative analysis because of their uniform sensitivity over a wide frequency band. They are used mainly for calibration because of their very delicate construction. Piezoelectric (PZT: lead zirconate titanate ceramics, and PVDF: polymer polyvinylidene fluoride) transducers are used for laboratory and field tests.

The main element in the acoustic emission transducer is the piezoelectric ceramic. An appropriately sized ceramic is placed in a metallic case which is electrostatically shielded and mechanically protected. An outlet is provided for necessary cabling. The sensor is placed in direct contact with the specimen or structure, and connected to the preamplifier.

10.1.2.4. Amplifiers and Signal Processors

Very few transducers are commercially available that incorporate the amplifier with the sensor. Below we note some details about the pre- and postamplifiers (or called simply, amplifier).

Acoustic emission preamplifiers have a relatively flat frequency response, between 20 KHz and 2 MHz, in the absence of bandpass filters. They generally have a fixed gain of either 40 or 60 dB. The postamplifier gives an additional noise rejection capability to the system—usually 20 dB. The sum of the gain of the pre- and postamplifier represents the total gain of the system.

The signal processor contains voltage-controlled gates, envelope processors, and logarithmic converters.

The transient recorder allows the study of the individual acoustic emission burst signals by digitizing the signal in real time and storing it in memory. From the memory the signal can be played back into an oscilloscope, a spectrum analyzer, or a computer for complex processing or examination of the waveform.

10.1.2.5. Signal Processing

The aim of acoustic emission signal processing is to locate the source of emissions. Acoustic emission from wooden materials is a random process. The signals are nonperiodic and contain many frequencies. However, these nonstationary signals can be characterized by their statistical properties, including:

- Mean square values, which describe the intensity of the signals
- Probability density functions, which show the probability that the signals will assume a value within some defined range at any instant in time
- Autocorrelation functions, which indicate the general dependence of the values of the signals at one

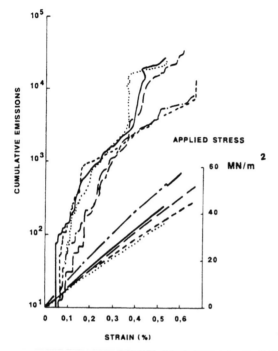

Figure 10.5. Cumulative acoustic emission counts vs. strain (%) and stress (MN/m⁻²) on several specimens of plywood. (From Ansell, M. P., *Wood Sci. Technol.*, 16, 35, 1982a. With permission.)

time on the values at another time
* Power density functions, which give the contribution of each frequency component to the total power

The acoustic emission signals produced by wood are similar to signals produced by other materials. The processing of acoustic emission signals is featured prominently in the literature [ASTM STP 571, Beattie (1983), Ceranoglu and Pao (1981), Crostack (1977), Egle and Trato (1967), Green (1980), Hsu et al. (1977), Mitrakovic et al. (1985), Muenow (1973), Pollock (1974), Sachse and Kim (1987), Sachse (1988), Simpson (1974), Stephens and Pollock (1971), Stone and Dingwall (1977), Weisinger (1980), Woodward (1976), Yamaguchi (1988), Ying et al. (1974)]. The simplest way to examine such an acoustic emission signal is to visually observe the waves from an oscilloscope and to set the gain so that the preamplifier noise is visible on the trace. In addition, verify:

* the influence of the geometry of the specimen on the received signals
* the effect of the coupling medium between the probe and the sample
* the electronic system used to select, filter, transmit, and amplify the signals.

Commonly, with modern systems incorporating digital computers, the operator counts the signals emitted during deformation and automatically plots the results as a function of strain, stress or other parameters (Figure 10.5). Furthermore, the amplitude distribution may be obtained as a plot of the number of events vs. the threshold amplitudes, etc. In more sophisticated analyses the acoustic emission signal is analyzed by fast Fourier transform (FFT) to determine frequency components or the power spectrum of the signal. This procedure provides a rational way for studying the inhomogeneous materials for which no two signals ever have exactly the same frequency components and for which the analysis of pulse dispersion is essential for the characterization of materials.

Difficulties (e.g., wave attenuation, geometric dispersion effect, scattering) relating to the source location can be avoided if a triangulation technique is used (Sachse, 1988).

10.1.2.6. Factors Affecting Acoustic Emission Responses from Wooden Materials

The principal advantages of acoustic emission testing over other forms of nondestructive testing are the wide volume surveyed, the real-time nature of the technique, and the capability of continously

Table 10.3. Factors that influence acoustic emission detectability in solid wood and wood composites

Factors resulting in	
High amplitude signals	**Low amplitude signals**
High strength	Low strength
High strain rate	Low strain rate
Low moisture content	High moisture content
High yield strength	Low yield strength
High anisotropy	Low anisotropy
High latewood proportion	Low latewood proportion
Narrow annual ring	Large annual ring
Low temperature	High temperature
Tension deformation	Compression deformation
Crack propagation	Radiation damage
Some natural defects	Impregnation
Long tracheids or fibers	Short tracheids or fibers
Basic crystalline structure	Content of lignin

monitoring structures. This technique is interesting mainly because it is nonlocalized, meaning that the receiver is not necessarily placed near the source or in the area under test. The environment in which acoustic emission tests are performed is an essential factor in determining the rate of emission. In comparison with vibration measurements or strain gauge measurements, the acoustic emission technique has also been proven effective. The rewards from using acoustic emission monitoring of wooden structures are considerable, but the difficulties are also great. The factors affecting the detectability of the acoustic emission response from wooden materials are summarized in Table 10.3.

A better understanding of the influence of the properties of wood on the acoustic emission parameters can be achieved by using the acousto-ultrasonic technique. While it is commonly accepted that the acoustic emission technique uses an external mechanical stress to induce the response of the specimen under test, the acousto-ultrasonic technique investigates the response of specimen when stimulated by an ultrasonic pulse. The technique is also called the acoustic stimulation technique, and was developed for the location of delaminations in wood-based composites and for the detection of defects in lumber such as decay, knots, voids, and cross-grain. The influence of species, moisture content, and type of transducer on acousto-ultrasonic parameters was studied by Lemaster and Dornfeld (1987).

10.2. ACOUSTIC EMISSION FOR THE STRUCTURAL EVALUATION OF TREES, SOLID WOOD, PARTICLEBOARD, AND OTHER WOOD-BASED COMPOSITES

Many interesting applications for acoustic emission testing have been developed in recent years in wood science. The objective of the technique is to measure parameters for specific properties of the material under test. If the selection of measured parameters is appropriate, they will correlate with material characteristics. Furthermore, the correlation relationship will produce calibration curves that can be used to verify the quality of the material. Practical applications were developed and documented for the following topics:

- Cavitation events in xylem for determining the hydraulic sufficiency or stress on woody stems
- Detection of fungi and termite activity in wood
- Fracture mechanics in solid wood and composites

10.2.1. CAVITATION

Cavitation is generally accepted as being detrimental to the water economy of plants and forest trees. In the last 20 years cavitation in forest trees was thoroughly studied by Milburn and Johnson (1966), Milburn and Crombie (1984), Milburn (1973a,b), Zimmermann and Milburn (1982), Zimmermann et al. (1980), Tyree and Dixon (1983), Zimmermann (1983), Tyree et al. (1984a, b), Tyree and Sperry (1989), Oertli (1990), Cochard and Tyree (1990), Cochard (1992), Jones et al. (1990), Robson (1990),

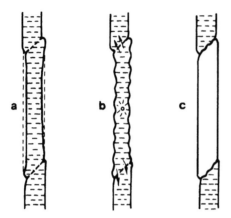

Figure 10.6. Model proposed by Milburn (1979) for cavitation of sap into a xylem-conducting element. (a) Under tension, sap strains the walls inward from dotted line. (b) Cavitation "bubble" forms and strained walls produce vibration detectable as a "click." Water exits to adjoining conduit. (c) Evacuated conduit gradually becomes air-filled by dissolution of gas from sap in walls. (From Milburn, J. A., *Water Flow in Plants*, Longman, London, 1979. With permission of the author.)

Salleo and Lo Gullo (1990), Pisante et al. (1990), and Sperry (1990). Cavitation is defined as the formation of one or more pockets of gas, or cavities (referred to as bubbles or voids) in a liquid. (Apfel, 1981; Vaughan and Leeman, 1989).

The hydraulic architecture of the xylem structure of plants and trees is based on xylem walls. Water is taken up from the soil by the roots and is transported to the leaves. In xylem bound water coexists with free water and water vapor and other gases located in the lumen and intercellular spaces. The storage of water in trees is determined by three main factors: elasticity of tissues, capillarity, and cavitation. The elasticity of tissues enables the plant body or xylem to shrink or swell, and is related to the pressure fluctuation of the ascending sap. Capillarity and water storage is dependent on the type and size of cells. The greatest storage capacity in a tree is observed in autumn, when the sap pressure is very low. Cavitation occurs when the pressure of the rising sap drops at a critical point and the embolism of conducting elements is observed. Cavitation pressure varies with species; e.g., very low values were measured for ash (-30 to -70 bar) and eucalyptus (-28 to -40 bar) (Zimmermann and Milburn, 1982). The mechanism for the formation of bubbles in conducting elements was described by Milburn (1979) in Figure 10.6 and by Zimmermann (1983), as an explosion of evaporated water in the vessel lumen. The increasing stress developed by bubbles in the conducting elements determines the oscillation of the tissues surrounding the cavitating zone. These vibrations produce pulses and consequently acoustic events in the sonic and ultrasonic ranges, depending on the size and elastic properties of vessels, tracheids, or fibers (Ritman and Milburn, 1988; Crombie et al., 1985).

The disruption of the continuity of sap in a small number of long, wide vessels may more seriously affect the flow of sap in the xylem than would the same number of narrow anatomical elements.

The efficiency of counting acoustic emission events was based on the assumption that all conducting elements cavitated during the stress period (Sandfor and Grace, 1985). Each event is considered to correspond to a single cavitation occurring in the sap circulating within the fibers, vessels, or tracheids. Sometimes total count measurements were performed simultaneously, with a more complex analysis of pulses given by the measurements of velocity and attenuation of the ultrasonic signals emitted during cavitation (Tyree and Sperry, 1989). For detecting acoustic emission activity in plants or in wood tissues, Tyree and Dixon (1983), Sandfor and Grace (1985), Poliszko et al. (1988), and Tyree and Sperry (1989) used piezoelectric transducers that had good sensitivity in the frequency range 100 kHz to 1 MHz and high sensitivity around 500 kHz.

The frequency response of acoustic emission signals is broad band (from kilo- to megahertz). The resonant anatomic elements may be identified through the cutoff frequencies, which are related to the dimensions of these elements. Large elements such as vessels can produce more acoustic events in the audible spectrum than fibers and other small elements that are expected to produce acoustic events in the ultrasonic range. Such effects may be useful for monitoring disruptive water stress in xylem tissues (Sandford and Grace, 1984), if each event is interpreted as a single cavitation occurring in the sap within the conducting element. Frequency analysis (FFT) was also used by Tyree and Sperry (1989) for the studies related to cavitation during the dehydratation of thuja, maple, and pine stems.

Analyzing the hydraulic sufficiency in xylem, Tyree (1989) divided the embolism cycle into five steps, characterized with specific duration as: water stress producing bubbles (μs), water vapor filled

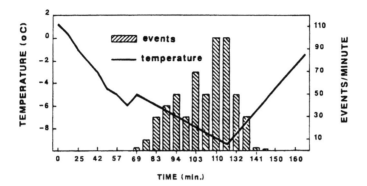

Figure 10.7. Influence of temperature on acoustic emission activity in Eucalyptus stem. (From Raschi, A. et al., *Ann. Sci. For.* 46 (Suppl.), 3425, 1989. With permission.)

conduits (ms), extension of vapors (minutes), embolism of the conduit (days to months), and the loss of hydraulic conductance.

The main factors inducing cavitation in trees could be:

- The hydraulic state of woody stems (dehydratation, water shortage, or water stress)
- Cycles of freezing and thawing
- Infection with fungi that parasitize the xylem cells

Ultrasound emission and water content in pine in relation to water stress periods were studied by Pena and Grace (1986). It was noted that during drought periods, the emitted ultrasound pulses occurred at 60 events per minute, while at the beginning of the experiment only 25 events per minute were registered. After rewatering, they found that "the plants that had been droughted failed to produce ultrasound emissions when water potential fell". These responses of the living xylem tissues suggested that an acoustic emission sensor could be used (Jones et al., 1989; Borghetti et al., 1989) as a detector for water stress in fruit trees or forest trees for the practical purpose of monitoring diurnal watering in the field.

Raschi et al. (1989) detected cavitation through acoustic emission (event number) measurements during cycles of freezing and thawing of coniferous (*Araucaria excelsa*) and broadleaved trees (*Eucalyptus occidentalis*) (Figure 10.7). Acoustic emission activity started at −4.5°C, prior to ice formation. The maximum in acoustic events was observed at the lowest temperature, which reached −8°C. Furthermore, a decrease in acoustic events was observed when the temperature rose. This technique is suitable for the study of the frost-cracking mechanism in trees, when the temperature falls from 0° to −25°C or more and the xylem tissue expands or swells. Acoustic emission testing was also used to follow microcrack formation during the volumetric swelling of wood in water (Poliszko et al., 1988). The radial direction is the main direction for the appearance of microcracks during mechanosorptive activity. Microcracks are induced by the simultaneous effects of strain and moisture gradient. In this experiment nonstationary conditions for acoustic emission activity were caused by the gradual penetration of water into the specimen and by swelling stresses. Locally, these stresses could exceed the strength of wood, inducing microcracks. It was demonstrated that the acoustic emission activity was related to the increase in swelling.

The third cited factor for inducing cavitation is the infestation of vessels with fungi. The model proposed by Zimmermann (1983) for the dynamics of cavitation and infestation is very attractive and is presented in Figure 10.8. This model is a reconstruction of a segment "of a stem of a seedling of *Ulmus americana*, injected in the axial scale with a spore suspension of the Dutch elm disease fungus *Ceratocystis ulmi*, 4 days before harvest of the stem" (Zimmermann, 1983). The nonconducting vessels are marked by an arrowhead.

10.2.2. DETECTING THE ACTIVITY OF BIOLOGICAL AGENTS

A survey of a number of publications on the detection of activity of biological agents in solid wood by acoustic emission testing reveals that this technique is not widely used. We note two main applications

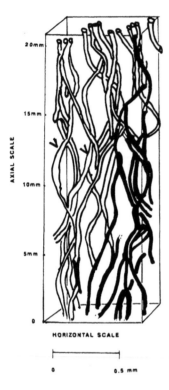

AXIAL SCALE

20mm

15mm

10mm

5mm

0

HORIZONTAL SCALE

0   0.5 mm

Figure 10.8. Model proposed by Zimmermann (1983) for cavitation in vessels of *Ulmus americana,* induced by fungus attack. White vessels are embolized; black vessels are living vessels. (From Zimmermann, M. H., *Xylem Structure and the Ascent of Sap.* Springer-Verlag, Berlin, 1983. With permission.)

for acoustic emission: (1) the detection of fungus activity producing wood decay and (2) the monitoring or detection of termite activity. The acoustic emission technique was used for the detection of fungal attack by Noguchi et al. (1985b, 1986, 1991, 1992), Imamura et al. (1991) in Japan, Beall and Wilcox (1987) in the U.S., and by Niemz et al. (1990) in Germany. Fujii et al. (1990) dealt mainly with the detection of termite activity in wood in Japan. For example, the incipient activity of white-rot fungus (*Coriolus versicolor* (Fr.) Quel.) can be detected on standard specimens submitted to bending stress. The decayed wood begins to generate acoustic emission signals at a lower stress level than sound wood. Relationships between the weight loss of a decayed specimen and cumulative acoustic emission counts are presented in Figure 10.9. It was noted that acoustic emissions measured at 150 kHz indicate decay that occurs at below 5% weight loss.

Beall and Wilcox (1987) detected brown-rot decay (*Poria placenta*) in radially compressed white fir wood using the acoustic emission technique. Figure 10.10a shows acoustic emission activity up to 2% compression, probably due to microcracks within the cell wall. The acoustic emission events of a sound white fir specimen, monitored at 100 dB gain, to about 5 and 1.5% compression is presented in Figure 10.10b. The curves for the decayed specimen are very different from those of the sound wood.

Data analysis allowed the deduction of regression coefficients. The logarithm of stress at 100 events vs. square root of mass loss ($r = 0.98$) and the logarithm of events per unit stress vs. the same mass loss ($r = 0.97$) are very sensitive parameters for gauging the level of fungus attack in solid wood.

The invasion of wood by termites was studied on western hemlock specimens, using a 150-kHz acoustic emission sensor at 70 dB gain and 0.1 V threshold. Acoustic emission activity was detected on specimens severely attacked as well as on the specimens exposed at a very early stage. The distribution of acoustic events corresponds to the region attacked (Figure 10.11).

10.2.3. ACOUSTIC EMISSION AND FRACTURE MECHANICS

One application of quantitative acoustic emission is to study the fracture process. The investigation of this process relies on several elements: the mechanical loading system, the deformation measurement system, and the broad-band acoustic emission system with displacement or velocity transducers and appropriate signal processing equipment.

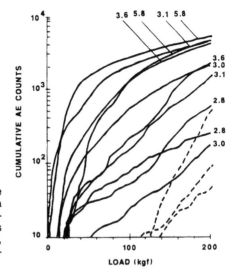

Figure 10.9. Relationships between the cumulative acoustic emission counts, load, and weight loss on a decayed western hemlock specimen tested in bending. (----) Sound wood; (——) decayed wood; numbers at top and right denote weight loss (g). (From Noguchi, M. et al., *Acoust. Lett.,* 9(6), 79, 1985b. With permission.)

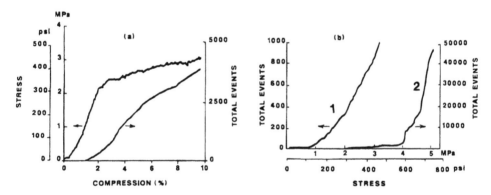

Figure 10.10. Radial compressive stress and corresponding acoustic emission parameter, on decayed (a) and sound white fir (b). (1) Acoustic emission activity at 1.5% compression; (2) acoustic emission activity at 5.0% compression. (From Beall, F. C. and Wilcox, W. W., *For. Prod. J.,* 37(4), 38, 1987. With permission.)

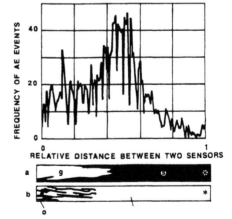

Figure 10.11. Distribution of acoustic emission events on a termite-attacked specimen. (a) X-ray image of the specimen; (b) image at the surface of the specimen with galleries; (*) position of the acoustic emission transducer; (g) galleries; (s) sound part of the specimen; (o) opening of galleries. (From Fujii, Y. et al., *For. Prod. J.,* 40(1), 34, 1990. With permission.)

10.2.3.1. In Solid Wood

Typical patterns of acoustic emission activity in solid wood under standard fracture mechanics tests were provided by Ansell (1982a), Sato et al. (1983, 1984), Miller (1963), Nakao et al. (1986), Niemz and Hansel (1987); Vautrin and Harris (1987), Knuffel (1988), Nakagawa et al. (1989), Niemz and Lühmann (1992), and Pentony and Porter (1964). A familiar example of emissions development due to flaw growth in the specimen under static test is given in Figure 10.5, in which the stress-strain diagram is related to an acoustic emission cumulative counts-stress graph. This is acoustic emission testing as it is performed in a typical laboratory situation.

From the various experimental investigations in acoustic emission over the past years (Vautrin and Harris, 1987) it may be said that softwoods are characterized by more intense acoustic activity than are hardwoods. Emissions were measured at very low levels of ultimate strength (5 to 15%) in longitudinal tension tests in softwoods. In longitudinal compression tests specimens produced less events than in longitudinal tension tests. For both tests the cumulative events increased linearly with strain to the proportional limit.

In static bending tests the behavior in solid wood specimens is more complex and the critical levels of damage, clearly related to the microstructural mechanism of ruin, are difficult to establish. The acoustic emission count-rate decreased with the increasing number of cycles in tensile cyclic loading. Sato et al. (1984) classified acoustical events as "slow" and "rapid". The first type is generated in the early stage of emission and is induced by microcracks, while the second type is determined by the ductile characteristics of the fracture process and is calculated as the difference between the total number of acoustic emission cumulative counts and estimated slow counts. Significant relationships were established between the slow or rapid counts and the fracture stress for Japanese coniferous species.

Ansell and Harris (1979) established relationships among toughness, acoustic emission counts, and fracture topography of softwoods. Fiber-matrix interfaces in solid wood contribute to the high toughness of this material. The slow count emissions are related to the "gradual opening of microflaws as the helically-wound cell wall reinforcement extends elastically within the matrix of hemicellulose and lignin. The rapid emissions are attributed to either interlaminar shear in planes of weakness, for example at ray cell-tracheid interfaces and earlywood-latewood interfaces, or to brittle failure of tracheids." However, the acoustic activity was related to very fine elements of anatomical structure such as the angle of microfibrils in cell walls and pits (Sato et al., 1983, 1986). It was supposed that the extension of microcracks observed on the cross-field of pits (between the rays and longitudinal cells) on radial walls of tracheids produced slow acoustic emission activity. In static bending the microcracks were also considered to be the cause of early events at 40 to 50% ultimate load. Early emissions also were used as a preliminary means for detecting flawed specimens.

In his pioneering work Ansell (1979, 1982a, b) presents a considerable amount of data on acoustic emission activity and on the topography of softwood fracture under tensile test. He related structural parameters (deduced from macroscopic and microscopic analyses) to acoustic emission characteristics under tension and impact. The behavior of three species (Parana pine, Scotch pine, and Douglas fir) was analyzed. The selection of species was determined by their heterogeneity, induced by the proportion of the latewood in annual rings (little latewood in Parana pine and marked transition between earlywood and latewood in Douglas fir).

Parana pine, which has a very uniform structure, generated emissions from around 20% of the fracture strain. Furthermore, the increase in the applied stress was approximately linearly related to the cumulative emissions, and the curves have a "fingerprints" shape. In contrast to this species Douglas fir emitted many acoustic emission counts in the very early stage of the testing. The linearity of the log cumulative emissions-strain is less pronounced than in the case of Parana pine.

The influence of anatomical structure on acoustic emission activity was demonstrated by Ansell (*op. cit.*) on specimens cut in different anisotropic planes or at an angle to them (e.g., specimens loaded in *LR* or *LT* planes or cut at different angles—15°, 30°, 45°—from the *L* axis in these planes). The profiles of cumulative counts vs. stress and strain at 0° on *LT* and *LR* planes are different probably because of the failure mechanism in each case. The *LT* specimens were observed to fail along the interface between latewood and earlywood via intrawall shear through a plane of earlywood tracheids, but the *LR* specimens cracked along the same interface by delamination more in the direction of cell growth. Measurements performed on specimens cut at an angle to the *LT* plane (15°, 30°, 45°) resulted in very different acoustic emission curves when compared to standard specimens. It was noted that with increasing grain angle, samples deformed with progressively fewer counts. The interfacial fracture takes place exclusively in earlywood and the fracture path is limited by the geometry of the test specimen. Similar results were reported by Ando et al. (1992).

Slow count rates are attributable to the gradual opening of microflaws in the cell wall (Ando et al., 1991). The increase in fracture activity is assigned to several factors, including interlaminar shear in planes of weakness (at ray and tracheid interfaces or at earlywood-latewood interfaces) or to brittle failure of tracheids. Using fracture micrographs Ansell and Harris (1979) observed that the fracture through the earlywood is smooth, unlike the fracture through the latewood, which is affected by the thicker-walled tracheids which act as barriers to crack propagation.

Emissions from earlywood, latewood, and solid wood are notably different. Cumulative acoustic emission from latewood and earlywood under tension in Scotch pine, plotted vs. strain and stress, is given in Figure 10.12. Some suggestions for emission mechanisms have been advanced:

1. In latewood a rapid accumulation of counts occurs at a low strain level (<10%), which is probably determined by microcracks from the tension in cell walls.
2. In earlywood no activity takes place below 10% fracture strain. Rapid jumps in emission are attributed to the tension and shear cracks of the walls or to flaw enlargement.
3. In both structures the sharp increase in counts corresponds to ultimate failure.
4. A dramatic change in the slope of logarithmic cumulative emissions for earlywood, latewood, and solid wood does not correspond to the deviation in the slope of the stress-strain characteristics.

From preceding discussions we can deduce that in static tension wood is able to absorb an increasing load, despite intermittent high-energy failure events. The acoustic emission technique, more so than the static tensile tests, is able to substantially contribute to the understanding of fracture mechanisms in softwoods in tension.

The behavior of Douglas fir in bending was described by Vautrin and Harris (1987). They used ring-down counting to examine structural mechanisms of failure. The mechanical testing was carried out in four-point bending at a low crosshead speed (0.5 mm/min) on standard specimens of 270 mm in length. Simultaneous examination of fracture mechanisms and acoustic emission curves established criteria for critical levels of damage in wood. Three regions were defined from acoustic emission curves (Figure 10.13):

Region I—below the "significance threshold".
Region II—the slope of the acoustic emission vs. displacement curve is reduced; a plateau region on the load displacement curve (A) was observed followed by the development of shear bands at 45° in the compression-ruined zone, observed by electronic microscopy.
Region III—a sharp increase in acoustic emissions and sudden drops in the rigidity of samples related to nonlinear mechanical behavior. Failure will occur when the slope of the load-count curve goes to zero. Simultaneous propagation of damage will take place as compression/shear or tension/shear occurs.

Vautrin and Harris (*op. cit.*) used the same acoustic emission technique to determine, via ring-down cumulative counting, the responses of beech and oak specimens to be load (Figure 10.13b, d). The curves are "fingerprints", similar to those for softwoods, with a sharp increase in acoustic emission counts at the failure load. In beech wood initial failure was observed on the first annual ring on the outer face of the specimen. The local failure initiated by tensile fracture of a group of cells was followed by shear failure at the earlywood-latewood interface. Different failure patterns were observed in oak. This species exhibited marked "fingerprint" paths in the acoustic emission curves corresponding to bursts of emissions generated by interlaminar shear in weak planes (e.g., ray cell-fiber and fiber-vessel interfaces). Brittle failure of fibers was also noted. The characteristics of the source of emission are unfortunately obscured by the dispersion of the acoustic waves propagating through the structure. The authors also stated that the cumulative acoustic emission events at the point of failure is nearly 10^5 counts. By comparing three species it seems that the softwood Douglas fir specimens generate more acoustic activity than the hardwoods, beech and oak.

An extension of the analysis adopted by Vautrin and Harris (1987) may be made to the nonlinear zone by using spectral analysis. Additional tests using multiple sensors could help locate the acoustic emission sources; however, the results reported are empirical and influenced by the particular experimental conditions. For further study, an active ultrasonic testing system could be used such as that suggested by Sachse and Kim (1987) for composites (referred to as the point-source/point-receiver) with an appropriate signal processing system.

234

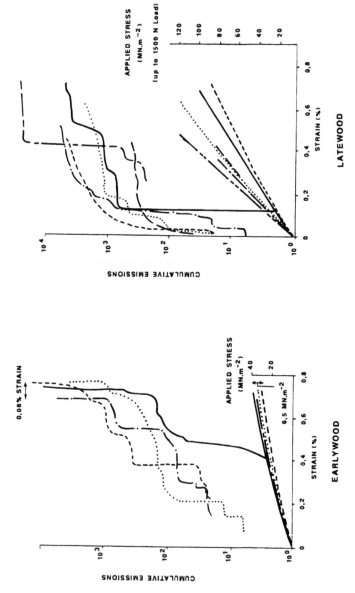

Figure 10.12. Acoustic emission from earlywood and latewood (TL specimens) under tension on Scotch pine. (From Ansell, M. P., in *Structural Uses of Wood in Adverse Environments.* Meyer, R. W. and Kellogg, R. M. (Eds. Van Nostrand Reinhold, New York, 451, 1982b. With permision.)

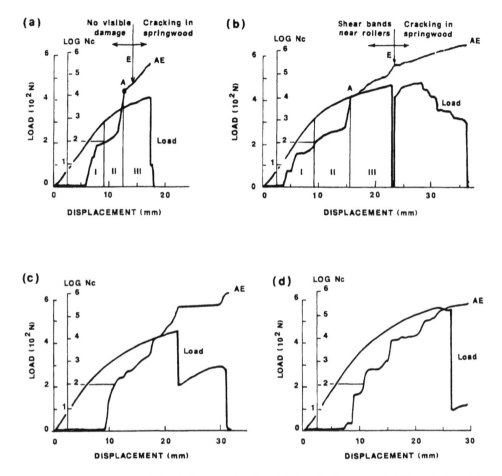

Figure 10.13. Acoustic emission in four-point bending. (a) Douglas fir sapwood, (b) Douglas fir heartwood, (c) beech, (d) oak. (From Vautrin, A. and Harris, B., *J. Mater. Sci.*, 22, 3707, 1987. With permission.)

10.2.3.2. In Wood Composites

As in solid wood, the acoustic emission phenomenon observed in wood composites (particleboards, plywood, laminates, modified solid wood with resins, etc.) is a random process. The acoustic emission signals are nonperiodic, containing many frequencies, and cannot be described by an explicit mathematical function. Between 1970 and 1990 many acoustic emission studies on wood composites were published (Porter et al., 1972; Morgner et al., 1980; Ansell, 1982b; Niemz et al., 1981, 1983, 1988, 1989; Reis and McFarland, 1986; Beall, 1985, 1986a,b; Patton-Mallory, 1988; Sims et al., 1977), and consequently, useful nondestructive techniques have emerged. The cumulative counts and rates of emissions were correlated with factors such as the applied stress, the ultimate stress, the total strain, and internal bonding.

Ansell (1982b) established acoustic emission-strain data for plywood and emphasized that this material behaved in a more predictable manner than did solid wood. The acoustic activity is smooth until failure, the strain is proportional to the total counts, and the experimental data are not very scattered.

Beech wood modified with polystyrene generates double the number of cumulative acoustic events, 1260, than does natural wood, 544, at the point of failure in bending. The density was modified from 715 to 780 kg/m^3 (Niemz and Paprzycki, 1988). The fracture topography in modified wood is ductile and single broken fibers are observed.

For particleboard and other similar composites under static load, the emissions begin at a much earlier stage of loading than for solid wood. This is probably due to the presence of resin as a principal

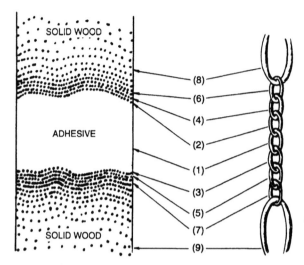

Figure 10.14. Adhesive joint as represented by Marra's model of the adherend-adhesive-adherend system. (1) Adhesive film; (2, 3) intra-adhesive boundary layer; (4, 5) adhesive-adherend interface; (6,7) adherend subsurface; and (8, 9) adherend proper. (From Blomquist, R. F. et al., *Adhesive Bonding of Wood and Other Structural Materials*, Pennsylvania State University, State College PA, 1984. With permission.)

component of the structure. Total events to failure correlate well with the resin content. Measurement of particleboards in flexural creep revealed microcracks in wood particles and in resin (Morgner et al., 1980). When particleboard, fiberboard, and oriented strandboard were cyclically loaded in bending, in increasing steps with relaxation to zero, the Felicity ratio was determined (Beall, 1985). This coefficient lay between 0.84 and 0.98 for a corresponding internal bond between 475 and 915 kPa. Acoustic emission activity induced by swelling was also observed in particleboard subjected to cyclic water-soak exposure (Beall, 1986a,b). It was shown that cumulative emissions could be a nondestructive predictor of the dimensional stability of specimens.

10.3. ACOUSTIC EMISSION FOR MONITORING TECHNOLOGICAL PROCESSES

We have already outlined the application of the acoustic emission technique in laboratory measurements. This section applies the physical principles described earlier to the monitoring of different technological processes such as drying, machining, and the assessment of structural integrity. We give only a brief introduction to the broad subject of acoustic emission and its application in wood technology. It is emphasized that the efficiency of monitoring systems is determined by the capability of sensors to detect and locate acoustic signals and to display reproducible data suitable for comparison with past and future records. Details of experimental design and associated data acquisition, processing, and display are described.

10.3.1. ADHESIVE CURING AND ADHESIVE STRENGTH

The overall strength of a laminate of wood and polymer is determined by the cohesive strength of both components. As suggested by Marra (1964), the model of the adherend-adhesive-adherend joint system is usually considered to be composed of a chain of nine link (Figure 10.14) that is symmetrical to the adhesive film. The intra-adhesive boundary layer is influenced by the adherend. The adhesive-adherend interface is the site of adhesive forces. The adherend subsurfaces are associated with the proper adherend. This model can assist in analyzing the failure of bonded joints using acoustic emission and ultrasonic techniques.

Beall (1987), Suzuki and Schniewind (1987), Yoshimura et al. (1987), Quarles and Lemaster (1988), Sato et al. (1989), Hwang et al. (1991); and Sato and Fushitani (1991) attempted to classify the known acoustic emission methods for the nondestructive testing of adhesive strength. They used the following headings:

- To determine the curing time, which depends on the adhesive properties of wood interfaces and the elastic properties of the polymer

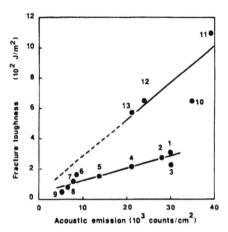

Figure 10.15. Fracture toughness and cumulative counts of acoustic emission per fracture area of Douglas cantilever beam for different adhesives. (1) Isocyanate, (2) Phenol-resorcinol, (3) urea formaldehyde with 20 parts filler, (4) isocyanate with 25 parts starch, (5) phenol-formaldehyde, (6) urea-formaldehyde, (7) starch with 50 parts isocyanate, (8) ammonium ligno-sulfonate with 25 parts formaldehyde, (9) ammonium lignosulfonate with 100 parts isocyanate, (10) epoxy with 200 μm glueline, (11) polyvinylacetate with 100 μm glueline, (12) emulsion polymer with 10 parts isocyanate, (13) phenol-formaldehyde with 10 parts polyvinylacetate. (From Suzuki, M. and Schniewind, A. P., *Wood Sci. Technol.*, 21, 121, 1987. With permission.)

- To establish the relationships between adhesive strength in joints and the acoustic emission parameters or ultrasonic velocity and attenuation
- To select the parameters most appropriate to predict nondestructively the strength of joints because this depends on many factors such as type of loading, wood species, the nature of adhesives, surface conditions, and moisture content

The fracture mechanics-acoustic approach proposed by Suzuki and Schniewind (1987) provides an example of a new direction for the investigation of adhesive bonds.

The most commonly measured acoustic emission parameters are total count and event rates. Figure 10.15 presents linear relations between acoustic emission counts and fracture toughness of a double cantilever Douglas fir beam, with rupture in mode I for joints bonded with different adhesives. The adhesive systems are in two groups, the traditional thermosetting synthetic resin adhesives, urea- or phenol-formaldehyde, (no. 1 to 9) with a correlation coefficient of $r = 0.934$, and the epoxy and polyvinylacetate systems (no. 10 to 13) with a correlation coefficient of $r = 0.787$. The second group produces more acoustic activity when compared to the brittle behavior of urea formaldehyde adhesives.

Beall (1987) noted that ultrasonic velocity and attenuation of an interface wave, containing longitudinal and shear modes, could be good indicators of the curing time of the adhesive. This statement is also connected with the fact that ultrasonic velocity measured with a particular adhesive is dependent on the cure cycle and on the correct mixing of hardener and resin, and is consequently strongly related to cohesive strength. In addition, as the adhesive cured the root mean square voltage increased.

Measurements on two layer adhesive bonds using root mean square voltage and ultrasonic velocities across and along the specimens were made by Bucur and Perrin (1988) to provide some evidence for the location of delaminations. The defective zone is characterized root mean square by voltage of 1.8 $\times$ 10^{-2} V/cm and an ultrasonic velocity of 1500 m/s (in radial symmetry direction of wood for a longitudinal wave), while measurements in the sound zone of 27 $\times$ 10^{-3} V/cm and 1896 m/s.

Perhaps the first spectroscopic testing of laminates was suggested by Yoshimura et al. (1987). In plywood compresses with a hot press counts vs. time and temperature were plotted during the cooling of specimens. The smooth slope of the curves in samples with delaminations between layers (Figure 10.16) were noted. Two spectra were presented to compare the defective specimen to the sound specimen. Spectral analysis of acoustic events identified the presence of the defect. In sound specimens the frequency range is between 10 and 40 kHz, while in delaminated specimens the acoustic activity is observed to be between 60 and 90 kHz.

Beall (1985, 1986) reported interesting results from acoustic emission testing for the determination of internal bond strength of wood-based composite panel materials. The quality of the adhesion between wood particles and resin in particleboards, medium density fiberboards, or oriented strandboards is generally expressed by the coefficient of internal bond strength, calculated in tension normal to the specimen face. The internal bond strength could be predicted on the basis of the event-load curves, as shown in Figure 10.17.

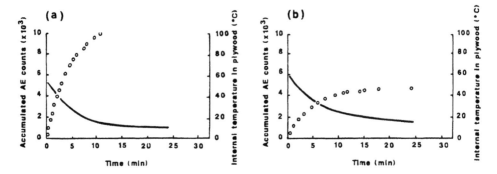

Figure 10.16. Accumulated acoustic emission counts in plywood during the cooling of specimens stressed in a hot press. (a) Sound specimen; (b) defective specimen. (From Yoshimura, N. et al., *Mokuzai Gakkaishi*, 33(8), 650, 1987. With permission.)

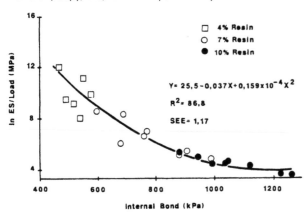

$$Y = 25,5 - 0,037X + 0,159 \times 10^{-4} X^2$$

$$R^2 = 86,8$$

$$SEE = 1,17$$

□ 4% Resin
○ 7% Resin
● 10% Resin

Figure 10.17. Relationship between the internal bond of different resin-content particleboards and the logarithm of events vs. load at failure. (From Beall, F. C., *For. Prod. J.*, 36 (7/8), 29, 1986a. With permission.)

10.3.2. FOR CONTROLLING DRYING OF LUMBER

Well-documented acoustic emission studies related to the drying of lumber (Quarles, 1992; Rice and Skaar, 1990; Sadanari and Kitayama, 1989; Niemz and Hansel, 1988; Noguchi et al., 1980, 1985, 1987; Wassipaul et al., 1986; Ogino et al., 1986; Okumura et al., 1986, 1987; Ogino et al., 1986; Honeycutt et al., 1985; Kitayama et al., 1985; Becker, 1982; Skarr et al., 1980; Kagawa et al., 1980; Poliszko et al., 1988; Rice and Kabir, 1992; Rice and Peacock, 1992) associated acoustic events and shrinkage as well as swelling and internal fractures induced by drying stresses. Acoustic emission systems "listen" to material by amplifying, filtering, recording, and analyzing signals emitted by wood during drying. Commonly, the measured parameters are: count rate, amplitude distribution of signals, energy levels and frequency spectra. These parameters are studied as a function of drying rate or moisture loss. Figure 10.18 shows a typical pattern of acoustic emission activity during the drying of oak lumber. Figure

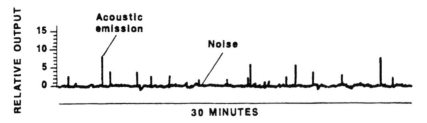

Figure 10.18. Acoustic emission activity (electrical output of instrumentation) during drying of oak lumber. (From Skaar, C. et al., *For. Prod. J.*, 30(2), 21, 1980. With permission.)

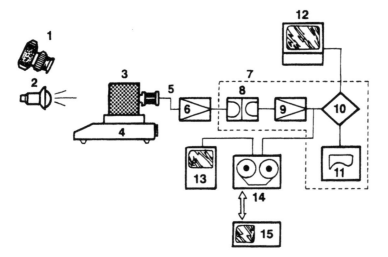

Figure 10.19. Schematic representation of acoustic emission set up used for drying Japanese larch. (1) Camera; (2) infrared lamp; (3) specimen; (4) balance; (5) AE transducer; (6) preamplifier; (7) checking monitoring; (8) filter; (9) main amplifier; (10) processor; (11) printer; (12) computer; (13) oscilloscope; (14) data recorder; (15) spectrum analyzer. (From Ogino, S. et al., *J. Acoust. Emission,* 5(2), 61, 1986. With permission.)

10.19 presents a schematic representation of the acoustic emission setup (Ogino et al., 1986) used for laboratory studies relating moisture loss of a $50 \times 70 \times 70$ mm Japanese larch specimen to the shrinkage and formation of checks. The drying conditions were controlled by an infrared lamp placed 30 cm from the drying surface of the specimen. The initial moisture content was 80%. Some of the technical characteristics of components shown in Figure 10.19 are

- Preamplifier, with a frequency response of 2 kHZ to 1 MHz, 40 dB gain
- Bandpass filter for signals in the range 10 kHZ to 1 MHz
- Amplifier, 30 dB, with a frequency response of 2 kHz to 1 MHZ
- Threshold voltage of 0.08 V and a maximum measured peak of 1 V

The accumulated energy (expressed as the square of the event amplitude) vs. moisture loss and drying rate is illustrated in Figure 10.20. The initiation of checking corresponds to the increasing rate

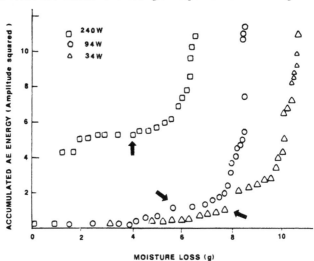

Figure 10.20. Accumulated acoustic emission energy (expressed as the square of the amplitude) vs. moisture loss for drying Japanese larch. √ indicates the initiation of checking. (From Ogino, S. et al., *J. Acoust. Emission,* 5(2), 61, 1986. With permission.)

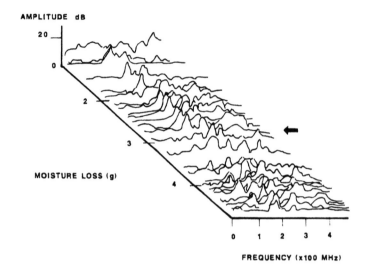

Figure 10.21. Frequency spectrum patterns of acoustic emission activity during drying when a 34-W infrared lamp was used. (From Ogino, S. et al., *J. Acoust. Emission,* 5(2), 61, 1986. With permission.)

of accumulation of acoustic emission energy. It was argued that the slow rate of accumulation of acoustic emission events is due to the capillary flow of free water in wood. Furthermore, when the fiber saturation point was reached, cell wall shrinkage produced surface tensile stresses and consequently checking of the structure.

Another interesting and very sensitive indicator of acoustic emission activity, for the analysis of different drying conditions, is the frequency spectrum (Figure 10.21), the shape of which reflects each

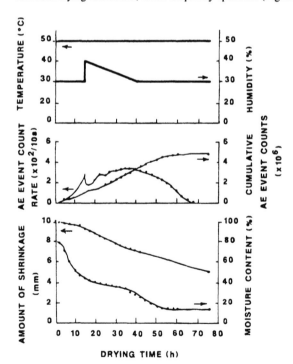

Figure 10.22. Temperature, relative humidity, count rate, and shrinkage vs. drying time of *Zelkova serrata.* (From Noguchi, M. et al., *For. Prod. J,* 37(1), 28, 1987. With permission.)

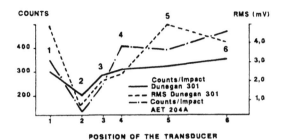

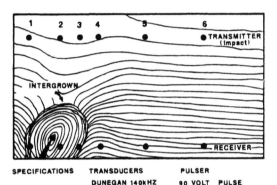

Figure 10.23. Acoustic emission parameters (counts; root mean square voltage) during the scanning of a Ponderosa pine board with an encased knot of 6 cm. (From Lemaster, R. L. and Dornfeld, D. A., *J. Acoust. Emission*, 6(3), 157, 1987. With permission.)

drying rate. The initiation of checking is connected to acoustic emission events having low frequency components (<200 kHz).

Establishing relationships among emission rate, wood moisture content, and drying conditions was the main task of a number of authors. As an example, Becker (1982) reported drying pine lumber that had a variable proportion of sapwood under severe conditions of relative humidity. Further acoustic events were generated by specimens containing 20 to 25% sapwood.

These advances demonstrate the potential for the development of automatic drying control systems through acoustic emission monitoring in commercial kilns (Kitayama et al., 1985; Noguchi et al., 1987). Figure 10.22 predicts check formation, when moisture content decreased under severe drying conditions (50°C dry-bulb temperature, 30% relative humidity, and 1.2 m/s air velocity, from green to air-dried conditions of a *Zelkova serrata* specimen). The increase of acoustic emission activity at 15 h of drying was followed by a sudden check. To avoid the development of this or any subsequent checks, the relative humidity was modified from 30 to 40%. A decreasing count rate was consequently observed.

Because the acoustic emission technique is intended to provide a useful tool for controlling the drying process for softwoods and hardwoods, Sadanari and Kitayama (1989) proposed that the count rate, the waveform, the mean square value of amplitude, the maximum of amplitude, and the frequency spectrum should be considered simultaneously. Despite the complexity of this approach, a full understanding of the interaction of wave parameters and electronic instrumentation details is probably the best approach to successfully monitoring drying control with acoustic emission sensing.

10.3.3. ACOUSTIC EMISSION AS A STRENGTH PREDICTOR IN TIMBER AND LARGE WOOD STRUCTURES

Acoustic emission testing was developed to be a strength predictor for timber and large structural components and an inexpensive, nondestructive alternative to visual or machine stress grading (Sato et al., 1990). Visual grading is based on the critical examination of flaws in individual timber members, while stress grading is based on the measurement of the modulus of elasticity, which is correlated with

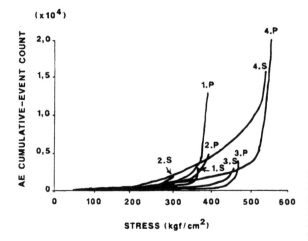

Figure 10.24. Air-dried western hemlock, lumber section 9 × 9 cm. Relationships between stress and acoustic emission counts during a bending test until failure. (P) Zone of timber sample with straight grain; (S) zone of timber sample with slope of grain; (1P) acoustic emission counts on specimen 1 zone P; (1S) acoustic emission counts on specimen 1 zone S. (From Nakagawa, M. et al., *Mokuzai Gakkaishi*, 35(3), 190, 1989. With permission.)

Table 10.4. Acoustic emission parameters and the slope of grain

Method	Correlation coefficient
Root mean square voltage using pulser	−97.4
Count rate using pulser	−97.6
Root mean square voltage using sine wave generator	−96.8
Total counts using slide hammer	−69.2

Note: Correlation coefficient (*r*) determined from second order polynomial least-squares regression.
From Lemaster, R. L. and Dornfeld, D. A., *J. Acoust. Emission*, 6(3), 157, 1987. With permission.

the modulus of rupture. The static bending test is the most appropriate test for the measurement of the elasticity of lumber modulus. Some softwood products are stress graded (timber, posts, stringers, beams, checking, some boards). Mechanical stress grading used in the U.S. and Canada is described in the *Wood handbook* (USDA Forest Products Laboratory, 1987), and has three basic components: "the mechanical sorting and prediction of strength through nondestructive determination of the modulus of elasticity, the assignment of allowable design stress based upon strength predictions and quality control". Grading machines are commonly designed to detect bending stiffness in an approximately 4 ft (1.20 m) span. However, the designation of the unique value of the modulus of elasticity does not furnish complete proof of structural element quality. To improve the mechanical prediction of lumber quality, visual detection of the size and type of defects (knots, checks, shakes, skips, splits, wane, and warp) is made. In addition, to check simultaneously for correct machine operation and visual observations, acoustic emission tests were developed.

The detection of an encased knot on a Ponderosa pine board is shown in Figure 10.23. Acoustic emission signals were induced by a stress wave which was generated by a hammer falling on the board. The counts per impact and the root mean square voltage were recorded for each impact position. The receiver transducer was 175 kHz, with a gain of 100 dB and a filter (0.125 to 2 MHz). Note that both acoustic emission parameters recorded decreased in the knot zone.

Nakagawa et al. (1988) reported results on three large specimens. The transducer located near the knot detected more counts than that located near the roller, probably because of the crack propagation along the slope of grain around the knot. On the other hand, Figure 10.24 strongly suggests that the timber sample without a knot and with the fiber direction parallel to the edges of the piece (no slope

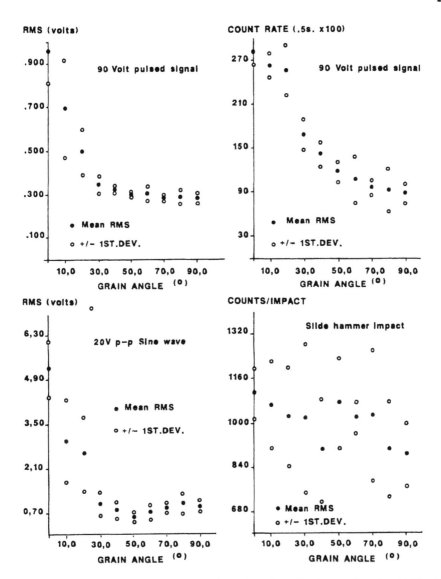

Figure 10.25. Acoustic emission parameters vs. the grain angle on Ponderosa pine, when (a, b) pulsed signal, (c) sine wave signal, and (d) hammer impact were monitored. (From Lemaster, and Dornfeld, D. A., *J. Acoust. Emission*, 6(3), 157, 1987. With permission.)

of grain) behaves similarly to small clear specimens for which, before ultimate failure, a sudden increase in the acoustic emission activity is observed.

The acoustic emission parameters are closely related to the grain angle (Table 10.4). The root mean square voltage and the count rate using a pulsed signal seem to give the most interesting experimental results. Figure 10.25 displays the root mean square and count rate vs. grain angle for different modes of excitation (pulsed signal, sinewave, or slide hammer impact). The standard deviation is smaller for pulsed signals. Root mean square voltage measurements with pulsed signals are prefered at grain angles of <30°. Count rate measurement is better at higher angles. The acoustic emission technique also provides information about the energy corresponding to acoustic events. Bearing in mind the proof

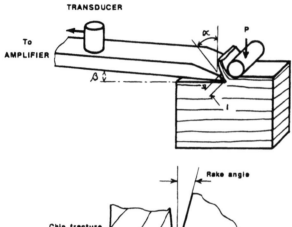

Figure 10.26. Cutting parameters. (α) Rake angle; (β) clearance angle; (1) contact length; (P) nosebar pressure. (From Lemaster, R. L. et al., *Wood Sci.*, 15(2), 150, 1982. With permission.

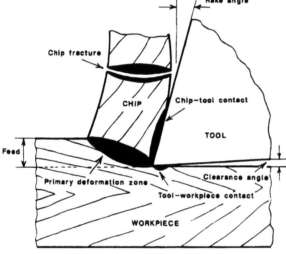

Figure 10.27. The stages of chip formation—plastic deformation of the shear zone before chip fracture, chip sliding on the tool rake face, and chip fracture—are sources of acoustic emission signals in wood cutting operations. (From Lemaster, R. L. et al., *Wear,* 101, 273, 1985. With permission.)

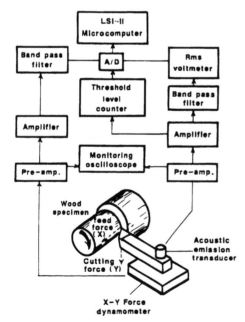

Figure 10.28. Device for measuring acoustic emission activity during wood specimen machining/turning. Machining parameters: depth of cut, 1.5 mm; cutting speed, 3 m/s; feed speed, 0.127 mm/rev; diameter, 20 cm; species, white fir. (From Murase, Y. et al., *Mokuzai Gakkaishi,* 34(3), 207, 1988. With permission.)

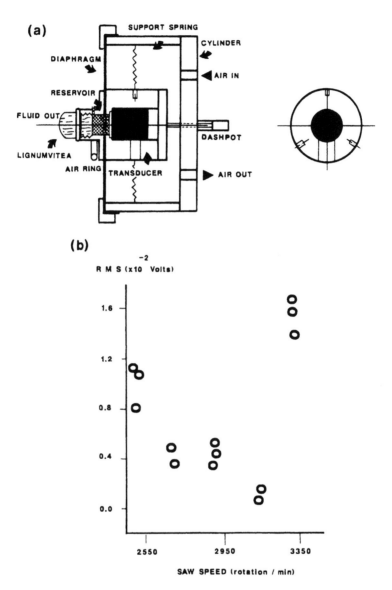

Figure 10.29. Device containing the wireless acoustic emission transducer for measurements on a circular blade. The position of the transducer is controlled by the air pressure (13.8 kPa) injected into the cylinder. Maximum rotation speed was 400 RPM and the speed at the tip of the blade was 510 m/min. (a) Diagrammatic representation of the setup. (b) Acoustic emission parameter. Root mean square vs. feed obtained on the experimental setup. (From Lemaster, R. L. and Dornfeld, D. A., *J. Acoust. Emission*, 7(2), 103, 1988. With permission.)

load/stress grading system for structural timber (34 × 100 mm section and 3.3 m long) Knuffel (1988) suggested the use of acoustic emission energy as a discriminant parameter to prevent damage during the sorting of boards. Based on the "take off" point (defined as corresponding to the sudden increases of the microfractures and of the acoustic emission activity, determined from the graph of stress vs. cumulative energy), the grading system can adjust the load to each board. This made the best testing condition suitable to each board.

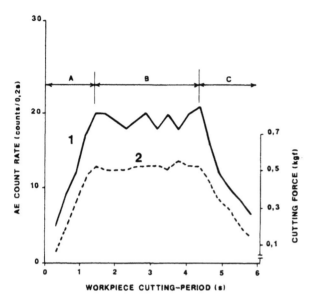

Figure 10.30. Acoustic emission count rate and cutting force sampled at a rate of 3.3 Hz for Japanese beech machining at a cutting speed of 3000 RPM and a workpiece feed rate of 2 m/min, in circular sawing. (1) Acoustic emission count rate; (2) cutting force; (A) increasing path; (B) contact path; (C) decreasing path. (From Tanaka, C. et al., *Holz. Roh. Werk.*, 48, 139, 1990. With permission.)

Porter et al. (1972) attempted to develop acoustic emission testing as an engineering tool for the prediction of the structure lifetime on flat sawn Douglas fir laminating stock profiled with finger joints. The most accurate prediction of failure (1.8%) was obtained when the load-count data were considered to be at the proportional limit of the stress-strain curve. However, Groom and Polensek (1987) investigated Douglas fir structural boards of nominal size (50 × 100 mm section and 3.6 m long) and concluded that the count rate and cumulative acoustic emission counts at various load levels appeared to be better predictors of the modulus of rupture than the stress at the proportional limit or the modulus of elasticity.

Biernacki and Beall (1993) demonstrated that the acoustic emission technique, combined with an ultrasonic technique based on the measurement of the "stress wave factor" (defined as "a product of pulse repetition rate, reset time of the counter and number of positive slope crossings of the waveform over a preselected threshold"), could be a powerful technique for the nondestructive evaluation of both laminated and solid wood.

Table 10.5. Regression equations established between workpiece feed rate and roughness or acoustic emission count rate, measured on specimens machined with circular saw

Species	Density (kg/m^3)	Regression equation (Y = A + Bx)	Correlation coefficient
Roughness (Y) and Workpiece Feed—Rate (X)			
Cryptomeria japonica	320	$Y = 5.56 + 2.04 X$	0.98
Picea jezoensis	390	$Y = 5.14 + 2.36 X$	0.97
Chamaecyparis obtusa	480	$Y = 4.54 + 0.64 X$	0.96
Quercus serrata	800	$Y = 3.38 + 1.04 X$	0.98
Zelkova serrata	640	$Y = 3.32 + 1.16 X$	0.99
Fagus crenata	600	$Y = 5.46 + 0.94 X$	0.99
Count Rates (Y) and Workpiece Feed; Rates (X)			
Cryptomeria japonica	320	$Y = 2.20 + 28.6 X$	0.99
Picea jezoensis	390	$Y = 24.0 + 34.2 X$	0.99
Chamaecyparis obtusa	480	$Y = 3.60 + 18.6 X$	0.99
Quercus serrata	800	$Y = 43.2 + 18.0 X$	0.98
Zelkova serrata	640	$Y = 2.60 + 18.4 X$	0.99
Fagus crenata	600	$Y = 21.4 + 16.2 X$	0.99

From Zhao, C. et al., *Mokuzai Gakkaishi*, 36(3), 169, 1990. With permission.

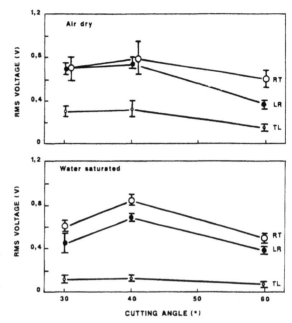

Figure 10.31. Relationships between the root mean square voltage and the cutting angle for spruce in air-dry conditions and saturated in water. (From Murase, Y. and Kawanami, S. *Mokuzai Gakkaishi.* 36(9), 717, 1990. With permission.)

10.3.4. SIGNALS IN WOOD MACHINING

Wood machining survey by acoustic emission technique is a relatively new nondestructive method (developed in the last decade). This technique links machining parameters such as tool geometry, cutting speed and workpiece feed rate with the power consumption during the process, forces acting on the cutting tool, tool wear, and roughness of the surface. When processing logs, lumber, or other wood products the cutting tool may be either in a stationary position (e.g., peeling, turning) or in a rotating position (e.g., circular saw). Also note that when the cutting tool is stationary, the main parameters (Figure 10.26) to be considered (because of their influence on the smoothness of the surface and on tool wear) are the rake angle and the clearance angle, the contact length between the tool and the workpiece (Lemaster et al., 1982).

The mechanism of chip formation is described in Figure 10.27 in three stages: the plastic deformation of the shear zone before chip fracture, the chip sliding on the tool rake face, and the chip fracture. This mechanism is a source of acoustic emission signals of overlapping, continuous burst-type signals (Lemaster et al., 1988). The peak of signals in wood machining was detected as being between 0.1 and 0.3 MHz.

The device commonly used for monitoring acoustic emission signals is presented in Figure 10.28.The acoustic emission transducer is coupled to the tool holder. The machining forces (vertical for cutting and horizontal for feed) were measured with strain gauges. The acoustic emission signal is processed as described in Figure 10.27 and the counts and the root mean square voltage are registered. When the tool is in the rotating position, the experimental procedure is more delicate. For this typical situation Lemaster and Dornfeld (1988) proposed an apparatus that allows the wireless acoustic emission transducer to be coupled to the operating sawblade (Figure 10.29). In this case acoustic emission counts and root mean square voltage vs. feed and chip thickness were recorded. The acoustic emission signals may be modified by varying the cutting parameters (rake angle, clearance angle, specimen temperature or species and roller bar pressure).

To achieve continuous monitoring of the sawing process using a circular saw some relationships were established between acoustic emission signals and cutting forces. Acoustic emission count rate and the cutting force follow a similar pattern, with three principal stages corresponding to increasing, constant, and decreasing paths (Figure 10.30).

The ability of the acoustic emission technique, especially the ring-down method, to monitor machining parameters such as roughness with a circular saw, was analyzed by Zhao et al. (1990) and Tanaka et al. (1990). Softwoods with densities ranging from 320 to 580 kg/m³ and hardwoods of densities of 280 to

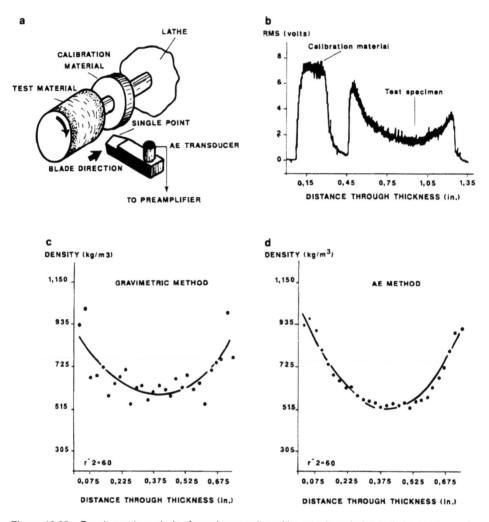

Figure 10.32. Densitometric analysis of wood composites with acoustic emission technique. (a) experimental setup; (b) root mean square voltage output for medium density particleboard; (c) acoustic densitometric profile compared with gravimetric densitometric profile in medium density particleboard. (From Lemaster, R. L. et al., *J. Acoust. Emission.* 7(2), 111, 1988. With permission.)

890 kg/m³, having ring and diffuse-porous structure, were analyzed. The list of species with corresponding density is given in Table 10.5.

Because absolute quantitative relationships between acoustic emission events and roughness of specimens were unavailable, correlations were sought which were based upon qualitative relationships between topographical variations of roughness and acoustic emission count rates or density (Table 10.5). The measured acoustic emission count rate (counts per 0.2 s) increased with increases in the workpiece feed rate, and was related to the density and the anatomical structure of the specimens. The roughness of softwoods and hardwoods of ring-porous structure decreased with increasing density. Hardwoods with diffuse-porous structure were insensitive to variations in density.

These experimental results suggest the development of an online anatomical control optimization system to produce a desired surface roughness. Indeed, it is also possible to survey tool wear (Lemaster et al., 1982, 1985) so that one may distinguish between a sharp or a worn tool using power spectra (Murase et al., 1988).

The level of acoustic emission signals is higher in hardwoods than in softwoods, and the level increased with increasing cutting speed. Moreover, an increase occurred in acoustic emission activity when the depth of the cut increased (Dornfeld et al., 1982). In the initial stage of cutting the root mean square voltage depends linearly on tool wear. When the blade became severely worn, the level of acoustic emission signals decreased significantly. This fact must be understood in relation to the long, continuous chip produced by a dull tool (Lemaster and Dornfeld, 1988). Murase et al. (1990a, b) studied acoustic emission activity during cutting as it related to the cutting angle, the moisture content of the specimen (noted *RT, LR, TL*), and the anisotropic plane in which the cutting process took place (Figure 10.31). The emission is more active when cutting against the grain in the *RT* plane rather than parallel in the *LR* and *TL* plane, probably because of the chip formation process. As expected, the level of acoustic emission activity decreased for water-saturated specimens.

A skillful utilization of the acoustic emission technique was described by Lemaster et al. (1988) for the determination of density profiles in wood composites (medium and high density particleboard and medium density fiberboard). The experimental setup is shown in Figure 10.32. The transducer was coupled to a single-point cutting tool. The specimen to be tested, a disk, was mounted on a lathe along with the calibration material of known density. The tool was moved across both the specimen and calibrator, in small layers (0.06 cm), and the root mean square voltage was recorded simultaneously. The chips produced by the tool for the entire thickness of the sample were weighed and the gravimetric profile was established. Bearing in mind the high correlation coefficient between the root mean square voltage and gravimetric density ($r^2 = 0.96$), it was stated that the method was consistent for all materials tested and the procedure was considered superior to the planing, sawing, or sanding tests commonly used for density profile measurements in composites.

Finally, when the acoustic emission technique is used for monitoring wood machining it is not surprising to learn (Lemaster et al., 1982) that acoustic emission phenomena are induced by the interaction of the tool and workpiece and by the formation of chip and check.

Based on the results discussed here it appears that the acoustic emission testing could be successfully used for the on-line control of tool wear, and consequently for the control of blade sharpness, or for the evaluation of sawing efficiency without having to halt production because of tool failure.

REFERENCES

Ando, K., Sato, K., and Fushitani, M. (1991). Fracture toughness and acoustic emission characteristics of wood. I. Effect of location of a crack tip in an annual ring. *Mokuzai Gakkaishi.* 37(2), 1129–1134.

Ando, K., Sato, K., and Fushitani, M. (1992). Fracture toughness and acoustic emission characteristics of wood. II. Effect of grain angle. *Mokuzai Gakkaishi.* 38(4), 342–349.

Ansell, M. P. and Harris, B. (1979). The relationship between toughness and fracture surface topography in wood and composites. I.C.M. 3, Cambridge, Vol. 3, 309–318.

Ansell, M. P. (1982a). Acoustic emission from softwoods in tension. *Wood Sci. Technol.* 16, 35–58.

Ansell, M. P. (1982b). Acoustic emission as a technique for monitoring fracture processes in wood. in *Structural Uses of Wood in Adverse Environments.* Meyer, R. W. and Kellogg, R. M. (Eds.). Van Nostrand Reinhold, New York, 451–466.

Apfel, R. E. (1981). Acoustic cavitation. in *Ultrasonics.* Edmonds, P. D. (Ed.). Vol. 19. *Methods of Experimental Physics.* Academic Press, New York, 356–411.

ASTM (1972). Acoustic Emission. STP 505. American Society for Testing and Materials, Philadelphia.

ASTM (1975). Monitoring Structural Integrity by Acoustic Emission. STP 571. American Society for Testing and Materials, Philadelphia.

Beall, F. C. (1985). Relationship of acoustic emission to internal bond strength of wood-based composite panel materials. *J. Acoust. Emission.* 4(1), 19–29.

Beall, F. C. (1986a). Effect of resin content and density on acoustic emission from particleboard during internal bond testing. *For. Prod. J.* 36(7/8), 29–33.

Beall, F. C. (1986b). Effect of moisture conditioning on acoustic emission from particleboard. *J. Acoust. Emission.* 5(2), 71–76.

Beall, F. C. and Wilcox, W. W. (1987). Relationship of acoustic emission during radial compression to mass loss from decay. *For. Prod. J.* 37(4), 38–42.

Beall, F. C. (1987). Acousto-ultrasonic monitoring of glueline curing. *Wood Fiber Sci.* 19(2), 204–214.

Beattie, A. G. (1983). Acoustic emission, principles and instrumentation. *J. Acoust. Emission.* 2(1/2), 95–129.

Becker, H. F. (1982). Schallemissionen während der Holztrocknung. *Holz Roh Werkst.* 40(9), 345–350.

Biernacki, J. M. and Beall, F. C. (1993). Development of an acoustic scanning system for nondestructive evaluation of wood and wood laminates. *Wood Fiber Sci.* 25(3), 289–297.

Blomquist, R. F., Christiansen, A. W., Gillespie, R. H., and Myers, C. E. (1984). Adhesive Bonding of Wood and Other Structural Materials. Materials Research Laboratory, Pennsylvania State University, State College, PA.

Borghetti, M., Raschi, A., and Grace, J. (1989). Ultrasound acoustic emission in water-stressed plants of *Picea abies* Karst. *Ann. Sci. For.* 46 (Suppl.), 346–349.

Brindley, B. J., Holt, J., and Palmer, I. G. (1973). Acoustic emission. III. The use of ring-down counting. *Non-Destruct. Test.* December, 299–306.

Bucur, V. and Perrin, J. R. (1988). Etude du Vieillissement à la Vapeur d'eau des Structures Lamellées-Collées par Ultrasons. INRA, Centre de Recherches Forestiéres de Nancy, France.

Castagnede, B., Sachse, W., and Kim, K. Y. (1989). Location of pointlike acoustic emission sources in anisotropic plates. *J. Acoust. Soc. Am.* 86(3), 1161–1171.

Ceranoglu, A. N. and Pao, Y. (1981). Propagation of elastic pulses and acoustic emission in a plate. *J. Appl. Mech. Trans. Asme.* I. (48), 125–132; II. (48), 133–138; III. (48), 139–147.

Cochard, H. and Tyree, M. T. (1990). Xylem dysfunction in *Quercus:* vessel sizes, tyloses, cavitation and seasonal changes in embolism. *Tree Physiol.* 6, 393–407.

Cochard, H. (1992). Vulnerability of several conifers to air embolism. *Tree Physiol.* In press.

Crombie, D. S., Milburn, J. A., and Hipkins, M. F. (1985). Maximum sustainable xylem sap tensions in Rhododendron and other species. *Planta.* 163, 27–33.

Crostack, H. A. (1977). Basic aspects of the application of frequency analysis. *Ultrasonics.* 15(6), 253–262.

Egle, D. M. and Tatro, C. A. (1967). Analysis of acoustic emission strain waves. *J. Acoust. Soc. A.* 41, 321–327.

Fujii, Y., Noguchi, M., Imamura, Y., and Tokoro, M. (1990). Using acoustic emission monitoring to detect termite activity in wood. *For. Prod. J.* 40(1), 34–36.

Grabec, I. (1980). Relation between development of defects in materials and acoustic emission. *Ultrasonics.* 18(1), 9–12.

Green, R., Jr. (1980). Basic wave analysis of acoustic emission. in *Mechanics of Nondestructive Testing.* Stinchcomb, W. W. (Ed.). Plenum Press, New York, 57–76.

Groom, L. and Polensek, A. (1987). Nondestructive prediction of load—deflection relations for lumber. *Wood Fiber Sci.* 19(3), 298–312.

Honeycutt, R. M., Skaar, C., and Simpson, W. T. (1985). Use of acoustic emission to control drying rate of red oak. *For. Prod. J.* 35(1), 48–50.

Hwang, G. S., Okumura, S., and Noguchi, M. (1991). Acoustic emission generation during bending strength tests for wood in shear. *Mokuzai Gakkaishi.* 37(11), 1034–1040.

Hsu, N. N., Simmons, J. A., and Hardy, S. C. (1977). An approach to acoustic emission signal analysis. Theory and experiment. *Mater. Eval.* 35(10), 100–106.

Imamura, Y., Fujii, Y., Noguchi, M., Fujisawa, K., and Yukimune, K. (1991). Acoustic emission during bending test of decayed wood. *Mokuzai Gakkaishi.* 37(11), 1084–1090.

Jones, H. G., Higgs, K. H., and Bergamini, A. (1989). The use of ultrasonic detectors for water stress determination in fruit trees. *Ann. Sci. For.* 46(Suppl.), 338s–341s.

Jones, H. G., Higgs, K. H., and Sutherland, R. A. (1990). Ultrasonic cavitation detection for irrigation scheduling. Abstr. Int. Workshop Analysis of Water Transport in Plants and Cavitation of Xylem Conduits. Abbazia di Vallombrosa, Firenze, Italy, May 29 to 31, 15.

Kaiser, J. (1953). Untersuchungen über das Auftreten von Geräuschen beim Zugversuch. (An investigation into the occurrence of noises in tensile tests). *Arch., Eisenhuettenwes.* 24, 43–45.

Kato, K., Tsuzuki, K., and Asano, I. (1971–1974). Studies on vibration cutting of wood. I. *Mokuzai Gakkaishi.* 17(2), 57–64; II. Cutting forces in longitudinal vibration cutting. 20(5), 191–195; III. Cutting forces in lateral vibration cutting. 29(9), 418–423; IV. The effect of fiber direction and annual ring on vibration cutting forces. 29(9), 424–429.

Kagawa, Y., Noguchi, M., and Katagiri, J. (1980). Detection of acoustic emission in the process of timber drying. *Acoust. Lett.* 3(8), 150–153.

Kitayama, S., Noguchi, M., and Satoyoshi, K. (1985). Automatic control system of drying Zelkova wood by acoustic emission monitoring. *Acoust. Lett.* 9(4), 45–48.

Knuffel, W. E. (1988). Acoustic emission as a strength predictor in structural timber. Holzforschung. 42, 195–198.

Kollmann, F. F. (1983). Holz und Schall—Theorie und Nutzanwendung. (Wood and sound—theory and practical utilization.) *Holz Zentralbl.* 109(14), 201–202.

Lemaster, R. L., Klamecki, B. E., and Dornfeld, D. A. (1982). Analysis of acoustic emission in slow speed wood cutting. *Wood Sci.* 15(2), 150–160.

Lemaster, R. L., Tee, L. B., and Dornfeld, D. A. (1985). Monitoring tool wear during wood machining with acoustic emission. *Wear.* 101, 273–282.

Lemaster, R. L. and Dornfeld, D. A. (1987). Preliminary investigation of the feasibility of using acousto-ultrasonics to measure defects in lumber. *J. Acoust. Emission.* 6(3), 157–165.

Lemaster, R. L. and Dornfeld, D. A. (1988). Apparatus for coupling an acoustic emission transducer to a rotating circular saw. *J. Acoust. Emission.* 7(2), 103–110.

Lemaster, R. L., Gasick, M. F., and Dornfeld, D. A. (1988). Measurement of density profiles in wood composites using acoustic emission. *J. Acoust. Emission.* 7(2), 111–118.

Lemaster, R. L. and Quarles, S. L. (1990). The effect of same-side and through-thickness transmission modes on signal propagation in wood. *J. Acoust. Emission.* 9(1), 17–24.

Liptai, R. G., Harris, D. O., and Tatro, C. A. (1972). An Introduction to Acoustic Emission. STP 505. American Society for Testing Materials, Philadelphia.

Lord, A. E., Jr. (1983). Acoustic emission—an update. in *Physical Acoustics.* Vol. 15. Masson, W. P. (ed.). Academic Press, New York, 295–360.

Lynnworth, L. C. (1989). *Ultrasonic Measurements for Process Control. Theory, Techniques, Applications.* Academic Press, New York.

Marra, A. A. (1964). Proc. Conf. Theory of Wood Adhesion. University of Michigan, Ann Arbor, July 26 to August 4, 1961, cited by Blomquist et al. (1983).

Milburn, J. A. and Johnson, R. P. C. (1966). The conduction of sap. II. Detection of vibrations produced by sap cavitation in *Ricinus xylem. Planta.* 69, 43–52.

Milburn, J. A. (1973a). Cavitation in *Ricinus* by acoustic detection. Induction in excised leaves by various factors. *Planta.* 110, 253–265.

Milburn, J. A. (1973b). Cavitation studies on whole *Ricinus* plants by acoustic detection. *Planta.* 112, 333–342.

Milburn, J. A. and Crombie, D. S. (1984). Sounds made by plants. *Bull. Aust. Acoust. Soc.* 12, 15–19.

Milburn, J. A. (1979). Detection of cavitation. *in Water Flow in Plants.* Longmans, London.

Miller, G. D. (1963). Sound generated by wood when under stress. *Can. Dept. For. Prod. Bull. Res. News.* 6, 6–7.

Mitrakovic, D., Grabec, I., and Sedemak, S. (1985). Simulation of AE signals and signal analysis systems. *Ultrasonics.* 23(5), 227–232.

Morgner, W., Niemz, P., Theis, K. (1980). Anwendung der Schallemission analyse zur Untertsuchung von Bruch-und Kriechvorgängen in Werkstoffen aus Holz. (The use of acoustic emission analysis for the investigation of fracture and creep in wood based materials.) *Holztechnologie.* 21(2), 77–82.

Muenow, R. A. (1973). Large scale applications of acoustic emission. *IEEE Trans. Sonics Ultrasonics.* 1, 48.

Murase, Y., Ike, K., and Mori, M. (1988). Acoustic emission monitoring of wood cutting. I. Detection of tool wear by AE signals. *Mokuzai Gakkaishi.* 34(3), 207–213.

Murase, Y., Ike, K., and Mori, M. (1988). Acoustic emission monitoring of wood cutting. II. Effect of slope angle of grain on AE characteristics. *Mokuzai Gakkaishi.* 34(3), 271–274.

Murase, Y. and Torihara, N. (1990a). Acoustic emission monitoring of wood cutting. III. Effect of cutting angle on AE characteristics in cutting parallel and perpendicular to the grain. *Mokuzai Gakkaishi.* 36(4), 269–275.

Murase, Y. and Kawanami, S. (1990b). Acoustic emission monitoring of wood cutting. IV. Effects of moisture content and species on acoustic emission characteristics in wood cutting. *Mokuzai Gakkaishi.* 36(9), 717–724.

Nakao, T., Tanaka, C., and Takahashi, A. (1986). Source wave analysis of large amplitude acoustic emission in the bending of wood. *Mokuzai Gakkaishi.* 32(8), 591–595.

Nakagawa, M., Masuda, M., and Nogucho, M. (1989). Acoustic emission in the bending of structural lumber containing knots. *Mokuzai Gakkaishi.* 35(3), 190–196.

Nichols, R. W. (Ed.). (1976). *Acoustic Emission.* Applied Science, London.

Niemz, P., Wagner. M., and Theis, K. (1983). Stand und Möglichkeiten der Anwendung der Schallemissionanalyse in der Holzforschung. (Use of acoustic emission in wood research.) *Holztechnologie,* 24(2), 91–95.

Niemz, P. and Hansel, A. (1988). Untersuchungen zur Ermittlung wesentlicher Einflussfaktoren auf die Schallemission von Vollholz und Holzwerkstoffen. (Determination of important factors influencing acoustic emission from wood and wood-based materials.) *Holztechnologie.* 29(2), 79–81.

Niemz, P. and Paprzycki, O. (1988). Deformation and failure of beech wood modified with styrene studied using SEM and acoustic emission analysis. *Holzforsch. Holzverwert.* 40(3), 48–50.

Niemz, P., Hansel, A., and Schweitzer, F. (1989). Untersuchungen zu ausgewåhlten Einflussgrössen auf die Schallemission von Vollholz und Holzwerkstoffen. *Holztechnologie.* 30(2), 70–71.

Niemz, P., Bemann, A., and Wagenfuhr, R. (1990). Untersuchungen zu den Eigenschaften von immissionsgeschådigtem Fichtenholz. Mechanische Eigenschaften und Schallemissionanalyse. *Holztechnologie.* 30(2), 70–71.

Niemz, P. and Lühmann, A. (1992). Application of the acoustic emission analysis to evaluate the fracture behaviour of wood and derived timber products. *Holz.Roh.Werkst.* 50, 191–194.

Noguchi, M., Kagawa, Y., and Katagiri, J. (1980). Detection of acoustic emission during drying of hardwoods. *Mokuzai Gakkaishi.* 26(9), 637–638.

Noguchi, M., Kagawa, Y., and Katagiri, J. (1983). Acoustic emission generation in the process of drying hardwoods. *Mokuzai Gakkaishi.* 29(1), 20–23.

Noguchi, M., Okumura, S., and Kawamoto, S. (1985a). Characteristics of acoustic emissions during wood drying. *Mokuzai Gakkaishi.* 31(3), 171–175.

Noguchi, M., Fujii, Y., and Imamura, Y. (1985b). Use of acoustic emission in the detection of incipient stages of decay in western hemlock wood. *Acoust. Lett.* 9(6), 79–81.

Noguchi, M., Nishimoto, K., Imamura, Y., Fujii, Y., Okumura, S., and Miyauchi, T. (1986). Detection of very early stages of decay in western hemlock wood using acoustic emission. *For. Prod. J.* 36(4), 35–36.

Noguchi, M., Kitayama, S., Satoyoshi, K., and Umetsu, J. (1987). Feedback control for drying *Zelkova serrata* using in-process acoustic emission monitoring. *For. Prod. J.* 37(1), 28–34.

Noguchi, M. (1991). Use of acoustic emission in wood processing and evaluating properties of wood. *Mokuzai Gakkaishi.* 37(1), 1–8.

Noguchi, M., Ishii, R., Fujii, Y., Imamura, Y. (1992). Acoustic emission monitoring during partial compression to detect early stages of decay. *Wood Sci. Tech.* 26, 279–287.

Oertli, J. J. (1990). The effect of cavitation on the status of water in plants. Abstr. Int. Workshop Analysis of Water Transport in Plants and Cavitation of Xylem Conduits. Abbazia di Vallombrosa, Firenze, Italy, May 29 to 31, 7.

Ogino, S., Kaiono, K., and Suzuki, M. (1986). Prediction of lumber checking during drying by means of acoustic emission technique. *J. Acoust. Emission.* 5(2), 61–65.

Okumura, S., Kawamoto, S., Nakagawa, M., Noguchi, M. (1986). Relationship between drying stresses and acoustic emission of wood. *Bull. Kyoto Univ. For.* No. 58, 251–259.

Okumura, S., Kawamoto, S., Mori, T., and Noguchi, M. (1986). Acoustic emission during drying of Japanese oak (*Quercus mongolica* var. *grosseserrata* [*Q. crispula*]). *Bull. Kyoto Univ. For.* No. 57, 300–307.

Okumura, S., Kiyotaki, T., and Noguchi, M. (1987). A few experiments on acoustic emission during wood drying. *Bull. Kyoto Univ. For.* No. 59, 283–291.

Pao, Y. H. (1978). Theory of acoustic emission. in *Elastic Waves and Nondestructive Testing of Materials.* Vol. 29. Pao, Y. H. (Ed.). American Society of Mechanical Engineers, New York, 107–128.

Patton-Mallory, M. (1988). Use of acoustic emission in evaluating failure processes of wood products. Proc. 1988 Int. Timber Eng. Vol. 1. Washington State University, Pullman, 596–600.

Pena, J. and Grace, J. (1986). Water relations and ultrasound emissions of *Pinus sylvestris* L. before, during and after a period of water stress. *New Phytol.* 103, 515–524.

Pentoney, R. E. and Porter, A. W. (1964). Acoustic Emission from Wood Under Applied Stress. Rep. 18th Annu. Meet. F.P.R.S., Chicago.

Pisante, M., Moretti, N., and Frisullo, S. (1990). Water relations and ultrasound acoustic emissions in Douglas fir seedlings infected with *Phomopsis occulta* and *Diplodia pinea*. Abstr. Int. Workshop Analysis of Water Transport in Plants and Cavitation of Xylem Conduits. Abbazia di Vallombrosa, Firenze, Italy, May 29 to 31, 43.

Poliszko, S., Molinski, W., and Raczkowski, J. (1988). Acoustic emission of wood during swelling in water. Proc. Eur. Cong. Mechanical Behaviour of Wood. Bordeaux, June 8 to 9, 331–340.

Pollock, A. A. (1970). Acoustic Emission from Solids Undergoing Deformation. Ph.D. thesis, University of London.

Pollock, A. A. (1973). Acoustic emission. II. Acoustic emission amplitudes. *Non-Destruct. Test.* October, 264–269.

Pollock, A. A. (1974). Acoustic emission. in *Acoustics and vibration progress*. Stephens, R. W. B. and Levinthall, H. C. (Eds.). Chapman & Hall, London.

Porter, A. W., El-Osta, M. L., and Kusec, D. J. (1972). Prediction of failure of finger joints using acoustic emission. *For. Prod. J.* 22(9), 74–82.

Quarles, S. L. and Lemaster, R. L. (1988). Current AE/AU research in solid wood and wood-based composites at the University of California, Forest Products Laboratory. Proc. Conf. NDT and Evaluation for Manufacturing and Construction. August 10 to 12, Urbana, IL.

Quarles, S. L. (1992). Acoustic emission associated with oak during drying. *Wood Fiber Sci.* 24(1), 2–12.

Raschi, A., Scarascia-Mugnozza, G. E., Valentini, R., and Vazzana, C. (1989). Detection of cavitation events upon freezing and thawing of water in stems using ultrasound techniques. *Ann. Sci. For.* 46 (Suppl.), 342s–345s.

dos Reis, H. L. M. and McFarland, D. M. (1986). On the acoustoultrasonic characterization of wood fiber hardboard. *J. Acoust. Emission.* 5(2), 67–70.

Rice, R. W. and Skaar, C. (1990). Acoustic emission patterns from the surface of red oak wafers under transverse bending stress. *Wood Sci. Tech.* 24, 123–129.

Rice, R. W. and Kabir, F. R. A. (1992). The acoustic response of three species of wood while immersed in three different liquids. *Wood Sci. Tech.* 26, 131–137.

Rice, R. W. and Peacock, E. (1992). Acoustic emission resulting from cyclic wetting of Southern yellow pine. *Holz Roh Werkst.* 50, 304–307.

Ritman, K. T. and Milburn, J. A. (1988). Acoustic emission from plants. *J. Exp. Bot.* 36(206), 1237–1248.

Robson, D. J. (1990). Development of a theory of freezing in conifer xylem. Abstr. Int. Workshop Analysis of Water Transport in Plants and Cavitation of Xylem Conduits. Abbazia di Vallombrosa, Firenze, Italy, May 29 to 31, 14.

Sachse, W. and Kim, K. Y. (1987). Point-source/point-receiver materials testing. in *Review of Progress in Quantitative Nondestructive Evaluation.* Vol. 6A. Thompson, D. O. and Chimenti, D. E. (Eds.). Plenum Press, New York, 311–320.

Sachse, W. (1988). The processing of acoustic emission signals. in *Progress in Acoustic Emission IV. Proc. 9th Int. Acoustic Emission Symp.* Japanese Society for Nondestructive Inspection, Kobe.

Sadanari, M. and Kitayama, S. (1989). Waveform analysis of acoustic emission generated in the wood-drying process. *Mokuzai Gakkaishi.* 35(7), 602–608.

Salleo, S. and Lo Gullo, M. A. (1990). Drought resistance strategies and vulnerability to cavitation of some Mediterranean scleriphyllous trees. Abstr. Int. Workshop Analysis of Water Transport in Plants and Cavitation of Xylem Conduits. Abbazia di Vallombrosa, Firenze, Italy, May 29 to 31, 19.

Sandford, A. P. and Grace, J. (1985). The measurement and interpretation of ultrasound from woody stems. *J. Exp. Bot.* 36, 298–311.

Sato, K., Noguchi M., and Fushitani, M. (1983). The characteristics of acoustic emission of wood generated during several types of loading. *Mokuzai Gakkaishi.* 29(6), 409–414.

Sato, K., Kamei, N., Fushitani, M., and Noguchi, M. (1984). Discussion of tensile fracture of wood using acoustic emissions. *Mokuzai Gakkaishi.* 30(8), 653–659 and 30(2), 117–123.

Sato, K., Honda, T., and Fushitani, M. (1986). Fracture toughness and acoustic emission in wood. in *Progress in Acoustic Emission III. Proc. 8th Int. Acoustic Emission Symp.* Japanese Society for Nondestructive Inspection, Kobe, 602–608.

Sato, K., Yamaguchi, K., Ando, N., and Fushitani, M. (1989). Detection of poor bond in plywood utilizing acoustic emission technique. in *Proc. 12th World Conf. Nondestructive Testing.* Boogaard, J., and van Dijk, G. M. (Eds.). Elsevier, Amsterdam, 1627–1629.

Sato, K., Takeuchi, H., Yamaguchi, K., Ando, N., and Fushitani, M. (1990). Lumber stress grading utilizing acoustic emission technique. *J. Acoust. Emission.* 9(3), 209–213.

Sato, K. and Fushitani, M. (1991). Development of nondestructive testing system for wood based materials utilizing acoustic emission technique. Proc. 8th Int. Nondestructive Testing of Wood Symp. Washington State University, Pullman, 33–43.

Simpson, W. A., Jr. (1974). Time-frequency domain formulation of ultrasonic frequency analysis. *J. Acoust. Soc. Am.* 56(6), 1776–1781.

Sims, D. G., Dean, G. D., Read, B. E., and Western, B. C. (1977). Assessment of damage in GRP laminates by stress wave emission and dynamic mechanical measurements. *J. Mater. Sci.* 12, 2329–2342.

Skarr, C., Simpson, W., and Honeycutt, R. M. (1980). Use of acoustic emission to identify high levels of stress during oak lumber drying. *For. Prod. J.* 30(2), 21–22.

Sperry, J. (1990). Causes and consequences of xylem embolism. Abstr. Int. Workshop Analysis of Water Transport in Plants and Cavitation of Xylem Conduits. Abbazia di Vallombrosa, Firenze, Italy, May 29 to 31, 13.

Stephens, R. W. B. and Pollock, A. A. (1971). Waveforms and frequency spectra of acoustic emission. *J. Acoust. Soc. Am.* 50(3, Part 2), 904–910.

Stephens, R. W. B. and Levinthall, H. C. (1974). *Acoustics and Vibration Progress.* Chapman & Hall, London.

Stone, D. E. W. and Dingwall, P. F. (1977). Acoustic emission parameters and their interpretation. *NDT Int.* April, 51–62.

Suzuki, M. and Schniewind, A. P. (1987). Relationship between fracture toughness and acoustic emission during cleavage failure in adhesive joints. *Wood Sci. Technol.* 21, 121–130.

Swindlehurst, W. (1973). Acoustic emission. I. Introduction. *Non-Destruct. Test.* June, 152–158.

Tanaka, C., Nakao, T., Takahashi, A., and Schniewind, A. P. (1990). On-line control of feed-speed in circular sawing. *Holz Roh Werk.* 48, 139–145.

Tyree, M. T. and Dixon, M. A. (1983). Cavitation events in *Thuja occidentaalis* L. *Plant Physiol.* 72, 1094–1099.

Tyree, M. T., Dixon, M. A., and Thompson, R. G. (1984a). Ultrasonic acoustic emission from the sapwood of *Thuja occidentalis* measured inside a pressure bomb. *Plant Physiol.* 74, 1046–1049.

Tyree, M. T., Tyree, E. L., and Johnson, R. (1984b). Ultrasonic acoustic emissions from the sapwood of cedar and hemlock. An examination of 3 hypotheses regarding cavitations. *Plant Physiol.* 75, 988–992.

Tyree, M. T. and Sperry, J. S. (1989). Characterization and propagation of acoustic emission signals in woody plants: towards an improved acoustic emission counter. *Plant Cell Environ.* 12(4), 371–382.

Tyree, M. T. (1989). Cavitation in trees and hydraulic sufficiency of woody stems. *Ann. Sci. For.* 46(Suppl.), 330s–337s.

USDA. (1987). *Wood Handbook. Wood as an Engineering Material.* Forest Products Laboratory, Madison, WI.

Vary, A. and Bowles, K. J. (1979). An ultrasonic-acoustic technique for nondestructive evaluation of fiber composite quality. *Polym. Eng. Sci.* 19(5), 373–376.

Vary, A. (1980). Ultrasonic measurement of material properties. in *"Research Techniques in Nondestructive Testing."* Vol. 4. Sharpe, R. S. (Ed.). Academic Press, London, 159–204.

Vaughan, P. W. and Leeman, S. (1989). Acoustic cavitation revised. *Acustica.* 69(3), 109–119.

Vautrin, A. and Harris, B. (1987). Acoustic emission characterization of flexural loading damage in wood. *J. Mater. Sci.* 22, 3707–3717.

Wassipaul, F., Vanek, M., and Maydhofer, A. (1986). Klima und Schallemissionen bei der Holztrocknung. (Climate and acoustic emissions during the drying of wood.) *Holzforsch. Holzverwert.* 38(4), 73–79.

Weisinger, R. (1980). Determination of fundamental acoustic emission signal characteristics. in *Mechanics of Nondestructive Testing.* Stinchcomb, W. W. (Ed.). Plenum Press, New York, 165–185.

Williams, R. V. (1980). *Acoustic Emission.* Adam Hilger Ltd., Bristol, England.

Woodward, B. (1976). Identification of acoustic emission source mechanisms by energy spectrum analysis. *Ultrasonics.* 14(5), 249–255.

Yamaguchi, K. (1988). Instrumentation and data processing for acoustic emission technology and applications. in *Proc. 9th Int. Acoustic Emission Symp.* Japanese Society for Nondestructive Inspection, Kobe.

Ying, S. P., Hamlin, D. R., and Tanneberger, D. (1974). A multichannel acoustic emission monitoring system with simultaneous multiple event data analysis. *J. Acoust. Soc. Am.* 55(2), 350–356.

Yoshimura, N., Shimotsu, M., Honjou, K., Kotaki, M., Ogasawara, Y., Okuyama, D., and Noto, F. (1987). Detection of starved joints in plywood by acoustic emission. *Mokuzai Gakkaishi.* 33(8), 650–653.

Zhao, C., Tanaka, C., Nakao, T., Takahashi, A., and Tsuzii, T. (1990). Relationship between surface finish qualities and acoustic emission count rate. *Mokuzai Gakkaishi.* 36(3), 169–173.

Zimmermann, M. H., Brown, C., and Tyree, M. T. (1980). *Trees, Structure and Function.* Springer-Verlag, Berlin.

Zimmermann, M. H. and Milburn, J. A. (1982). Transport and storage of water. in *Encyclopaedia of Plant Physiology,* New Series, Vol. 12B. Lange, O. L., Nobel, P. S., Osmond, C. B., and Ziegler, H. (Eds.). Springer-Verlag, Berlin, 135–151.

Zimmermann, M. H. (1983). *Xylem Structure and the Ascent of Sap.* Springer-Verlag, Berlin.

Chapter 11

High-Energy Ultrasonic Treatment for Wood Processing

11.1. MODIFICATIONS INDUCED IN ANATOMIC STRUCTURE BY HIGH-ENERGY ULTRASONIC WAVES

Modifications induced in wood by high energy ultrasonic waves were reported by Parpala et al. (1970, 1974) and by Filipovici et al. (1970). The parameters of the ultrasonic pulse used by Parpala et al. for the treatment of standard fir specimens (20 × 20 × 300 mm), immersed in water at 25°C for 45 min were 1 MHz frequency, 1.1 kV, 100 mA intensity, and 5.5×10^4 W/m^2 energy. Beech specimens were treated in an ultrasonic welder at 18 kHz frequency and 25 W/cm^2 energy during a period of time varying from 30 s to 6 min. The moisture content of specimens was 40%. Analyzing the scanning electronic micrographs of transversal anatomic sections published by the authors, it was observed that in fir, the middle lamella of the earlywood is the most affected by the treatment. The cavitation phenomenon acts as a lignin softening agent, reducing the cohesion between the microfibrils and the layers of the cellular wall. The walls of the vessels in beech were ruined. Dislocations of the S$_2$ layer generated oriented slip-lines, with a characteristic step probably related to the value of the microfibril angle. Subsequent to the structural modifications induced by the ultrasonic treatment, mechanical properties were modified. Reductions from 20 to 80% of Young's modulus measured under static bending test and of the toughness coefficient were reported.

11.2. HIGH-ENERGY ULTRASONIC TREATMENT, A TOOL FOR WOOD PROCESSING

Ultrasonic processing of materials, using high-energy ultrasound, with high amplitude and displacement up to 150 μm, was the subject of studies by Rozenberg (1973) and Rooney (1981). The advantages of ultrasonic techniques have been demonstrated for the ultrasonic welding of metals and plastics and the machining of hard, brittle materials, cemented carbides, etc. The development of these techniques was based on understanding the nonlinear phenomena of high-amplitude ultrasonic waves propagation and the quantification of nonlinear properties of materials. Developments in the ultrasonic processing of wood began with the empirical determination of some parameters selected by the experimenters. No reference was made to nonlinear properties of the medium. The lack of a theoretical approach to this technique may partially explain the impossibility of efficiently solving practical problems.

11.2.1. DRYING

There is no doubt that selecting a drying process is an important economic decision for the lumber manufacturer. Recognizing the relative difficulty of drying Australian eucalyptus, Kauman (1964) proposed the use of ultrasound for inducing cavitation and reducing collapse during drying via impregnation with bulking or wetting agents. His approach was limited to laboratory testing, and was based on the assumption that the effect of the surface forces developed during drying could be offset by the nucleation of bubbles inside the lumen of the cells.

Neylon (1978) also attempted to avoid collapse in the eucalyptus by heating wood with ultrasonic energy. He tested boards (25 × 100 × 300 mm) with a commercially available welder, which produced energy at 750 W and 20 kHz. The rise in temperature at 150°C and the contact pressure between the horn and the specimen at 140 kPa determined high shrinkage of the tested specimen, and consequently its degradation. To defeat this effect Neylon proposed a combination of preultrasonic treatment with drying at low temperatures. No commercially viable techniques have been reported since.

11.2.2. DEFIBERING

Disintegration of cellulosic fibers (Morehead, 1950) is thought to be one of the first practical applications of the high-energy ultrasonic technique for wood material.

Of all the characteristics of the anatomic elements of wood, cell length is without question the most important because of its strong relationship to wood quality. Cell length—tracheids length in softwoods

Table 11.1. Anatomical characteristics of different species (Istas and Raekelboom, 1970)

Species	Density (kg/m³)	Anatomic characteristics		
		Length (mm)	Diameter (μm)	Wall (μm)
Acer pseudoplatanus	482	0.86	23.0	4.0
Alnus glutinosa	468	1.09	25.3	5.7
Betula verrucosa	484	1.27	23.7	5.7
Quercus robur	568	1.12	20.1	2.3
Castanea sativa	461	1.02	22.8	3.7
Fagus sylvatica	579	1.32	20.8	8.0
Picea abies	370	3.61	42.3	5.0
Pinus strobus	318	2.95	50.1	4.6
Pinus sylvestris	431	3.60	50.0	5.8
Populus tremula	332	1.31	32.3	3.6
Pseudotsuga menziessii	470	2.46	32.2	5.6
Tilia cordata	385	1.15	37.8	4.7

and fiber and vessel element length in hardwoods—is understood in production operations as "fiber length", when referring to both tracheids and fibers. For the sake of simplicity the term "fiber length" is used here. It is generally admitted that fiber length varies greatly both within and among trees, is strongly genetically controlled (Zobel and van Buijtenen, 1989), and affects the quality of manufactured products or mechanical pulps and papers (Clark, 1978). Table 11.1 lists some dimensional characteristics of the anatomical elements of different wood species.

Preparing wood fibers for length measurement by maceration of solid wood specimens and filtration of slurries is indeed one of the oldest methods invented using wood chemistry. When only limited sizes of wood samples are available, the low-consistency slurries must be concentrated first. For this purpose, ultrasound can be used to facilitate concentration or aggregation of fibers in suspension in order to separate fibers, parenchyma cells, and vessels for the examination of the microstructural distribution of lignin in those elements (Kutsuki and Higuchi, 1978) or for other fine analyses. The efficiency of the treatment depends on the frequency and power of the ultrasonic signal, the viscosity of the fluid medium, and the size and density of the particles. To avoid filtration, Wang and Micko (1985) placed the vials containing fiber slurries in an ultrasonic cleaner (frequency, 55 kHz and intensity, 0.34 W/cm²). The fiber suspensions coalesced into clumps and formed a ring that can be easily removed, eliminating the need for filtration. Coniferous specimens required immersion for several seconds in the ultrasonic bath, but short-fiber species require longer treatment periods.

Taneda (1972) used the ultrasonic treatment (frequency, 400 kHz and energy, 150 W) on *Tilia japonica* fibers and chips and on wood products (veneer, plywood, hardboards) to greatly accelerate the grafting of various monomers (methylmethacrylate, styrene, etc.) onto the fine structure of wood. Monomer-modified new wood products showed high resistance to chemicals, water, and abrasion. Physicomechanical characteristics such as hardness, dimensional stability and machinability are improved as compared to natural products. Figure 11.1 shows the effects of ultrasonic irradiation on various factors of graft copolymerization of styrene by hydrogen peroxide in wood. The maximum rate of polymerization was attained at 150 W ultrasonic power and a frequency of 400 kHz. An important acceleration of grafting on wood fibers was observed via ultrasonic irradiation in the early stages of polymerization. The acceleration process at maximum seems to correspond to 3 W/cm² for the ultrasonic intensity.

Fiber length measurement associated with the ultrasonic technique was the object of an interesting acousto-optic technique developed by Dion and co-workers (1982, 1987, 1988) and Garceau et al. (1989). The device (Figure 11.2) itself can measure a fiber length distribution of several thousand fibers within a few seconds when a stationary, 150-kHz ultrasonic field acts on a dilute pulp suspension. Under the influence of the radiation pressure induced by the transducers (T_1 and T_2) in the cavity containing pulp in suspension, the fibers are reoriented in the ultrasonic field parallel to the direction of wave propagation, at different characteristic times, depending on their length and distribution. During ultrasonic excitation the characteristic time is monitored using a collimated beam of incoherent light (S). This device, based on the original idea of using ultrasound, completed the series of apparatus for fiber length measurements developed in the 1980s by Swedish and French laboratories (Janin and Ory, 1984).

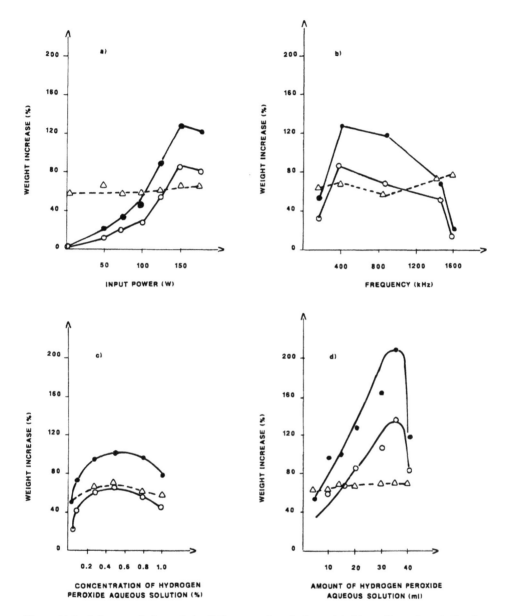

Figure 11.1. Influence of ultrasonic irradiation on various factors related to graft copolymerization by hydrogen peroxide of *Tilia japonica* wood fibers. (a) 400 kHz, 0.5% hydrogen peroxide, 20 ml, styrene, 5 ml, 70°C, 3 h; (b) 150 W, 0.5% hydrogen peroxide, 20 ml, styrene, 5 ml, 70°C, 3 h; (c) 400 kHz, 150 W, 0.5% hydrogen peroxide in aqueous solution, 15 ml, styrene, 5 ml, 70°C, 3 h; (d) 400 kHz, 150 W, 0.5% hydrogen peroxide, styrene, 5 ml, 70°C, 3 h. (●) Weight increase; (o) grafting; (△) graft efficient (From Taneda, K., *Hokkaido For. Prod. Res. Inst.*, Rep. 59, 1972. With permission).

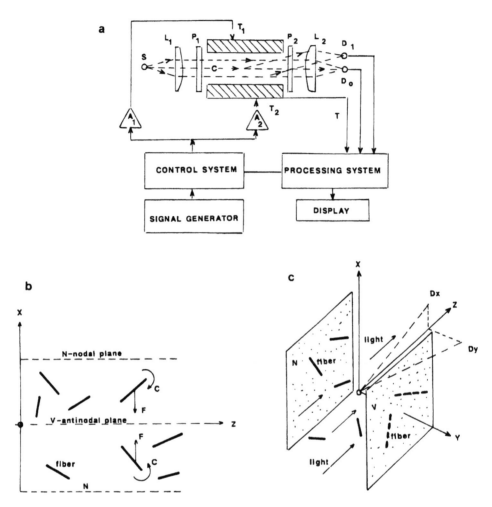

Figure 11.2. Ultrasonic device for fiber length measurement. (a) Measurement system: T_1, T_2, transducers; C, cavity containing fibers in suspension; A_1, A_2, amplifiers; S, source of light; L_1, L_2, optical lens; P_1, P_2, polarizers; D_0, D_1, photodetectors for measurement of transmitted and scattered light intensity. (b) Relative position of fibers in the ultrasonic stationary field. N, nodal plane, where the velocity is minimum and where the fibers are aggregated or stratified by the acoustic pressure; V, antinodal plane, where the amplitude is greatest; F, force acting on fibers; C, torque. (c) Light scattering by the suspension in Dx direction, parallel to the V plane, and in Dy direction, perpendicular to the V plane. (From Dion, J. L. et al., *Acustica.* 65, 284, 1988. With permission.)

The following optical principle forms the basis for this technique. The pulp suspension passes through a capillary tube interrupting a light beam. This interruption generates pulses proportional to the fiber length when the duration is measured or to the fiber diameter when the intensity is considered.

On the assumption that fibers are isotropic infinite cylinders, Habeger and Baum (1983) reported some interesting theoretical investigations about the fundamental mechanism that governs ultrasonic propagation in fiber slurries. Ultrasonic velocity and attenuation were measured and an apparatus was designed for the characterization of fiber suspension. Better quantitative results have been predicted for synthetic fiber system than for wood fibers. Indeed, the use of the high-energy ultrasonic treatment could cause one to question the changes in morphology or the chemical properties of wood fibers. When a 1-h treatment and 5 to 6 W/cm^2 were applied to suspensions of dissolving pulp, Dolgin et al. (1968) reported several modifications: slightly reduced (5 to 8%) degree of polymerization of cellulose; slightly reduced (5 to 12%)

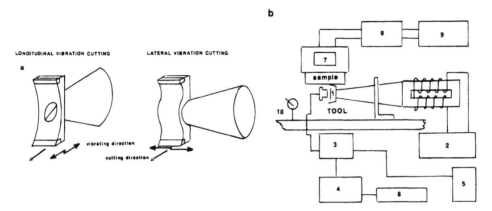

Figure 11.3. Ultrasonic device for cutting wood in longitudinal and transversal directions. (a) Cutting direction and vibrating direction. (b) Ultrasonic device. (1) Tool; (2) ultrasonic oscillator; (3) displacement device; (4) preamplifier; (5) oscilloscope; (6) frequency counter; (7) load cell; (8) strain amplifier, (9) mV recorder; (10) dial gauge. (From Kato, K. et al., *Mokuzai Gakkaishi.* 17(2), 57, 1971. With permission.)

pentosan and lignin content of pulp; 50% reduction in the consumption of chloride for bleaching. However, when relatively low energy is used, the treatment is unbiased by the morphology and chemical properties of fibers (Wang and Micko, 1985).

Labosky and Martin (1969) and Clark (1978) reported on the enhancement of some properties of paper (breaking length, burst) obtained from ultrasonically beaten pulps. Clark noted that "the differences between ultrasonic treatment and mill beating are so extreme as to rule out the satisfactory use of ultrasonic treatment for laboratory purposes".

11.2.3. CUTTING

Ultrasonically cutting solid wood or composites is a relatively new process and little information is available. The need to introduce new machining methods, especially when cutting composite products, is determined by two main factors: electrical energy required and tool wear. Ultrasonic cutting offers advantages over conventional methods such as improvement of the smoothness quality of the machined surface (producing little kerf or sawdust), no deformation due to cutting forces, no burned surfaces, and little tool wear. The disadvantages are the slow cutting rate and the reduction of penetration rate with depth, which in the case of wood composites are strongly related to the type of resin or the size and the shape of the particles of the board.

The studies of Kato and Asano (1971, 1974) recognized the effectiveness of ultrasonically cutting wood. A 20-kHz and a 200-W oscillator is incorporated in the equipment presented in Figure 11.3. Manufacture of the tool horn was oriented to longitudinal cutting as well as to crosscutting. Experiments were done using hinoki (*Chamaecyparis obtusa* Endl.) and buna wood (*Fagus crenata* Blume) under air-dried conditions. The tool was fixed on the ultrasonic vibrator; the cut depth was 0.2 mm; the amplitude of the knife was 15 to 30 μm; the feed speeds were 1.2, 2.6, and 5.2 m/s; and the clearance angle was 5°.

The principal end products of ultrasonic cutting were improvement of the smoothness of the machined surface, reduction of the cutting force (Figure 11.4), and reduction of the coefficient of friction between the tool and the specimen (Table 11.2). The results are encouraging and could serve as a guide for developments in special applications justified by the special properties needed for the manufactured products.

11.2.4. PLASTICIZING EFFECT

It is well known that heat, moisture, various organic liquids, and aqueous solutions of urea of liquid ammonia behave as plasticizers for wood. Based on chemical plasticization, many practical procedures were patented for facilitating the bending of wood for furniture parts or other objects, but the general use of plasticizers has been limited because of the high cost. The search for ways to improve the plasticizing procedure for bending was extended to the ultrasonic technique (Filipovici et al., 1970, 1972). They injected high ultrasonic energy (18 kHz, 25 W/cm^2) in beech at 40% moisture content, causing the cavitation phenomena to

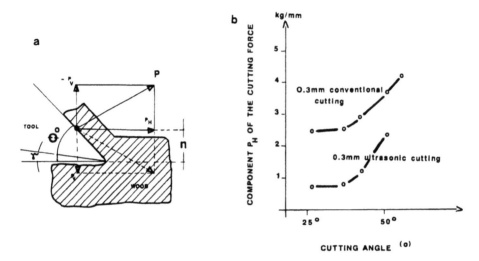

Figure 11.4. Relationships between cutting angle and cutting force in conventional and ultrasonic cutting parallel to grain in hinoki. (a) Cutting parameters. P_H, parallel tool force; P_V, normal tool forces; Θ, cutting angle; n, depth of cut; γ, clearance angle. (b) Comparison between ultrasonic cutting force components and conventional cutting parallel to the grain in hinoki. (From Kato, K. et al., *Mokuzai Gakkaishi.* 17(2), 57, 1971. With permission.)

Table 11.2. Calculated coefficient of friction in cutting parallel to the grain

Depth (mm)	Cutting angle								
	50°			42°			36°		
	c	l	t	c	l	t	c	l	t
	Hinoki								
0.20	0.80	0.65	0.50	0.95	0.65	0.20	1.50	0.80	0.10
0.30	0.70	0.50	0.35	0.90	0.65	0.20	1.20	0.55	0.15
0.40	0.70	0.50	0.30	0.90	0.5	0.20	1.10	0.65	0.15
0.50	0.65	0.45	0.25	0.90	0.45	0.15	1.10	0.60	0.40
	Buna								
0.20	1.10	0.35	0.60	0.80	0.60	0.15	1.00	0.55	0.30
0.30	1.10	0.30	0.55	0.90	0.45	0.10	0.95	0.50	0.25
0.40	0.85	0.35	0.40	0.80	0.35	0.10	0.80	0.50	0.25
0.50	0.85	0.30	0.35	0.80	—	0.10	0.85	—	0.10

Note: c = coefficient of friction in conventional cutting; l = coefficient of friction in longitudinal cutting; t = coefficient of friction in transversal cutting;
From Kato, K. et al., *Mokuzai Gakkaishi*, 17(2), 57, 1971. With permission.

occur simultaneously with the rise in the temperature of the sample. After that, the samples were easily bent in a special device. Published results did not show whether this method may be applied economically for commercial bending of furniture parts.

11.2.5. REGENERATION EFFECT ON AGED GLUE RESINS
Glue resins used in the wood industry are supplied as powders to be added to solvents, or as reactive, thick, viscous liquids or films. The adhesive performance of resins is strictly related to the conditions of storage and preparation of adhesives. The urea-formaldehyde resins are very popular in the particleboard industry. Aging processes occur during resin storage, manifested by the increase in viscosity and the incompatibility with water. The structural modification of resins produces perturbation in their pumping or container clean-

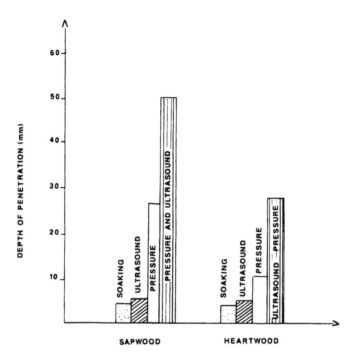

Figure 11.5. Influence of ultrasonic irradiation on the penetration effect of creosote into spruce. Histograms of the depth of penetration at 50°C on samples of 20 × 30 × 55 mm, when the following ultrasonic parameters were used: power, 1000 W; frequency, 22 kHz; irradiation time, 32 min. (From Kurjatko, S. and Marcok, M., *IAWA Bull.*, 11(2), 129, 1990. With permission.)

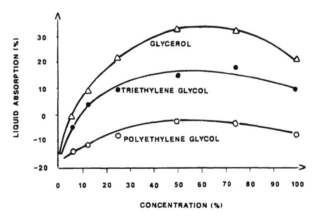

Figure 11.6. Effect of ultrasonic treatment (40 kHz and 500 W) on the absorption of hydrophilic organic liquids by *Pinus radiata*. (From Borgin, K. and Corbett, K., *Wood Sci. Technol.*, 4, 189, 1970. With permission.)

ing. Laboratory tests of the regeneration via ultrasound of aged glue urea-formaldehyde resins (Proszyk and Pradzynski, 1990) were encouraging.

Ultrasonic exposure of resin samples to various vibration amplitudes (10, 20, 30 μm) in an disintegrator at 21 kHz for 5 to 30 min enables full regeneration of sample properties.

11.3. IMPROVING WOOD PRESERVATION USING ULTRASOUND

Ultrasonic energy can be generated by different methods such as air jets, piezoelectricity, and magnetostriction. In order to improve the resistance of wood species to degrading agents, impregnation methods utilizing

Table 11.3 Retention of liquids (kg/m³) under ultrasonic treatment (200 kHz and 500 W) (data from Amemiya and Siriban, 1968)

Test Piece	Liquid	Time Minute	Ultrasonic	Hot (30°C) Dipping	Ratio
Veneer, 4 mm	Water	30	175	127	1.38
Veneer, 3 mm	Water	30	220	157	1.45
Veneer, 4 mm	Wolman[a]	5	97	109	0.89
		30	164	127	1.29
	Creosote	5	52	61	0.85
		30	72	67	1.08
Plywood, 5 mm	Wolman	5	27	29	0.93
		30	47	45	1.04
	Creosote	5	22	20	1.10
		30	29	24	1.21

[a]Wolman = Wolman salt 2% water solution; specimens on sheets of 10 × 10 cm on lauan or almon (shorea almon), a Japanese species at 13 to 15% moisture content.

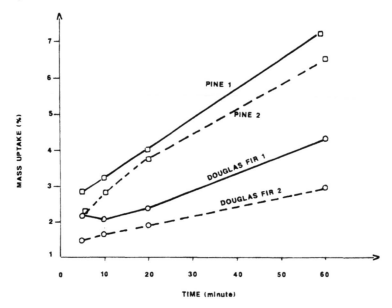

Figure 11.7. Relationships between the time of exposure to ultrasonic treatment and the mass uptake of different species impregnated with an aqueous solution of chromated copper arsenate at atmospheric pressure. (1) Treatment with ultrasound; (2) treatment without ultrasound. (From Avramidis, S., *Wood Fiber Sci.*, 20(3), 397, 1988. With permission.)

preservative liquids under ultrasonic energy were developed. These liquids could penetrate solid wood under the influence of the capillary pressure gradient or diffusion. The level of liquid retention or of depth of penetration depend on the anatomical structure of the wood, properties of the fluid, and conditions of application (temperature, pressure). The use of ultrasonic energy to improve liquid penetration in wood was proposed by Amemiya and Siriban (1968), Borgin and Corbett (1970), Walters (1977), Avramidis (1988), Kurjatko and Marcok (1990), and Lam et al. (1992).

The effect of ultrasound on liquid penetration is the result of the acoustic cavitation phenomena—the formation of bubbles that grow and collapse and generate shock waves which act on pit membranes. A consequent increase in fluid penetration and retention is observed. Chemical reactions also occur during cavitation (Sliwinski, 1974), and are essential for measuring the temperature or concentration of preservative liquids during treatment. Kurjatko and Marcok (1990) reported that optimum liquid penetration was measured in preservative pitch oil in spruce after ultrasound (22 kHz) and pressure (0.65 MPa) were used simultaneously. Such treatment could be positive or negative, depending on the possible chemical links with wood chemical structure (e.g., introducing one or several hydrophilic groups into straight hydrocarbon chains) and on the damage produced in the anatomical structure. Very often the bordered pits are destroyed and the margo microfibril broken. The penetration of creosote into sapwood is greater than that into the heartwood and is much more effective when pressure (0.65 MPa) is used simultaneously with ultrasound.

The absorption of the preservation liquids could be expressed as a weight modification of the specimen during the treatment, or as a percentage of the oven-dry weight. Figure 11.6 describes the effect of ultrasonic treatment on the absorption of various liquids that have a strong hydrophilic character by *Pinus radiata*, after 20 min of immersion. The weight increase of the specimen is due probably to capillary penetration and diffusion into the cell wall; the weight decrease is produced by the removal of wood extractives. It seems that concentrations of about 50 to 60% showed an optimum affinity for wood. The effect of the ultrasonic treatment was expressed as the weight ratio between the absorption of the untreated and treated samples. Amemiya and Siriban (1968) reported that the ultrasonic treatment in hardwoods accelerates the penetration of liquids in thin specimens and increases the retention of preservatives when compared to the hot dipping (30°C) treatment (Table 11.3). Figure 11.7 shows the effect of the ultrasonic treatment on Douglas fir and pine when a chromated copper arsenate solution is used. After 60 min in a laboratory-grade cleaner at 20°C, under 50 to 55 kHz, the best absorption is obtained in pine.

The most desirable research results are ultimately the efficiency of the high-energy ultrasonic treatment on large-size specimens with sapwood and heartwood as poles, structural lumber, or joinery products under atmospheric pressure.

REFERENCES

Amemiya, S. and Siribian, F. R. (1968). Research on wood preserving treatment. II. Effect of dipping with supersonic waves on the treatment of wood. *Bull. Gov. For. Exp. Stn.* 212, 167–178.

Avramidis, S. (1988). Experiments on the effect of ultrasonic energy on the absorption of preservatives by wood. *Wood Fiber Sci.* 20(3), 397–403.

Borgin, K. and Corbett, K. (1970). Improvement of capillary penetration of liquids into wood by use of supersonic waves. *Wood Sci. Technol.* 4, 189–194.

Clark, J. D. A. (1978). *Pulp Technology and Treatment for Paper.* Miller Freeman Publishing San Francisco.

Dion, J. L., Valade, J. L., and Law, K. N. (1988). Evolution d'une suspension de fibres dans un champs ultrasonore stationnaire. *Acustica.* 65, 284–287.

Dion, J. L., Garceau, J. J., and Baribeault, R. (1987). Preliminary test results of a new acousto-optical process for online characterization of fiber size. *Tappi J.* 70(4), 161–162.

Dion, J. L., Malutta, A., and Cielo, P. J. (1982). Ultrasonic inspection of fiber suspension. *J. Acoust. Soc. Am.* 72, 1524–1526.

Dolgin, G. L., Akim, L. E., and Vinogradova, L. I. (1968). Changes in the chemical composition of cellulosic fibers during treatment with ultrasound. *Tr. Leningrad Tekhnol. Inst. Tsellyul. Bumazh. Prom.* 21, 156–163.

Filipovici, J., Mihai, D., and Mihai, S. (1972). Possibilities of plant modification in the bending of chair frames. *Bull. Univ. Brasov Ser. B,* 14, 285–293.

Filipovici, J., Mihai, D., Mihai, S., Dragan, O., and Ciovica, D. (1970). Studies on the bending of beech wood in ultrasonic field. *Ind. Lemnului.* 21(3), 88–91.

Filipovici, J., Mihai, D., Mihai, S., and Staicu, L. (1970). Microfractographic study of beech wood plasticized in high energy ultrasonic field. *Ind. Lemnului.* 21(12), 441–444.

Garceau, J. J., Dion, J. L., Brodeur, P., and Luo, H. (1989). Acousto-optical fiber characterization. *Tappi J.* 72(8), 171–173.

Habeger, C. C. and Baum, G. A. (1983). Ultrasonic characterization of fiber suspensions. in *Trans. 7th BPBIF Fundamental Research Symp.* Role of Fundamental Research in Papermaking. Vol. 1. Brander, J. (Ed.). Cambridge, England, 277–308.

266

Istas, J. R. and Raekelboom, E. L. (1970). Etude Biométrique, Chimique et Papetiére de Quelques Essences. Publ. No. 5. Administration de la Donation Royale, Arboretum Gégraphique de Tervuren, Belgium.

Janin, G. and Ory, J. M. (1984). A high-speed individualized fiber length measurement device. *Tappi J.* 67(4), 136–137.

Kato, K., Tsuzuki, K., and Asano I. (1971). Studies on the vibration cutting of wood. I. *Mokuzai Gakkaishi.* 17(2), 57–65.

Kato, K. Asano, I. (1974). Studies on the vibration cutting of wood. II. Cutting forces in longitudinal vibration cutting. *Mokuzai Gakkaishi.* 20(5), 191–195.

Kato, K. and Asano, I. (1974). Studies on the vibration cutting of wood. III. Cutting forces in lateral vibration cutting. *Mokuzai Gakkaishi.* 20(9), 418–423.

Kato, K. and Asano, I. (1974). Studies on the vibration cutting of wood. IV. The effect of fiber direction and annual ring on vibration cutting forces. *Mokuzai Gakkaishi.* 20(5), 424–429.

Kauman, W. (1964). Cell collapse in wood. *Holz Roh Werkst.* 22(5), 185–196; *ibid.* 22(12), 465–472.

Kurjatko, S. Marcok, M. (1990). Effect of ultrasound on the penetration of preservative oil into spruce wood (Abstr.). *IAWA Bull.* 11(2), 129.

Kutsuki, H. and Higuchi, T. (1978). Formation of lignin of *Erythrina crista-galli. Mokuzai Gakkaishi.* 24(9), 625–631.

Laboski, P., Jr. and Martin, R. E. (1969). Properties of paper obtained from ultrasonically and mechanically beaten pulps. *Wood Sci.* 1(3), 183–192.

Lam, F., Avramidis, S., and Lee, G. (1992). Effect of ultrasonic vibration on convective heat transfer between water and wood cylinders. *Wood Fiber Sci.* 24(2), 154–160.

Morehead, F. F. (1950). Ultrasonic disintegration of cellulose fibers before and after acid hydrolysis. *Textile Res. J.* 20, 549–553.

Neylon, M. (1978). Ultrasonic drying of eucalyptus—a preliminary study. *Aust. For. Ind. J.* 44(5), 16–18.

Parpala, V., Pasternac, A., and Paraschiv, N. (1974). Influence of ultrasound on the structure of *Abies alba* wood. Rev. *Padurilor-Ind. Lemnului.* 25(3), 68–70.

Parpala, V., Popescu, A., Filipovici, J., Pasternac, A., and Terschak, H. (1970). Der Einfluss der Uberschalle auf die physisch-mechanischen Eigenschaften des Buchenholzes. *Bul. Inst. Politeh. Brasov Ser. B.* 12, 355–358; *ibid.* 11, 387–393.

Proszyk, S., and Pradzynski, W. (1990). Use of ultrasound for regeneration of aged glue urea-formaldehyde resin. *Holzforschung.* 44(3), 173–176.

Rooney, J. A. (1981). Nonlinear phenomena. in *Methods of Experimental Physics.* Vol. 19. *Ultrasonics,* Edmonds, P. E. (Ed.). Academic Press, New York, 299–353.

Rozenberg, L. D. (Ed.). (1973). *Physical Principles of Ultrasonic Technology.* Plenum Press, New York.

Sliwinski, A. (1974). Chemical aspects of ultrasonics. in *Acoustic and Vibration Progress.* Vol. 1. Stephens, R. W. B. and Leventhall, H. G. (Eds.). Chapman & Hall, London, 85–164.

Taneda, K. (1972). Graft copolymerization of vinyl monomers onto wood material by initiators. *Hokkaido For. Prod. Res. Inst. Rep.* 59.

Walters, C. S. (1977). Effects of surfactant and ultrasonic energy on the treatment of wood with chromated cooper arsenate. *Holzforschung.* 31(4), 112–115.

Wang, I. C. and Micko, M. M. (1985). An improved method of preparing wood fiber for fiber length determination. *Tappi J.* 68(7), 106–107.

Zobel, B. J. van Buijtenen, J. P. (1989). *Wood Variation. Its Causes and Control.* Springer-Verlag, Berlin.

Appendix

List of Wood Species Cited in This Book

Abies alba **Mill.**—silver fir
Abies sachalinensis **Mast.**—todomatsu—Mayr's fir
Acer campestre **L.**—common maple
Acer macrophyllum **Pursh**—bigleaf maple
Acer platanoides **L.**—maple (Norway maple)
Acer pseudoplatanus **L.** (of curly grain)—curly maple
Acer rubrum **L.**—red maple, soft maple, swamp maple
Acer saccharum **Marsh.**—sugar maple, hard maple, rock maple
Aesculus hippocastanum **L.**—horse chestnut
African blackwood—*Dalbergia melanoxylon* Guill. et Perr.
alder—*Alnus* spp.
Alnus glutinosa **(L) Gaertn.**—black alder
American beech—*Fagus grandifolia* Ehrh.
Araucaria K. Koch—Parana pine
ash—*Fraxinus* spp.
Australian species
Austrian pine—*Pinus nigra* Arn.
balsa—*Ochroma lagopus* Sw.
beech—*Fagus sylvatica* L.
Betula **spp.**—birch
Betula maximowicziana **Reg.**—makaba—Japanese red birch
Betula phatyphylla **Sukat.**—japanese birch
Betula schmidtsii **Regel**—onoore kanba
Betula verrucosa **Ehrh.**—European birch, common birch
birch—*Betula* spp.
black wood—Dalbergia melanoxylon Guill. et Perr.
bois d'amourette—snakewood—*Piratinera guianensis* Aubl.
boxwood—*Buxus microphylla* Sieb. et Zucc.
Brazilian rosewood—*Dalbergia nigra* Fr. All.
buna wood—*Fagus crenata* Bl.
Buxus sempervirens **L.**—European boxwood
Caesalpinia brasiliensis **Sw.**—pernambuco
Canadian red cedar—*Tsuga heterophylla* Sarg.—western hemlock
Canadian rock maple—*Acer saccharum* Marsh
canne de Provence
Carpinus japonica **Bl.**
Castanea sativa **Mill.**—European chestnut
cedar—*Cedrus* spp.
cedar of Lebanon—*Cedrus libani* Laws.
Cercidiphyllum japonicum **Sieb. et Zucc.**—katsuram—Katsura tree
Cerocarpus ledifolius **Nutt**—mountain mahogany
Chamaecyparis obtusa **Sieb. et Zucc.**—hinoki—Japanese cypress
chestnut—*Castanea* spp.
cocuswood—*Brya buxifolia* Urb.
Cordia trichotoma **Arrab.**—louro pardo, or freijo
Cornus controversa **Hemsl.**
Cryptomeria japonica **D. Don.**—sugi—japanese cedar
curly maple—*Acer pseudoplatanus*, of curly grain

Dalbergia **spp.**—palisander
Dalbergia cearensis **Ducke**—kingwood
Dalbergia latifolia **Roxb.**—Indian rosewood
Dalbergia nigra **Fr. All.**—Brazilian rosewood
Diospyros **spp.**—ebony
Douglas fir—*Pseudotsuga menziesii* (Mirb.) Franco
ebony—*Diospyros* spp.
Engelmann spruce—*Picea engelmannii* Engelm.—mountain spruce
Eucalyptus occidentalis—eucalypt
European plane—*Platanus acerifolia* Willd.
Fagus crenata **Blume.**—buna—Japanese beech
Fagus sylvatica **L.**—beech
Ferreira spectabilis **Fr. All.**—Mill.
fir—*Abies alba* Mill.
Fraxinus japonica **Bl.**
Fraxinus mandshurica **Rupr.**—tamo—Japanese ash
Fraxinus sapaethiana **Lingelsh**—shioji—Japanese ash
Fraxinus **spp.**—ash
Guilandia echinata **Spreng.**—Brazil wood pernambuco
green ash—*Fraximus* spp.
hazelfichte—spruce, resonance spruce
hazeltree—*Corylus avellana* L.
hickory—*Carya tomentosa* Nutt.
hinoki—*Chamaecyparis obtusa* Sieb. et Zucc.
Honduras rosewood—*Dalbergia stevensonii* Standl.
hornbeam—*Carpinus betulus* L.
horse chestnut—*Aesculus hippocastanum* L.
Huon pine—*Dacrydium franklinii* Hook.
Indian rosewood—*Dalbergia latifolia* Roxb.
Intsia bakeri **Prain.**—*Intsia palembanica* Miq.—merbau—mirabow
Juglans sieboldiana **Maxim.**—Japanese walnut
lancewood
Larix leptolepis **Gord.**—karamatsu—Japanese larch
lauan (**white lauan**)—*Pentacme contorta* Merr. et Rolfe
Liriodendron tulipifera **L.**—tulip tree
limba—*Terminalia superba* Engl. et Diels
Macassar ebony—*Diospyros celebica* Bakh.
Machaerium villosum **Vog.**—jacaranda pardo
Magnolia obovata **Thumb.**—magnolia—honoki
mahogany—*Dalbergia* spp.
makoro—*Tieghemella heckelii* Pierre.—*Dumoria heckelii* A. Chev.)
Manilkara bidentata **1. (A. Chev.)**—bulletrie—bulletwood
mansonia—*Mansonia altissima* A. Chev.
maple—*Acer platanoides* L.
maritime pine—*Pinus pinaster* Ait.
matoa—*Pometia pinnata* Forst.
Mauritius ebony
merbau—mirabow—*Intsia bakeri* Prain—*Intsia palembanica* Miq.
metasequoia—*Metasequoia glyptostroboides* Hu et Cheng.
mizunara (japonese oak)—*Quercus mongolica* Fisch.
mountain mahogany—*Cerocarpus ledifolius* Nutt.
Nigerian obeche—*Triplochiton scleroxylon* K. Schum.
oak—*Quercus* spp.
Ochroma lagopus **Swartz**—balsa
onoore kanba—*Betula schmidtii* Regel
Oregon pine

Parana pine
Paulownia tomentosa **Steud.**—kiri—royal paulownia
pernambuco—brazil wood—*Guilandia echinata* Spreng.
Picea abies (*Picea excelsa* Link) **Karst.**—European spruce
Picea engelmannii **Parry ex. Engelm.**—silver spruce
Picea glehnii **Mast.**—(akaezomatsu)—Hokkaido spruce
Picea jezoensis **Carr.**—kurozomatsu—Yeso spruce
Picea rubens **Sarg.**—red spruce
Picea sitchensis (**Bong**) **Carr.**—Sitka spruce
Pinus densiflora **Sieb. et Zucc.**—Japanese red pine—akamatzu
Pinus radiata **D. Don.**—radiata pine, Monterey pine
Pinus **spp.**—pine
Pinus strobus **L.**—Weymouth pine, yellow pine
Pinus sylvestris **L.**—Scots pine
Piptadenia macrocarpa **Benth**
Piratinera guianensis **Aubl.**—bois d'amourette—snakewood
Platanus acerifolia **Willd.**—European plane
Pometia pinnata **Forster.**—matoa
Ponderosa pine—*Pinus ponderosa* Dougl.
Populus **spp.**—poplar
Populus tremula (**tremuloides**) **Michx.**—trembling aspen
privets—*Ligustrum vulgare* L.
Prunus jamasakura **Sieb.**
Prunus glandulosa thunbergii **Parl.**—chinese aspen
Prunus verecunda—Japanese cherry tree
Pseudotsuga menziessii (**Mirb**) **Franco**—Douglas fir
Queensland maple—*Flindersia pimenteliana* F. v. M.
Queensland walnut—*Beilschmiedia tawa* Benth.
Quercus cerris **L.**—turkey oak
Quercus mongolica **Fish.**—Japanese oak
Quercus myrsinaefolia **Blume**—Japanese oak
Quercus pedunculata **Ehrh.**—*Quercus robur* L.—European oak
Quercus petraea **Liebl.**—sessile oak
Quercus robur **L.**—*Quercus pedunculata* Ehrh.—European oak
Quercus serrata **Thunb.**—Japanese oak
red spruce—*Picea rubens* Sarg.
rosewood—*Dalbergia* spp.
Romanian pine—commercial name of spruce resonance wood; botanical name *Picea abies*
sapele—*Entandrophragma cylindricum* Sprague
sassafras—*Sassafras albidum* Nees
Scotch pine—*Pinus sylvestris* L.
silver maple—*Acer saccharinum* L.
Sitka spruce—*Picea sitchensis* Carr.
southern pine—*Pinus* spp. or *Pinus elliotti* Engelm.
spruce pine—*Pinus glabra* Walt.
sugi—*Cryptomeria japonica* D. Don
Swartzia **spp.**
Tasmanian myrtle—*Nothofagus cunninghamii* Oerst.
teak—*Tectona grandis* L.
Thuja **spp.**—arbor vitae
Tilia cordata **Mill.**—small leaved European limetree
Tilia Japonica **Simonkai**—Japanese lime
Tsuga heterophylla **Sarg.**—western helmlock
tulip tree—yellow poplar—*Liriodendron tulipifera* L.
walnut—*Juglans* spp.

wawa (**Ghana**)—Ghana obeche—*Triplochiton scleroxylon* K. Schum.
Western hemlock—*Tsuga heterophylla* Sarg.—beitsuga
white spruce—*Picea engelmannii* Engelm.
***Zelkova serrata* Makino.**—(Japanese elm)—keyaki
Zollernia illicifilia

INDEX